P9-DTY-522

Progress in SEPARATION AND PURIFICATION

Volume 4

A Wiley-Interscience Series

Progress in SEPARATION AND PURIFICATION

VOLUME 4

Edited by

EDMOND S. PERRY

Research Laboratories
Eastman Kodak Company
Rochester, New York

and

CAREL J. VAN OSS

Department of Microbiology
Immunochemistry Laboratory
State University of New York at Buffalo
Buffalo, New York

Wiley-Interscience, A Division of John Wiley & Sons, Inc.
New York • London • Sydney • Toronto

Library of Congress Catalog Card Number: 67-29539

ISBN 0-471-68101-6

Printed in the United States of America

10 9 8 7 6 5 4 3 2 1

Preface to Volume 4

Progress in Separation and Purification was launched in 1968; Volume 2 appeared in 1969 and Volume 3 in 1970. Volume 4 continues our original objective of having a forum for special papers to provide the expert as well as the newcomer with a current awareness of the progress being made in this expanding segment of science.

This volume, like Volumes 1 and 3, contains articles on a variety of subjects within the field of separation and purification. Again, the authors present their particular expertise in a manner to benefit both the novice and the expert. Volume 4 differs somewhat from the first three volumes in the manner of printing. To combat simultaneously the continuously increasing costs and delays that accumulate through the classical type-setting, galley proofing, and printing processes, it was decided to use the "coldtype" printing process for Volume 4.

We extend our sincere gratitude to the contributors who made Volume 4 possible.

EDMOND S. PERRY
CAREL J. VAN OSS

Rochester, New York
Buffalo, New York
June 1971

Preface to Volume 1

Through the ages man has been aware of the importance of separating the valuable from the less valuable in nature's mixtures. This trait has persisted; it is difficult to single out an area of science of a science-based industry where separations and purifications do not play an important role. Yet, in spite of this heritage, only within recent times has the science of separations been recognized in its individuality and has finally been accorded independent identity. The beginning of this era is difficult to pinpoint, but it is approximately the time of the second world war. Certainly, the science and technology generated in the struggle to separate the fissionable isotopes, to isolate and purify marketable quantities of antibiotics, the discovery of gas-phase chromatography, and the industries based on solid-state phenomena were important factors in leading the way for the science of separations and purifications as we know it today.

In the intervening twenty-five to thirty years, there has occurred an extensive development in this field. The literature has grown immensely; it is still expanding at a rapid rate for both the science and the technology of the subject, and a forum is needed for specialists as well as those new to the field or those working in related areas to keep up with the development. *Progress in Separation and Purification* is devoted to this purpose. Its broadest objective is to provide its readers with a high degree of current awareness on the progress being made in this large and complex field. We hope that the Series will help the practitioner to keep abreast of the ever growing literature by providing him with authoritative summaries on significant new developments and critical evaluations of new methods, apparatus, and techniques. The organization and condensation of the literature which is now dispersed throughout the chemical, biological, and nuclear sciences will also help to bring to the science and technology of separations its rightful status.

Fast and expeditious reporting are particular objectives of this Series. We plan to publish volumes at intervals commensurate with the procurement of articles. Manuscripts will be processed as received, and a volume will be issued when sufficient material has been assembled. The choice of subject and the manner in which ideas and opinions are expressed are essentially left to the discretion of the authors. Our only request of them is to render a service which will be of value to the reader and to the science and technology of separation and purification.

The nine articles in this first volume fall within the spirit of this liberal policy. Each author is an acknowledged leader in the field and has written on a subject with which he has had intimate experience. The random order of appearance of the articles in the volume has helped to expedite publication. We shall be happy to receive suggestions and recommendations for improving the service this Series purports to provide, and we invite inquiries for publication in the Series from authors.

My sincere gratitude goes to the authors of this first volume who were willing to embark with me on this new venture. Special thanks are due Dr. Arnold Weissberger, who suggested this undertaking and has provided advice and counsel in getting it under way.

EDMOND S. PERRY

May 1968

Contributors to Volume 4

P. E. Barker, *University of Aston, Birmingham, England*

W. D. Betts, *Coal Tar Research Association, Cleckheaton, Yorks, England*

U. A. Th. Brinkman, *Department of Analytical Chemistry, Free Reformed University, Amsterdam, The Netherlands*

George B. Cline, *Department of Biology, College of General Studies, University of Alabama at Birmingham, Birmingham, Alabama*

G. W. Girling, *Benzole Producers Ltd., Watford, Herts, England*

Sun-Tak Hwang, *Department of Chemical Engineering, University of Iowa, Iowa City, Iowa*

Karl Kammermeyer, *Department of Chemical Engineering, University of Iowa, Iowa City, Iowa*

Alexander Kolin, *Department of Biophysics, School of Medicine, University of California, Los Angeles*

Stephen J. Luner, *Department of Biophysics, School of Medicine, University of California, Los Angeles*

H. W. Quinn, *Dow Chemical of Canada, Limited, Sarnia, Ontario, Canada*

Norman H. Sweed, *Department of Chemical Engineering, Princeton University, Princeton, New Jersey*

Contents

Progress in SEPARATION AND PURIFICATION

Volume 4

A Wiley-Interscience Series

PERMEABILITY AS A PHENOMENOLOGICAL COEFFICIENT

Sun-Tak Hwang and Karl Kammermeyer
Department of Chemical Engineering
University of Iowa
Iowa City, Iowa 52240

I. INTRODUCTION

With the advent of gaseous diffusion as a large scale separation process and the promising developments in water purification by reverse osmosis, membrane permeation has become rather popular as a separation and purification process. A wide variety of membranes is used in many different types of unit operations and research experiments. Invariably, one encounters the term "permeability."

Permeability is, at least historically, a natural definition to express the degree of permeation of a component through a certain material, usually in the form of a membrane. In many cases, the basic mechanism of permeation is molecular diffusion. In others, some other mechanisms such as hydrodynamic flow, capillary action or electrokinetic flow are the cause of permeation. For simple cases, the use of a diffusivity is obvious and sufficient in expressing the permeation rate. However, in general, the complexity and multiplicity of permeation mechanisms call for a quantity which tells the overall degree of permeation, that is, permeability. Another factor, which adds to the complexity of the term permeability, is the fact that it is used in so many different disciplines and research fields. It seems, at times, that there is hardly any communication among investigators in different fields of study. Numerous publications report all sorts of permeability data in the literature. However, it is impossible to make valid comparisons of the data and utilize them without a proper reference to basic definitions.

II. VARIOUS DEFINITIONS OF PERMEABILITIES

A. Molecular Diffusivity

If the permeation is due to molecular diffusion alone, then permeability is equivalent to diffusivity, which is a property of a given system. Provided that the concentrations of the diffusing species in the membrane are known, Fick's first law is adequate to describe such a system:

$$F = DA \frac{C_1 - C_2}{L} \qquad [1]$$

B. Overall Permeability

Frequently, however, the actual concentrations at the very surface or the inside of the membrane are not known, or are very difficult to measure experimentally. This requires a new definition of a phenomenological coefficient, namely permeability, in terms of measurable quantities, that is, outside concentrations or pressures:

$$F = Q_2 A \frac{\Gamma_1 - \Gamma_2}{L} \qquad [2]$$

or

$$F = Q_3 A \frac{P_1 - P_2}{L} \qquad [3]$$

Here, Γ represents the concentration of bulk fluid outside of the membrane while C represents the concentration within the membrane.

The permeability definitions given by Equations [2] and [3] may be used regardless of the actual transport mechanism. In fact, the mechanism does not need to be known for the measurements and for the calculation of permeabilities by these equations.

If the permeation mechanism is not solely due to diffusion, the following definition results, using the actual concentrations:

$$F = Q_4 A \frac{C_1 - C_2}{L} \qquad [4]$$

This permeability will have the same units as those of diffusivity, but it should never be called or confused with true diffusivity. One may call it "effective diffusivity."

In the case of a porous membrane, another permeability is defined (1) to take porosity into account:

$$F = Q_5 A \epsilon \frac{\Gamma_1 - \Gamma_2}{L} \qquad [5]$$

C. Darcy's Law of Permeability

Darcy's Law is widely used in the petroleum industry and in soil mechanics. Here, the flow coefficient is divided by the viscosity of the fluid in order to isolate the properties of the porous medium in its permeability (2).

$$F = \frac{Q_6 A}{\mu} \frac{P_1 - P_2}{L} \qquad [6]$$

D. Comparison

By any means, the definitions given above are not exhaustive. One can make countless combinations of units in the expression of permeability. For instance mass flow rate may be used as well as volume flow rate, or inches instead of centimeters for the thickness of the membrane, etc. However, these are considered as mere unit conversions.

On the other hand, the interconversion between the above given permeabilities may or may not be so simple. Sometimes it can be meaningless. For example, the gas phase permeation through a porous membrane may be expressed by any of the permeabilities defined by Equations [2], [3], [5] and [6]. And yet, the liquid phase permeability through a nonporous membrane can hardly be expressed by those given in Equations [3], [5] and [6]. Even though it could, one can find no physical meanings attached to the permeability.

Summing up, permeability is a phenomenological quantity; hence, it depends upon the system and the experimental conditions. It is, in general, not a property of the membrane as diffusivity is. Depending on the structure of the membrane or mechanism of permeation, one can also define special permeabilities as shown in the next section.

III. PERMEABILITIES FOR MICROPOROUS MEMBRANES

In this section, the permeabilities and diffusivities for flow in microporous membranes are discussed in order to illustrate how the theoretical interpretations and modeling of the flow mechanisms affect the definition of permeability. It is also possible to establish interrelationships between different flow coefficients. This would resolve the difficulty arising in the comparison of reported permeability data in the literature.

A survey of the literature on gas permeation at low surface coverage reveals little uniformity in the presentation of the gas and surface permeability data. Hence, the following discussions will be limited to the Knudsen regime for gas phase flow and to the region of low surface coverage for surface flow.

Gas phase flow in the Knudsen regime occurs where the mean free path of the molecules is greater than the diameter of the pore. Under these conditions, the gas molecules are transported with little intermolecular collisions, but collide mainly with the pore walls. Although the gas phase flow is then well described by the Knudsen law, it becomes difficult to represent a porous medium by a geometrical model in order to directly apply Knudsen's work. However, a number of models (4,5,6) have been proposed as a basis for calculating the Knudsen diffusivity, D_K, for various porous media. Such models try to take into account factors of tortuosity, porosity, etc.

In addition, several theoretical interpretations (7,8,9,10,11, 12,13) have evolved from the considerable evidence supporting the existence of transport by surface flow in porous media. Since a surface diffusivity, D_s is then often calculated for the many corresponding models, it should be emphasized that these surface diffusivities are only valid for a particular model and are often in a range where they are strongly dependent on surface coverage. Therefore both the model and amount of surface coverage must be considered when comparing reported surface diffusivities. The results discussed here are all in the region of low surface coverage.

Unfortunately, the experimental results of permeation are expressed as either diffusivities or permeabilities, with permeabilities being expressed in both units of cm^2/sec and (std.cc)(cm)/(sec)(cm^2)(cmHg). It should also be noted that the definition of "permeability" is usually the same as the definition of "permeability coefficient," and this is also the case with "diffusivity" and "diffusivity coefficient."

It is therefore pertinent to review these basic definitions used in the study of permeation of gases through porous media, to establish a relationship between them, and then to outline a method of calculating "diffusivities" from "permeability data." Since it is beyond the scope of the present discussion to perform a comprehensive evaluation of the proposed theories in gas diffusion, only some of the reported experimental results are considered in order to illustrate the relationship between "permeabilities" and "diffusivities" in the region of Knudsen flow and low surface coverage.

A. Gas Phase

The permeability of gases and vapors through microporous media is usually considered to be the sum of gas phase flow, F_g and surface flow, F_s. Using the concept that the two flow processes take place in parallel, the steady state flow (4,10) is:

$$F = F_g + F_s \ . \qquad [7]$$

Applying the analogy of Fick's first law of diffusion, the steady state flow rate can be expressed by

$$F = -D_t \ A \frac{dC_t}{dx} \ , \qquad [8]$$

and similarly for the gas phase

$$F_g = -D_g \ A \frac{dC_g}{dx} \ . \qquad [9]$$

Since the gas phase flow is in the Knudsen regime, a Knudsen diffusivity, D_K is often calculated for a given geometrical model and is defined by (14):

$$F_g = -D_K \ A' \ \frac{dC'_g}{dx} \ . \qquad [10]$$

Knudsen showed that if a straight cylindrical capillary is very long relative to its deameter, the following expression (4,5,6,14) is obtained for the Knudsen diffusivity in the Knudsen flow region:

$$D_K = \frac{4r}{3} \sqrt{\frac{2 \ R \ T}{\pi \ M}} \qquad [11]$$

It should be noted that the following relationships exist for the straight cylindrical capillary model:

$$A' = \varepsilon \ A \qquad [12]$$

$$C'_g = \frac{C_g}{\varepsilon} \qquad [13]$$

in which ε is the porosity of porous medium. When Equations [12] and [13] are substituted into Equation [10]:

$$F_g = -D_K \ A \frac{dC_g}{dx} \qquad [14]$$

and $D_K = D_G$, if the straight cylindrical capillary model represents the actual porous medium. However, usually this is not the case and Equation [11] is modified by more complicated models where Equations [12] and [13] are not applicable. In general, Equation [10] then becomes (10,15):

$$F_g = -D_K \ A \ G_1 \ \frac{dC_g}{dx} \ , \qquad [15]$$

where G_1 is a dimensionless geometrical factor.

Since the Knudsen diffusivity is a constant and is defined by the geometry of the model, it can be related to the effective gas diffusivity by (8,15):

$$D_g = G_1 D_K \quad . \qquad [16]$$

Thus, the calculations of D_K will depend on the model selected while D_g can be determined directly from the appropriate experimental data.

When the outside concentrations or the pressures are used instead of the concentrations within the porous membrane, the following permeabilities are defined: for total flow,

$$F = -Q_t A \frac{dP}{dx} \qquad [17]$$

or

$$F = -K_t A \frac{d\Gamma}{dx} \quad ; \qquad [18]$$

and similarly for the gas phase flow,

$$F_g = -Q_g A \frac{dP}{dx} \qquad [19]$$

or

$$F_g = -K_g A \frac{d\Gamma}{dx} \quad . \qquad [20]$$

When F_g is taken to be Knudsen flow, it can be expressed (10) as:

$$F_g = -\frac{A G_2}{\sqrt{2\pi MRT}} \frac{dP}{dx} \qquad [21]$$

Therefore, the gas phase permeability, Q_g, for the Knudsen regime can be obtained by

$$Q_g = \frac{G_2}{\sqrt{2\pi MRT}} \qquad [22]$$

B. Adsorbed Phase

Applying Fick's diffusion equation for two-dimensional steady state flow, the flow rate for the adsorbed phase in terms of surface concentration per unit area, C_s', becomes (5):

$$F_s = -W D_s \frac{dC_s'}{dx} \quad , \qquad [23]$$

where D_s is the surface diffusivity, and is assumed to be constant for the region of low surface coverage. In order to obtain Equation [23] in terms of measurable quantities, it is transformed into three-dimensional units. The surface concentration, C_s', is expressed as concentration per unit volume of porous medium by using the surface area per unit volume of the porous medium, S_v, such that:

$$C_s = C_s' S_v \quad . \qquad [24]$$

Combining Equations [23] and [24], the following three-dimensional expression is obtained:

$$F_s = -D_s \left(\frac{W}{S_v}\right) \frac{dC_s}{dx} \quad . \qquad [25]$$

In order to obtain the same form as Equation [8], the macroscopic cross sectional area, A, is included in Equation [25] so that:

$$F_s = -A\, D_s \left(\frac{W}{AS_v}\right) \frac{dC_s}{dx} \quad . \qquad [26]$$

If the quantity $\left(\frac{W}{AS_v}\right)$ is represented by a geometric factor G_3, then:

$$F_s = -AG_3D_s \frac{dC_s}{dx} \quad . \qquad [27]$$

An "apparent" surface diffusivity, D_s', is often defined as (8):

$$F_s = -A\, D_s' \frac{dC_s}{dx} \quad , \qquad [28]$$

and is related to D_s by:

$$D_s' = G_3D_s \quad . \qquad [29]$$

Calculation of the true surface diffusivity, D_s, can only be done by assuming a model and cannot, therefore, be explicit as also previously shown in the calculation of the Knudsen diffusivity, D_K from the effective gas diffusivity D_g. If, for an example, the straight cylindrical capillary model is again considered, then:

$$W = A\, S_v \quad , \qquad [30]$$

yielding $G_3 = 1$. Equation [27] becomes:

$$F_s = -A\, D_s \frac{dC_s}{dx} \qquad [31]$$

and D_s is equal to D_s' for this particular model.

In consideration of the above definitions, it is necessary that all authors clearly define all their reported diffusivities and corresponding models if the Knudsen diffusivity, $\underline{D}_K$, and actual surface diffusivity $\underline{D}_s$, are calculated.

The surface permeabilities are defined similarly to the gas phase permeabilities.

$$F_s = -Q_s A \frac{dP}{dx} \qquad [32]$$

or

$$F_s = -K_s A \frac{d\Gamma}{dx} \quad . \qquad [33]$$

C. Diffusivities from Permeabilities

The actual concentrations in the porous medium are very difficult to measure directly because they are the result of pressures or concentrations existing outside of the membrane. Expressing the equation of state for the gas phase as

$$\Gamma = \frac{P}{zRT} \quad , \qquad [34]$$

the following relationships are observed:

$$K_t = zRT\, Q_t \qquad [35]$$

$$K_g = zRT\, Q_g \qquad [36]$$

$$K_s = zRT\, Q_s \qquad [37]$$

1. Effective Gas Diffusivity

The relationship between $\underline{D}_g$ and $\underline{Q}_g$ is obtained from Equations [9] and [19].

$$D_g = Q_g \frac{dP}{dC_g} \quad . \qquad [38]$$

Since $\underline{C}_g$ is the concentration of transported species in the gas phase per unit volume of the porous medium, it can be related to pressure by:

$$C_g = \frac{\epsilon P}{zRT} \quad . \qquad [39]$$

It should be noted that Equation [39] ignores the void volume reduction due to the adsorbed species since the reduction is usually very small at low surface coverage. Taking the derivative of pressure, $\underline{P}$, with respect to the concentration, $\underline{C}_g$ in Equation [39]:

$$\frac{dP}{dC_g} = \frac{zRT}{\epsilon} \qquad [40]$$

and substituting into Equation [38], the effective gas diffusivity can be calculated using the following expression:

$$D_g = Q_g \frac{zRT}{\epsilon} \quad . \qquad [41]$$

2. Apparent Surface Diffusivity

The surface permeability is then obtained using Equations [7], [19] and [32]

$$Q_s = Q_t - Q_g \quad . \qquad [42]$$

The relationship between Q_s and D_s' is obtained from Equations [28] and [32] for the case where D_s' is constant:

$$D_s' = Q_s \frac{dP}{dC_s} \quad . \qquad [43]$$

In order to obtain $\frac{dP}{dC_s}$, the following relationships were developed. Since the adsorption isotherm is generally plotted as y (in std. cc. of adsorbate per gram of adsorbent) versus pressure, the slope of any given pressure can be defined as:

$$\alpha = \frac{dy}{dP} \quad . \qquad [44]$$

In the Henry's law region of the isotherm, which should hold at low surface coverage, the slope is constant and thus α is independent of pressure in this range. Since C_s is related to y by:

$$C_s = \rho y \quad . \qquad [45]$$

Equations [44] and [45] yield:

$$\frac{dP}{dC_s} = \frac{1}{\alpha\rho} \quad . \qquad [46]$$

Thus, the apparent surface diffusivity can be related to the surface permeability for the Henry's law region by substituting Equation [46] into Equation [43] to obtain:

$$D_s' = Q_s \frac{1}{\alpha\rho} \qquad [47]$$

3. Total Diffusivity

The total effective diffusivity can be related to the total

permeability by combining Equations [8] and [17] to yield:

$$D_t = Q_t \frac{dP}{dC_t} \qquad [48]$$

Since C_t can be expressed as:

$$C_t = C_g + C_s \quad , \qquad [49]$$

the following equation is obtained:

$$\frac{dP}{dC_t} = \frac{1}{\frac{dC_g}{dP} + \frac{dC_s}{dP}} \quad . \qquad [50]$$

Thus, by combining Equations [40], [46], [48] and [50] the relationship between D_t and Q_t is given by:

$$D_t = Q_t \frac{1}{\frac{\epsilon}{zRT} + \alpha\rho} \quad . \qquad [51]$$

D. Comparisons

Utilizing the equations developed in the preceding section, various diffusivities are calculated and listed in Table I from the permeability data of Hwang (10,11) and Huckins (16). The adsorption isotherm data reported by Barrer (4) are used to calculate the diffusivities shown in the table. These values are compared with those reported by Barrer (4). A comparison of the apparent surface diffusivities are slightly higher than those calculated from the data of Hwang and of Huckins. However, this is not surprising since the apparent surface diffusivities reported by Barrer are based on transient state measurements while Hwang's and Huckins' measurements were at steady state. Barrer (17) has shown that surface diffusivities for some alumina-silica and carbon plugs were about twice as large for the transient state cases as compared with the steady-state conditions.

The gas permeabilities calculated from the Knudsen flow constant reported by Hwang (10,11) are compared with some values reported by Barrer (17) in Table II. This table also includes a comparison of effective gas diffusivities calculated from Hwang's (10,11) gas permeabilities and the effective gas diffusivities reported by Barrer (4). It should be noted that Barrer's gas diffusivities are based on the conventional assumption of no surface diffusion for helium, which is incorrect. Hwang's gas diffusivities were calculated from a constant which was determined when helium surface diffusion is taken into account. A sample calculation is given in the Appendix.

E. Recommendation

The main objective of the foregoing analysis was to clarify a confusing situation existing in the reporting of diffusivity and

Table I

Comparison of calculated and reported diffusivities in the region of low surface coverage for vycor glass.

Gas	Temp. ^oK	$Q_t \times 10^6$ Reported by Hwang (10,11)	Huckins (16)	$D_g \times 10^5$ Calculated from Hwang's Constant (10,11)	$D'_x \times 10^5$ Calculated for Hwang (10,11)	Huckins (16)	Reported by Barrer≠ (4)	$D_t \times 10^5$ Calculated for Hwang (10,11)	Huckins (16)	Adsorption Isotherm Slope α from Barrer(4)
N_2	273	6.03	6.10	125	7.64	8.00	----	26.8	27.1	0.0125
"	290	5.78	5.92	128	9.95	11.0	23.35*	34.5	35.4	0.00884
"	323	5.38	5.61	136	15.0	17.9	30.53	49.7	51.8	0.00514
"	343	5.18	5.44	140	18.9	23.4	34.50	59.9	62.9	0.00381
O_2	273	5.62	5.97	117	7.01	8.88	----	25.0	26.6	0.0125
"	290	5.40	5.79	120	8.87	11.7	19.84*	31.3	33.6	0.00919
"	323	5.04	5.49	127	13.7	19.3	26.41	45.6	49.6	0.00529
"	343	4.86	5.33	131	17.3	25.1	31.29	54.8	60.0	0.00396
Ar	273	4.97	5.19	104	6.54	7.84	----	23.9	24.9	0.0114
"	290	4.77	5.04	108	8.16	10.2	20.00*	29.4	31.1	0.00850
"	323	4.46	4.77	113	11.7	15.6	25.19	40.4	43.2	0.00529
"	343	4.31	4.63	117	14.2	19.5	29.16	47.4	50.9	0.00410
H_2	273	20.4	21.2	466	150.	188.	----	350.	364.	0.00142#
"	298	19.5	20.3	487	197.	246.	----	395.	411.	0.00104#
"	313	19.1	19.8	499	233.	290.	----	423.	439.	0.00086#
CH_4	273	8.61	8.55	165	8.38	8.17	----	25.4	25.2	0.0201
"	294	8.06	8.24	171	10.9	11.7	18.93	34.0	34.8	0.0135
"	323	7.47	7.86	179	14.3	17.3	23.96	46.1	48.5	0.00872
"	343	7.14	7.63	185	17.7	22.7	26.41	56.6	60.5	0.00645
C_2H_6	294	8.56	----	125	3.16	----	3.96*	6.17	----	0.0901
"	323	7.39	----	131	5.87	----	7.02	12.4	----	0.0375
"	343	6.80	----	135	8.07	----	9.00	17.8	----	0.0235

* 292°K.

\# Adsorption isotherm data from Graham (18).

≠ Obtained by dividing all values in Table 8 of original article (4) by a tortuosity factor of 6.55.

Table II

Comparison of effective gas diffusivities and gas permeabilities for vycor glass at T = 292°K.

	Data taken from Barrer and Barrie (4)		Data taken from Barrer (17)		Calculated data from a constant given by Hwang (10,11)	
	(b)	(c)	(d)	(c)	(e)	(f)
Gas	$D_g \times 10^4$	$Q_g \times 10^5$	$D_g \times 10^4$	$Q_g \times 10^5$	$D_g \times 10^4$	$Q_g \times 10^5$
He	42.8	1.571	---	---	34.09	1.176
Ne	18.7	0.686	---	---	15.18	0.524
H_2	53.7	1.975	---	---	48.21	1.663
Ar	13.0	0.477	13.5	0.496	10.79	0.372
N_2	16.0	0.587	16.2	0.595	12.88	0.444
O_2	14.5	0.532	15.1	0.554	12.05	0.416
Kr^a	9.33	0.343	9.4	0.345	7.48	0.256
CH_4^a	20.2	0.742	21.4	0.786	17.11	0.586
C_2H_6	17.2	0.631	---	---	12.45	0.429

Note: Units of $\underline{D}_g$ and $\underline{Q}_g$ are listed in Nomenclature Section.

<u>Key</u>:

a Krypton and methane data for T + 294°K

b Calculated from Barrer and Barrie (4) who reported a $\underline{K}_K$ which is equal to $(D_g \epsilon)$, where $\epsilon = 0.298$

c Calculated from $\underline{D}_g$ by Eq. [41], $Q_g = D_g\left(\frac{\epsilon}{zRT}\right)$

d Reported by Barrer (17) as $\underline{K}_g$* which is the same as $\underline{Q}_5$ for gas phase

e Calculated from $\underline{Q}_g$ by Eq. [41], $D_g = Q_g\left(\frac{zRT}{\epsilon}\right)$

f Calculated from Knudsen flow constant, $Q_g\sqrt{MT} = 4.02 \times 10^{-4}$ as given by Hwang (10)

permeability data for microporous media. For ease in making comparisons of data it is necessary that authors report permeabilities in one of the forms defined in this section. In addition, if diffusivities are also desired, then the effective gas, apparent surface, and total diffusivities should be calculated, as they are based only on the concept of unhindered parallel flow and are independent of any pore model. Further, the relationship between these diffusivities and any other reported diffusivities should be clearly defined.

IV. PERMEABILITIES FOR LIQUID SYSTEMS THROUGH PLASTIC MEMBRANES

The previous section discussed the variation of permeability with the mechanism of mass transfer within the membrane. Additionally the permeability is also affected by the conditions outside a membrane.

A. Relationship between Permeability and Diffusivity

It is generally true that there are always some resistances to permeation at the interfaces between the bulk fluids and the membrane. These may be due to the presence of physical boundary layers near the membrane, to sorption and desorption processes, or both. Therefore, it is possible to propose the following general model for membrane permeation.

A hypothetical concentration profile and a schematic view across the membrane are shown in Figure 1. It is assumed that both sides

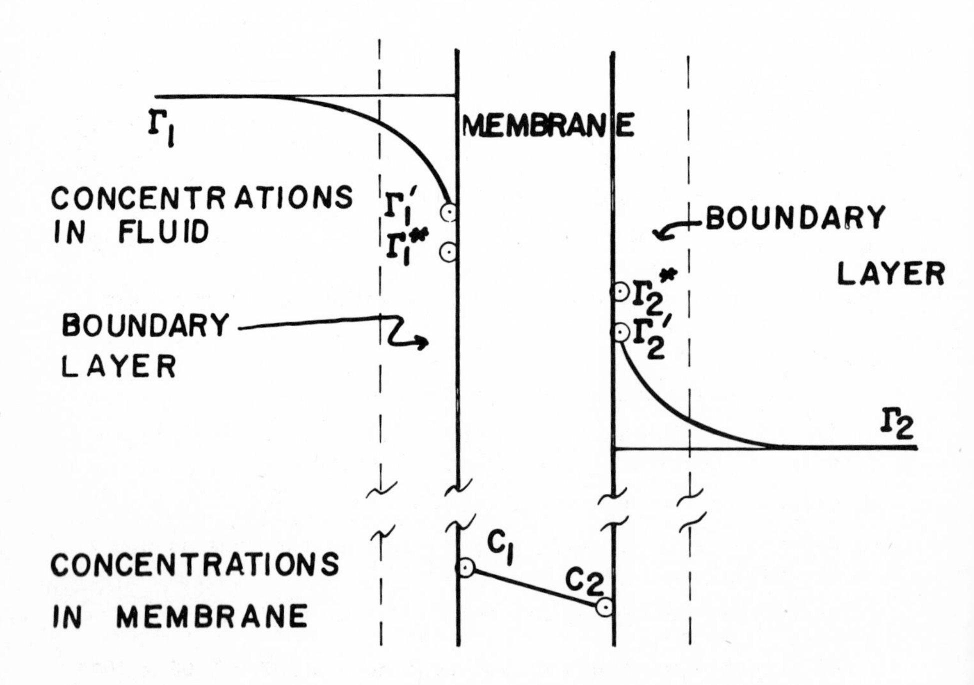

Figure 1. Concentration profile across the membrane.

of the membrane are covered by thin layers of immobile fluid, which would give an extra resistance to permeation. The mass transfer through such a system consists of the following stepwise processes:

1. Diffusion through the boundary layer.
2. Sorption into the membrane.
3. Diffusion through the membrane.
4. Desorption out of the membrane.
5. Diffusion through the boundary layer.

Each step represents a resistance to the gas transport of different magnitudes. For practical purposes, however, the resistances of some steps are negligible in comparison with those of others. In the case of gas-phase permeation, processes 1 and 5 are not involved, and the resistances due to the steps 2 and 4 may even be negligible. However, for liquid phase permeation, a large boundary resistance may result from steps 1, 2, 4 and 5. The presence of these boundary resistances reduces the available driving force for diffusion inside the membrane.

Using the definition of overall permeability given by Equation [2], the steady-state flow equation is written as:

$$F = QA\left(\frac{\Gamma_1 - \Gamma_2}{L}\right). \qquad [52]$$

If the diffusivity of the membrane is independent of concentration, the same steady-state flow rate for the inside of the membrane can also be expressed by:

$$F = DA\left(\frac{C_1 - C_2}{L}\right). \qquad [53]$$

Should diffusivity be concentration-dependent, an average diffusivity can be used.

$$F = \bar{D}A\left(\frac{C_1 - C_2}{L}\right) \qquad [54]$$

where:

$$\bar{D} = \int_{C_2}^{C_1} D dC/(C_1 - C_2) \quad . \qquad [55]$$

Because it is very difficult to separate steps 2 and 4 from steps 1 and 5 experimentally, it is convenient to lump the resistances of steps 1 and 2 in a group, and steps 4 and 5 in another. Then one can write the flow equation for one side of the membrane:

$$F = A\left(\frac{\Gamma_1 - \Gamma_1^*}{r_1}\right) \qquad [56]$$

and for the other side

$$F = A\left(\frac{\Gamma_2^* - \Gamma_2}{r_2}\right) \qquad [57]$$

where $\underline{r}_1$ and $\underline{r}_2$ are film resistances including resistances of sorption and desorption if they exist. The fictitious quantities, Γ_1^* and Γ_2^*, are the concentrations which would have produced the inside concentrations $\underline{C}_1$ and $\underline{C}_2$ respectively under equilibrium conditions. If a linear isotherm (Henry's Law) is applicable,

$$C_1 = S\,\Gamma_1^* \qquad [58]$$

$$C_2 = S \Gamma_2^* \quad . \tag{59}$$

Combining Equations [56], [57], [58] and [59] with Equation [53] and solving for F,

$$F = \frac{DS(\Gamma_1 + \Gamma_2)}{DS(r_1 + r_2) + L} \quad . \tag{60}$$

Comparing Equation [60] with Equation [52], the following is obvious:

$$Q = \frac{DSL}{DS(r_1 + r_2 + L)} \quad . \tag{61}$$

This equation tells exactly how the observed permeability changes as the thickness of a membrane varies. Also, it shows that the film resistance could be significant when the diffusivity of the membrane is large, or when the thickness of a membrane is small. If there is no such film present, then the film resistance simply becomes zero, and the observed premeability reduces to the familair form.

$$Q = DS \quad . \tag{62}$$

This equation has been used widely in many systems. However, it is clear from Equation [61] that Equation [62] holds only in a special case. Furthermore, Equation [61] illustrates the fact that permeability is a phenomenological coefficient rather than a property of a given system as a given in Equation [62]. A change of an outside condition, such as film resistance or membrane thickness, alters the value of permeability. Therefore, in general, a comparison of two permeabilities for the same system but at different experimental conditions may not be neaningful.

B. Dissolved Oxygen Permeation through Membrane

The studies by Yasuda (19,20) and Robb (21) on permeation of oxygen dissolved in water present an interesting comparison of gas phase permeability with the permeability of dissolved oxygen in the aqueous phase. Among many polymer membranes employed, silicone rubber showed a marked difference from other membranes in the magnitude of its permeability. Therefore, the system of silicone rubber membrane and dissolved oxygen would be quite suitable for testing the validity of Equation [61]. There are two reason for this. First, a stagnant film would exist at the surface of membrane to give the film resistance. Second, the overall permeation rate is so great that such an interface resistance may not be negligible in comparison with the diffusion resistance in the membrane.

1. Effect of Thickness

A recent study (22) showed that the observed permeability of dissolved oxygen through a silicone rubber membrane depends on the membrane thickness. In both, steady-state (circles) and unsteady-state (hexagons) measurements, the observed permeability increased with increasing membrane thickness and approached asymptotically a limiting value as shown in Figure 2. The analysis of the experimental data can best be done by plotting the inverse permeability

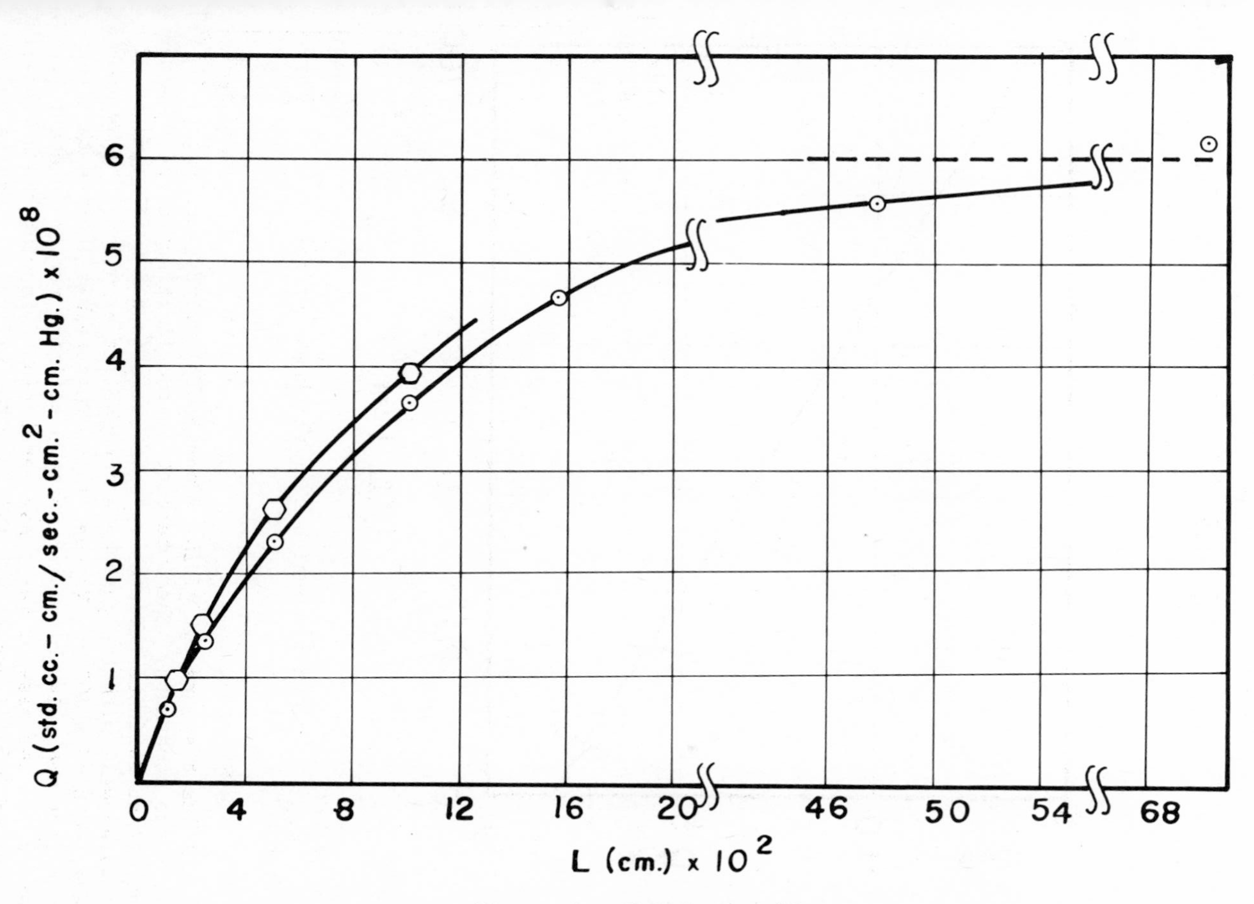

Figure 2. Thickness effect.

against the inverse thickness as in Figure 3. From Equation [61], a linear relationship is obtained between the inverse permeability and the inverse thickness.

$$\frac{1}{Q} = \frac{1}{DS} + (r_1 + r_2)\frac{1}{L} \quad . \qquad [63]$$

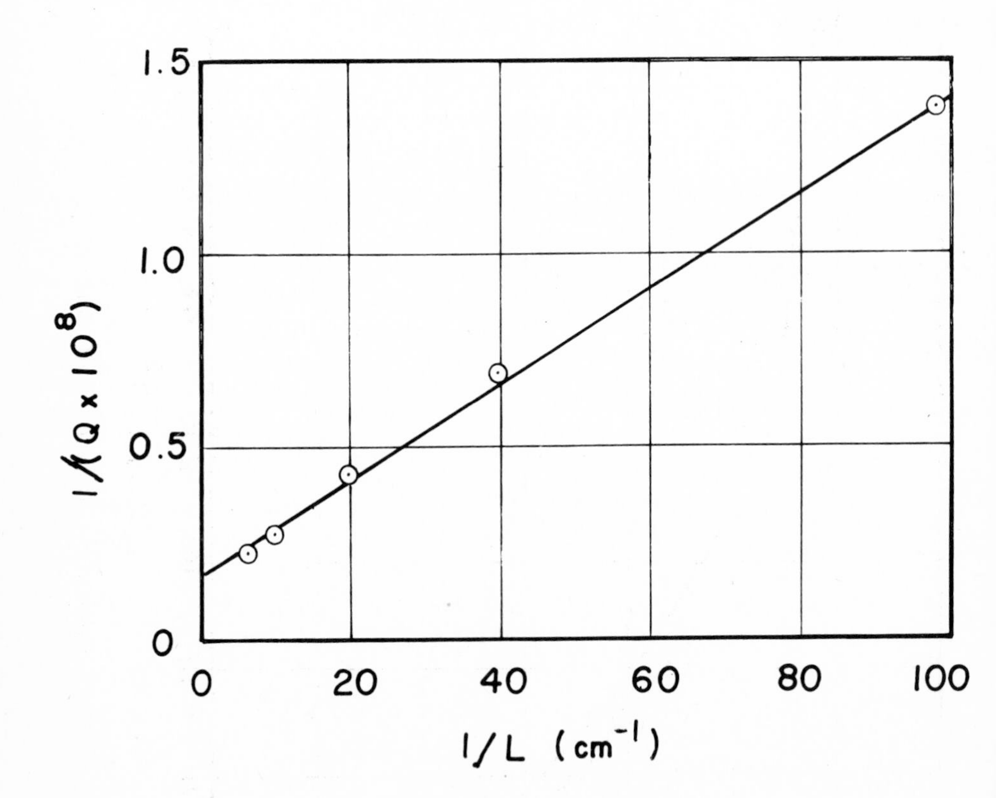

Figure 3. Analysis of thickness effect .

A least-square fit of the experimental points into Equation [63] gives two constants. The intercept is the limiting value of permeability at infinite thickness. This value is 6.0×10^{-8}, which compares well with the gas-phase permeabilities 6.2×10^{-8} by Robb (21) and 6.65×10^{-8} by Yasuda (19). This result implies that the transport mechanism of oxygen through silicone rubber is about the same as in the case of gas-phase permeation. Knowing the solubility of oxygen in silicone rubber (21,23), the diffusivity can be calculated by Equation [62]:

$$D = \frac{Q_\infty}{S} = \frac{6.0 \times 10^{-8}}{0.31 \times \frac{1}{76}} = 1.47 \times 10^{-5}\ cm^2/sec \ .$$

This value is in very good agreement with those in the literature, 1.6×10^{-5} by Robb (21) and 1.2×10^{-5} by Barrer (23).

Another constant obtained from Figure 3 is the slope of the straight line. This is the resistance of the boundary layers. Here, one can make several interesting calculations. For a 4-mil thick membrane, the film resistances amount to 88.7% of the total resistance. This means that the available concentration drop across the membrane was only 11.3% of the total concentration drop. Referring to Figure 4, the bulk concentrations were Γ_1 = 14.6 cm Hg and Γ_2 = 5.86 cm Hg. The values of $\underline{\Gamma}_1^*$ and $\underline{\Gamma}_2^*$ were calculated by Equations [56] and [57] as 10.73 and 9.73 cm Hg respectively. Then $\underline{C}_1$ and $\underline{C}_2$ could be calculated as 0.0438 and 0.0397 (std. cc/cc) respectively. Knowing the actual concentrations, the diffusivity can be calculated again by

$$D = Q \frac{\Gamma_1 - \Gamma_2}{C_1 - C_2} = (0.725 \times 10^{-8}) \frac{8.74}{0.0041} = 1.54 \times 10^{-5} \frac{cm^2}{sec} .$$

This value also checks the previous estimate. Another interesting point is that one can estimate the thickness of the boundary layers if the resistance is due to the diffusion of oxygen through water. The diffusivity of oxygen in water is 2.6×10^{-5} cm /sec (24). The thickness of an immobile boundary layer was calculated as 5.95×10^{-3} cm.

2. Effect of Concentration Level

The same study (22) on the permeation of dissolved oxygen concludes that the permeability is also a function of the concentration level employed. Here, the concentration level means the arithmetic average of the concentrations on the two sides of membrane. The permeability data at steady-state increased with increasing concentration level as shown in Figure 5. This effect is believed due to the change in boundary resistance.

3. Effect of Temperature

It is well known that permeability is dependent upon temperature. Usually it follows the Arrhenius type relationship:

$$Q = Q_o e^{-E_Q/RT} . \qquad [64]$$

The case of dissolved oxygen permeation through a silicone rubber membrane also conformed to this exponential function as shown in Figure 6. From the slope of the straight line, the activation energy was calculated as 2.65 Kcal/g-mole.

C. Liquid Permeation through Membrane

It has long been suspected that the concentration gradient within the membrane may not be constant and uniform throughout the entire membrane. Some attempts (25,26,27) were made to calculate concentration gradients for polymeric films based on a model. However, the published information on the actual concentration profile within a membrane is limited to one example (28).

Another important assumption, which is frequently made without

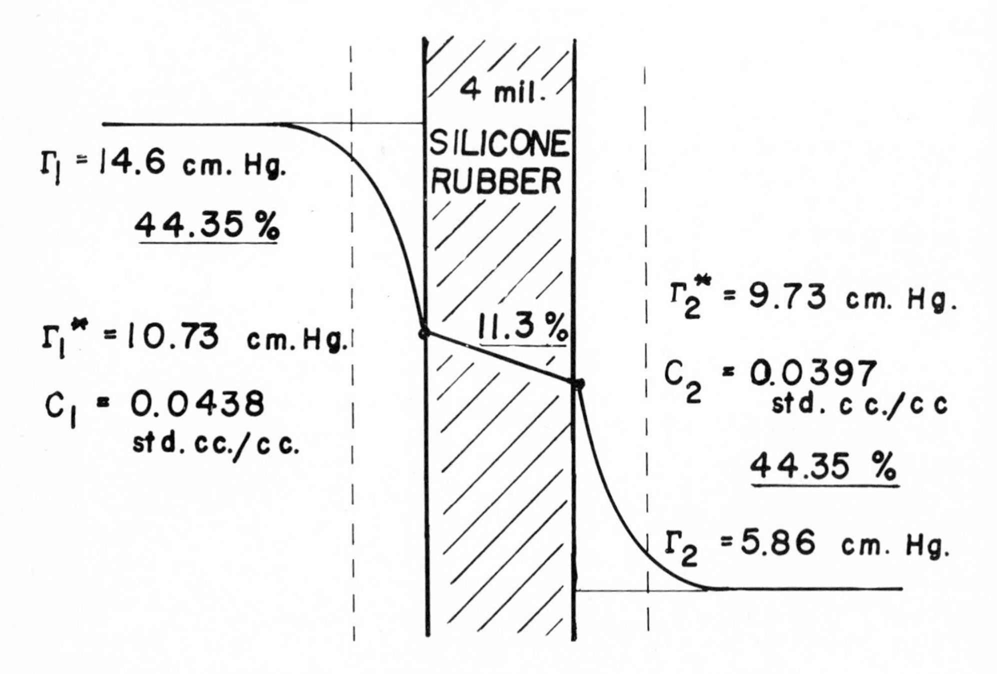

Figure 4. Concentration profile of oxygen across 4 mil silicone rubber.

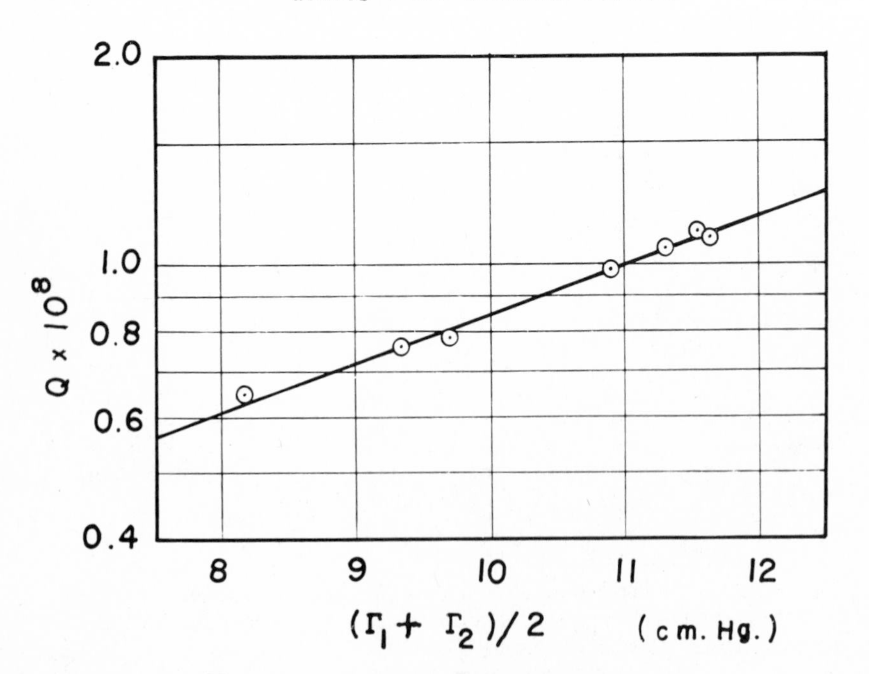

Figure 5. Concentration effect.

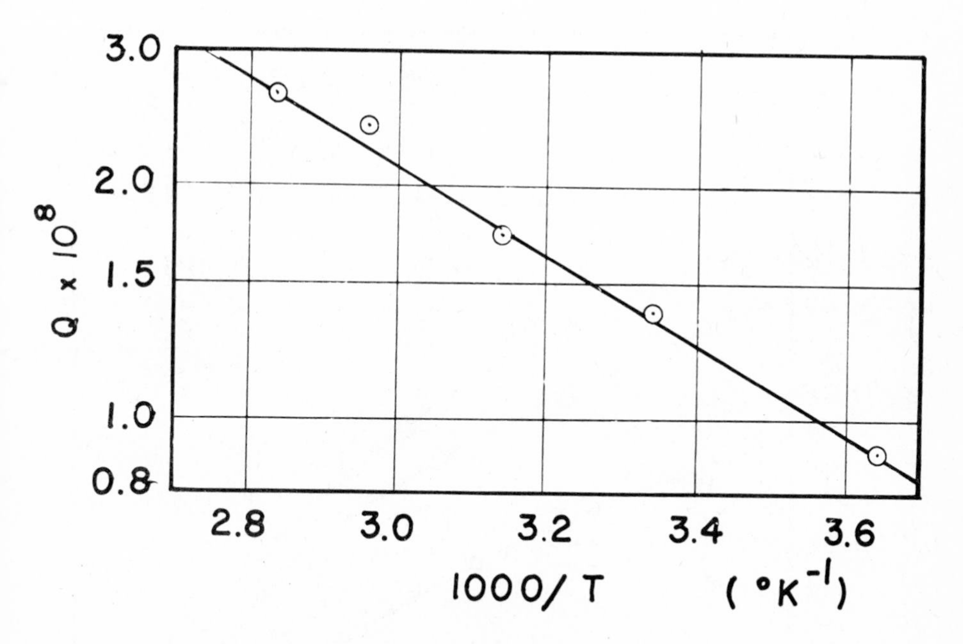

Figure 6. Temperature effect.

any evidence, is that the concentrations of permeate at the surfaces of membrane are the values of static equilibrium with the bulk concentrations.

The above two questions are well explored in a recent study by Kim (28) where the behavior of single components was investigated. A simple technique was developed to prepare stacks of thin films for the measurement of actual concentration profiles in the membrane. These sandwiched films were then employed in a steady-state permeation of pure liquid. One side of the membrane was in direct contact with the liquid and vacuum was applied on the other side. After a run was completed, each layer of film was peeled off carefully, then immediately placed into a weighing bottle to measure the weight gain. The weight gain in each film represented the average content of permeate in that section of a membrane.

A typical result is shown in Figure 7 for the system of nylon 6-water at 35°C. The actual concentration profile is not a straight line, instead it is a curve. Perhaps the most significant point is the fact that the interface concentration deviates more from the equilibrium concentration, $\underline{C}^*$, as the membrane becomes thinner.

This can be easily interpreted as follows: When the membrane is thick, the diffusional resistance within the membrane becomes sufficiently large so that the interfacial resistance may be neglected. In the case of pure component permeation, the interfacial resistance is only due to sorption or desorption processes, because

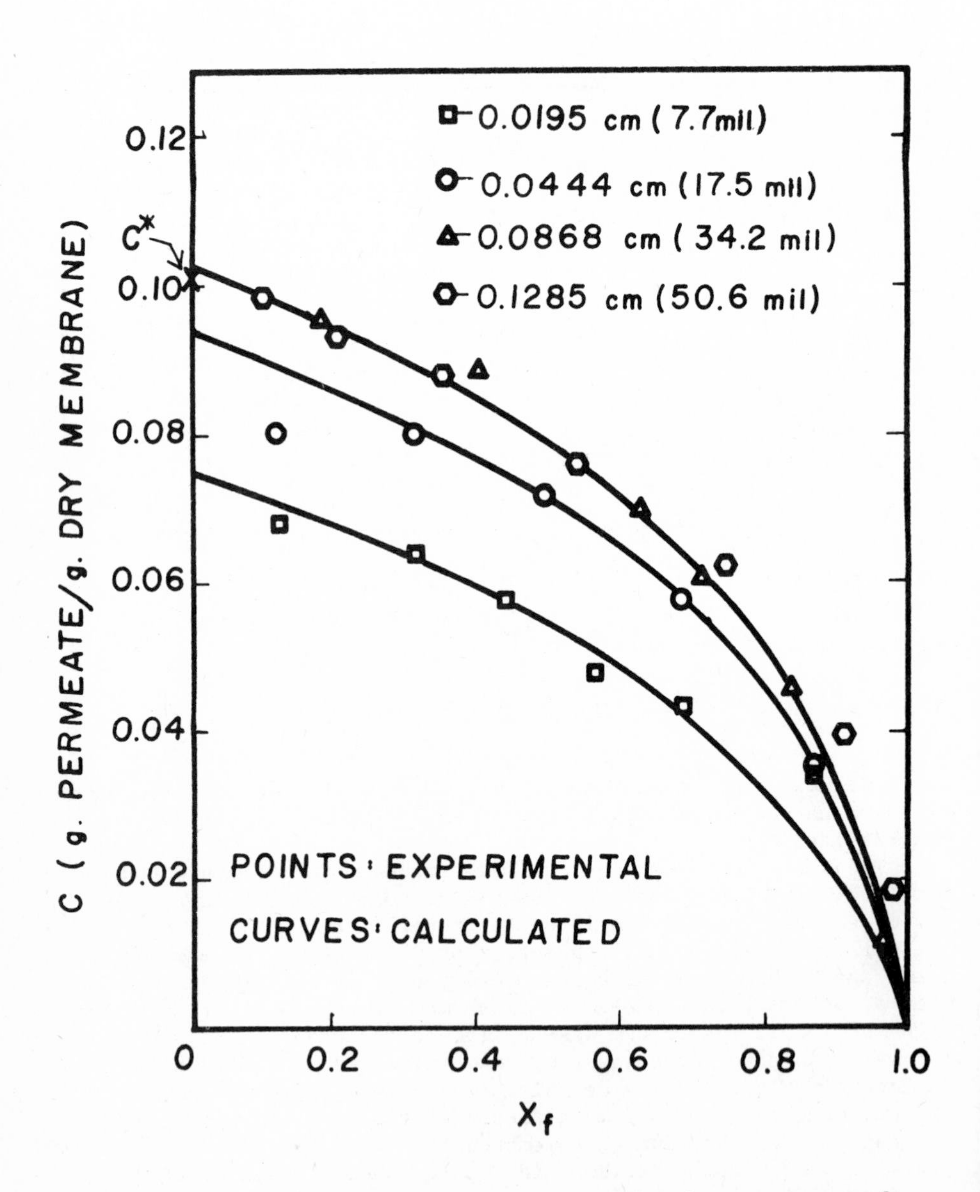

Figure 7. Concentration profile of nylone 6 - water at 35°C for variation of total membrane thickness.

there are no boundary layers to speak of. Therefore, the total permeability is mainly due to the diffusion process, and the surface concentration is virtually the equilibrium concentration, C^*. However, as the membrane becomes thinner, the diffusional resistance decreases while the interfacial resistance remains unchanged. Consequently, the influence of the interfacial resistance to the total permeability becomes greater. Thus, the surface concentration cannot reach the equilibrium value. The important parameter is the relative magnitude of the respective resistances and not their absolute values.

Some anomalous temperature effects were also discussed in the above work (28). Speculation is that the relative magnitudes of interfacial resistance and diffusional resistance may change as temperature varies.

V. CONCLUSIONS

Except in a few simple cases, permeation of matter through a membrane is not completely understood. Numerous models and mechanisms of permeation have been proposed and utilized with varying degrees of success. Among many definitions of permeabilities appearing in the literature, some fundamental ones were selected and reviewed.

Permeabilities that are defined on the basis of the mechanisms of permeation (like the surface permeability, Q_s, in microporous media) depend to a great extent on the interpretation of the permeation process. Usually, this kind of permeability is contingent upon what goes on inside the membrane.

The overall permeability tells what happened actually under a given set of experimental conditions. However, it should be remembered that this kind of permeability is dependent upon the outside conditions of the membrane (like the boundary layer resistances in dissolved oxygen permeation).

Thus, it seems appropriate to say that permeability is a phenomenological coefficient. This implies that the value of permeability depends upon the particular experimental environment chosen. In general, permeability is not a property of a membrane, while diffusivity is such a property.

VI. APPENDIX

SAMPLE CALCULATION

From Huckins (16):

Q_t = 5.92 x 10^{-6} (std. cc) (cm)/(sec)(cm^2)(cmHg) for N_2 at 290°K.

M = 28

T = 290°K

A. Calculation of Q_g and Q_s:

Using the value $Q_g \sqrt{MT} = 4.02 \times 10^{-4}$ (see [10])

$$Q_g = \frac{4.02 \times 10^{-4}}{\sqrt{(28)\ (290)}} = 4.46 \times 10^{-6} \text{ (std.cc)(cm)/(sec)(cm}^2\text{)cmHg)}$$

from Equation [42]:

$$Q_s = Q_t - Q_g = (5.92 - 4.46) \times 10^{-6} = 1.46 \times 10^{-6}$$
(std.cc)(cm)/(sec)(cm^2)(cmHg)

B. Calculation of D_g and D_s:

Using Equation [41]:

$$D_g = Q_g \left(\frac{zRT}{\epsilon}\right) = (4.46 \times 10^{-6}) \frac{(1)\ (0.278)\ (290)}{(0.28)}$$

$$D_g = 128 \times 10^{-5} \frac{cm^2}{sec.}$$

D_s' is calculated from Equation [47] and the slope of the isotherm as given by Barrer (1) is

$$\alpha = 0.8844 \times 10^{-2} \text{ cc/(gm)(cmHg)}.$$

$$D_s' = Q_s \left(\frac{1}{\alpha\rho}\right) = (1.46 \times 10^{-6}) \frac{1}{(0.8844 \times 10^{-2})\ (1.5)}$$

$$D_s' = 11.0 \times 10^{-5} \text{ cm}^2\text{/sec}$$

C. Calculation of D_t

Using Equation [51]:

$$D_t = Q_t \frac{1}{\frac{\epsilon}{zRT} + \alpha\rho}$$

$$D_t = (5.92 \times 10^{-6}) \frac{1}{\frac{0.28}{(1)\ (0.278)\ (290)} + (0.8844 \times 10^{-2})\ (1.5)}$$

$$D_t = 35.4 \times 10^{-5} \text{ cm}^2\text{/sec.}$$

VII. NOMENCLATURE

A Cross sectional area of the membrane, cm^2.

A' Total cross sectional area of pores for the porous membrane, cm^2.

C Concentration of permeate in the membrane.

C_1 Concentration of permeate at the upstream side surface.

C_2 Concentration of permeate at the downstream side surface.

C_g Concentration of permeate in gas phase based on unit volume of porous membrane, std. cc/cc.

C'_g Concentration of permeate in gas phase based on unit volume of pore space, std. cc/cc.

C_s Concentration of adsorbed phase based on unit volume of porous membrane, std. cc/cc.

C'_s Surface concentration of adsorbed phase based on unit area of adsorbent surface, std. cc/cm^2.

C_t Total concentration of permeate based on unit volume of porous membrane, std. cc/cc.

C^* Equilibrium concentration in the membrane with the bulk fluid.

D Diffusivity in the membrane, cm^2/sec.

$\overline{D}$ Average diffusivity in the membrane, cm^2/sec.

D_g Effective gas diffusivity, cm^2/sec.

D_k Knudsen Diffusivity, cm^2/sec.

D_s Surface diffusivity, cm^2/sec.

D_t Total effective diffusivity, cm^2/sec.

E_Q Activation energy

F Total flow rate of permeate at steady-state, std. cc/sec.

F_g Flow rate of gas phase at steady state, std. cc/sec.

F_s Surface flow rate at steady state, std. cc/sec.

G_1, G_2, G_3	Geometrical factor of porous membrane.
K_g	Gas phase permeability, cm^2/sec.
K_s	Surface permeability, cm^2/sec.
K_t	Total permeability, cm^2/sec.
L	Thickness of membrane, cm.
M	Molecular weight.
P	Pressure, cm Hg.
P_1	Upstream pressure, cm Hg.
P_2	Downstream pressure, cm Hg.
Q	Permeability, (std.cc)(cm)/(cm^2)(sec)(cmHg).
Q_o	Constant defined by Equation (64).
Q_2, Q_3, Q_4 Q_5, Q_6	Permeability
Q_∞	Permeability for a membrane of infinite thickness, (std. cc)(cm)/(cm^2)(sec)(cm Hg).
Q_g	Gas phase permeability, (std.cc)(cm)/(cm^2)(sec)(cmHg).
Q_s	Surface permeability, (std.cc)(cm)/(cm^2)(sec)(cmHg).
Q_t	Total permeability, (std.cc)(cm)/(cm^2)(sec)(cmHg).
R	Gas constant
r	Capillary radius, cm.
r_1, r_2	Resistance of boundary layer, (cm^2)(sec)(cmHg)/std.cc.
S	Solubility of permeate in the membrane, std.cc/(cc)(cmHg).
S_v	Surface area per unit volume of porous membrane, cm^2/cc.
T	Temperature, °K.
W	Curcumference of pores, cm.
x	Coordinate in the direction of permeation, cm.

x_f Fraction of membrane thickness, x/L.

y Concentration of adsorbed species per unit mass of porous membrane, std. cc/gm.

z Compressibility.

α Slope of adsorption isotherm, std. cc/(gm)(cmHg).

Γ_1 Concentration or partial pressure of permeate upstream from membrane.

Γ_2 Concentration or partial pressure of permeate downstream from membrane.

Γ_1^*, Γ_2^* Concentrations or partial pressures in equilibrium with C_1, C_2 respectively.

ε Porosity of porous membrane: cc. of porous volume/cc of total volume of membrane.

μ Viscosity of permeate.

ρ Apparent density of porous membrane, gm/cc.

Where units are not given, it is understood that any set of consistent units is to be used. Thus, for example, no units are shown for the concentration of permeate in the membrane, $\underline{C}$, because it could be expressed as grams permeate per unit volume of membranes, or gram moles per unit volume, etc.

VIII. REFERENCES

1. R. M. Barrer, Proc. Brit. Ceramic Soc., 5, 21 (1965).
2. P. C. Carman, "Flow of Gases Through Porous Media", Academic Press, New York, P. 1 (1956).
3. R. E. Collins, "Flow of Fluids Through Porous Materials", Reinhold Publishing Corp., New York, p. 10 (1961).
4. R. M. Barrer, and J. A. Barrie, Proc. Roy. Soc. A., 213, 250 (1952).
5. R. D. Present, "Kinetic Theory of Gases", McGraw-Hill, New York, p. 60 (1958).
6. P. C. Carman, "Flow of Gases Through Porous Media", Academic Press, New York, p. 79 (1956).
7. P. C. Carman, and P. Malherbe, Proc. Roy. Soc. A., 203, 165 (1950).
8. P. C. Carman, and F. A. Raal, Proc. Roy. Soc. A., 209, 38 (1951).
9. R. A. W. Haul, Angew. Chem. 62, 10 (1950).
10. S.-T. Hwang, and K. Kammermeyer, Can. J. Chem. Eng., 44, 82 (1966).
11. S.-T. Hwang, and K. Kammermeyer, Separation Sci., 1(5), 629 (1966).
12. S.-T. Hwang, and K. Kammermeyer, Separation Sci., 2(4), 555 (1967).
13. S.-T. Hwang, A.I.Ch.E. J., 14(5), 809 (1968).
14. M. Knudsen, Ann. Physik., 28, 75 (1909).
15. E. E. Peterson, "Chemical Reaction Analysis", Prentice-Hall, Inc., Englewood Cliffs, N. J., p. 115 (1965).
16. H. E. Huckins, and K. Kammermeyer, Chem. Engr. Progr., 49, 180 (1953).
17. R. M. Barrer, A.I.Ch.E. - I. Chem. Eng. Symp. Ser. No. 1, 112 (1965).
18. J. B. Graham, "Adsorption of Gases and Vapors on Porous Glass", M. S. Thesis, University of Iowa, Iowa City, Iowa (1958).
19. H. Yasuda, and W. Stone, Jr., J. Polymer Sci. Al., 4, 1314 (1966).
20. H. Yasuda, J. Polymer Sci. al., 5, 2952 (1967).
21. W. L. Robb, Research Lab Bulletin, General Electric Co., Winter 1964-65, p. 7-8.
22. S.-T. Hwang, T. E. S. Tang, and K. Kammermeyer, J. Macromolecular Sci.-Phys. In print.
23. R. M. Barrer, and H. T. Chio, J. Polymer Sci. Part C, 111 (1965).
24. C. R. Wilke, Chem. Eng. Progress, 45, 218 (1949).
25. N. N. Li and R. B. Long, A.I.Ch.E. Journal, 15(1), 73 (1969).
26. N. N. Li, R. B. Long, and E. J. Henley, I & EC, 57, 18 (1965).
27. R. B. Long, I & EC Fundamentals 4(4), 445 (1965).

28. S. N. Kim and K. Kammermeyer, preprint 29d, A.I.Ch.E. 67th National Meeting, Atlanta, Georgia, February 15-18, 1970.

CONTINUOUS COLUMN CRYSTALLIZATION

W. D. Betts
Coal Tar Research Association,
Cleckheaton, Yorks, England.

G. W. Girling
Benzole Producers Ltd.,
Watford, Herts, England.

I. INTRODUCTION

Fractional crystallization of materials either from a melt or from solution has both technical and economic advantages over other means of separation and purification. It may, for example, prove effective with mixtures whose vapor-liquid equilibrium properties are such as to make it difficult to obtain separation by fractional distillation. Compared with the latter process, crystallization uses less energy, since latent heats of fusion are usually 1/3 to 1/2 of latent heats of vaporization, and it requires smaller equipment, as vaporization usually involves a 300 times increase in volume.

This chapter is concerned with column crystallizers in which separation is enhanced by counter-current contacting of crystals and melt under adiabatic conditions. Such columns effect either the recovery of the higher melting point component from a binary system forming solid solutions in a purity consistent with several theoretical stages of purification, or the separation of a major component from admixture with one or more eutectogenic components at a purity consistent with a high degree of detachment of mother-liquor from the crystal phase. Although the use of columns has been long established for processes involving counter-current mass-transfer between two liquids or between liquid and gas, as in distillation, the development of analogous processes for liquid-solid systems is less well advanced. After briefly reviewing the various systems that have been proposed for continuous fractional crystallization detailed consideration will be limited to those in which the counter-current movement is induced by the action of a rotor, and in particular to designs based on the prototype devised by Schildknecht, using a rotating helix or Archimedean screw.

II. PRINCIPLES OF SEPARATION

The degree of separation that can be obtained by a single stage crystallization is dependent on the solid-liquid equilibrium for the mixture. The various forms of equilibrium that occur may be classified into two main types which are illustrated by the phase diagrams shown in Figure 1. The numerous variations on these basic types are described in the standard texts and have been discussed by Wolten & Wilcox (1).

Cooling of a eutectogenic mixture (Figure 1a) of composition $\underline{x}$ produces crystals of component B at temperature t_1. As the temperature is reduced, further B crystallizes and the liquid-phase composition follows the equilibrium line until the eutectic point $\underline{e}$ is reached. Crystals of composition $\underline{e}$ are then obtained until the whole system becomes solid. In principle, a mixture of this type may be separated into one pure component, and material of eutectic composition. In practice, the crystals are contaminated by adherent or occluded mother liquor and this must be removed in order to obtain the pure component. The adherent liquid may be removed by washing the crystals with a solvent or pure melt, but removal of occluded liquid may necessitate re-melting and recrystallization. Examples of mixtures forming eutectics are: benzene with many aliphatic and aromatic hydrocarbons; xylene isomers; water/alcohol; and water/salts.

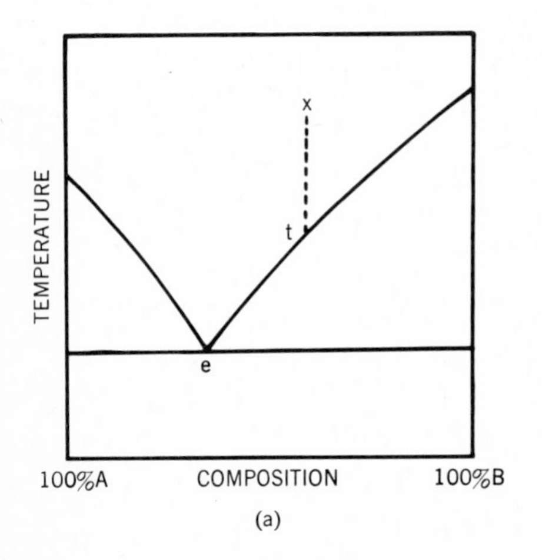

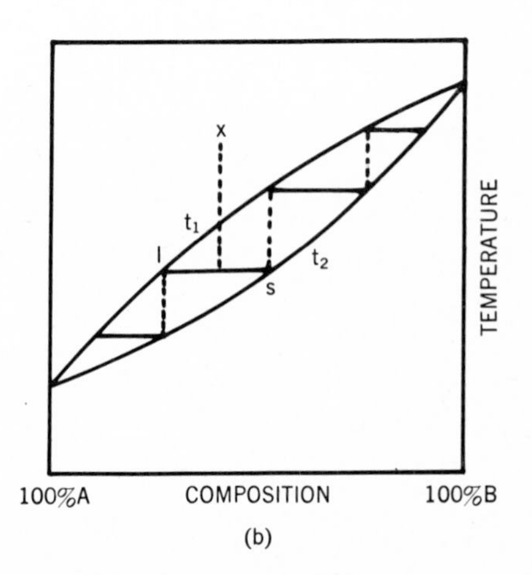

Figure 1. Phase diagrams (a) eutectic system (b) mixed-crystal system.

In a system forming a series of solid solutions (Figure 1b), cooling a mixture of composition $\underline{x}$ produces mixed crystals at temperature $\underline{t_1}$, and if cooling is continued to $\underline{t_2}$, the system at equilibrium will contain crystals of composition $\underline{s}$ and liquid of composition ℓ. In order to obtain further separation of the two components it is necessary to segregate the solid and liquid phases; the solid is then melted and partially re-frozen whilst the liquid undergoes further partial freezing. Repetition of these operations with separation of the phases at each stage, as indicated by the steps on the phase diagram, produces fractions rich in components A and B respectively. Given sufficient stages a system of the type illustrated may be separated into the two pure components, but with systems in which the freezing point curve has a maximum or minimum the ultimate separation is limited, as with eutectic systems, to one pure component and a mixture of constant freezing-point. Examples of systems forming mixed crystals are: benzene/thiophen; napthalene/thionaphthene; phenol/o-cresol.

As with other systems involving inter-phase mass-transfer, the most effective mode of operation for fractional crystallization involves counter-current contacting of the two phases, and the degree of separation can be increased by refluxing pure material to the separation system. A column crystallizer provides the means for effecting these operations with a minimum of equipment and without the need for repeated handling of the solid phase. It comprises three basic sections which may be contained in a single vessel:

(i) a cooling section to produce crystals from the liquid mixture;

(ii) a heating section to melt crystals to provide liquid for reflux; in a continuously operated system a stream of liquid pure material is also withdrawn from the melter;

(iii) an adiabatic purifying section in which a large part of the separation is effected.

The crystallizer also requires means of transporting crystals from the cooling section to the melter via the purifying section, and of promoting mass-transfer within the latter.

When treating a mixed-crystal system, an axial temperature gradient is established in the purifying section. Crystals are transported from the point of their formation to a higher temperature zone, where they then melt, with the formation of an equivalent quantity of crystals of higher melting-point i.e. richer in the higher-melting component of the mixture. With a eutectic system the gradient in the purifying section is small but the crystals pass through liquid of increasing purity so that the adherent mother liquor is replaced by pure material. With both systems, the end result is the production of fractions enriched in the lower- and higher-melting components respectively at the extremities of the column.

III. COLUMN CRYSTALLIZER DESIGNS

A. Non-Screw Systems

Numerous designs for counter-current column crystallizers have been proposed, but only a few of these have become established either as laboratory tools or in production units. The simplest approach to counter-current operation is to cause crystals to traverse a vertical column from the cold to the hot end by gravity, but many workers, including the present authors, have been unable to obtain significant separations by this means. One apparatus of this type uses a rotating perforated blade to remove crystals from the walls of a cooled section of column so that they fall through a column of liquid to a heated section at the base (2). The use of a stream of falling glass beads to assist transport has also been proposed, the beads being cooled so that a thin layer of solid material becomes frozen onto them. At the base of the column the solid is melted and part of the melt is displaced up the column to contact the material on the descending beads (3).

Combinations of gravity, entrainment and hydraulic pressure have been used to effect crystal transport in processes for water desalination (4, 5, 6) and recovery of p-xylene from mixed xylene feedstocks. In the Institut Français du Pétrole process, crystals of xylene are generated in one column by direct contact with a refrigerant, pass into the base of a conical column, and are forced to the top of this by the hydraulic pressure of the mother liquor. The crystals are removed mechanically from the top of the column for melting, and part of the melt is returned to wash the crystals in the top of the column (7, 8, 9).

Sedimentation columns employing systems of baffles and scrapers to promote inter-phase contacting have also been devised. Matz has described a column containing horizontal plates; one segment of each plate is open and the openings in successive plates are staggered. Crystals falling from one plate to the next are swept through the open segment by a rotating blade mounted just above each plate (10). In a process devised by I.C.I. a crystal slurry is fed to the top of a column containing several vertical rotating shafts fitted with horizontal blades. Crystals pass down this column, through a rising stream of liquid, into the conical base and thence to a melter. Part of the melt is returned to the column as reflux (11).

Other fully compartmented columns have been described with heat-exchange systems to effect melting and refreezing between each section of the column (12, 13). Another process uses an almost horizontal column having an outer cylindrical shell which is divided into compartments on its lower side and cooled. Crystals are removed from the cooled sections by paddles on a rotating inner cylinder, which is kept relatively warm. The crystals fall onto the inner cylinder and melt; the resulting liquid falls into the next higher compartment of the outer shell while uncrystallized material overflows the baffle into the next lower compartment (14).

The Phillips crystallization process uses a reciprocating piston or a liquid pulsing system for crystal transport and has been widely adopted for large-scale recovery of p-xylene and other materials. The development of this process, since the concept of counter-current contacting of crystals with the melt was introduced by Arnold (15), has been reviewed by McKay (16) and by Albertins *et al* (17). Crystals obtained in a scraped-surface chiller are fed to one end of a column and are forced through it by the action of a piston, or more usually by pulsing the liquid in the column, to form a relatively compact bed. The crystals are melted by a heater in the end of the column and part of the melt is withdrawn, while the remainder is displaced back through the crystal bed. The counter-current contacting of cold crystals with the reflux liquid results in partial re-freezing of the latter and simultaneous displacement of impurities. The mother-liquor is removed through filters at the inlet end of the column. Another pulsed column system has been described by I.C.I. (18), who have also developed a cyclic process using a column with a long and narrow spiral path for material movement. The mixture in the column is first partially frozen and the remaining liquid is then displaced towards one end of the column. This is warmed to melt some of the crystals, and the liquid phase is then displaced towards the other end of the column. Feed is supplied to the center of the column; pure and impure fractions respectively are withdrawn from the ends during the appropriate liquid displacement parts of the cycle (19).

B. Screw Systems

Whilst some of the devices already described utilized screw conveyors to move crystal slurry from one section of the equipment to another, e.g. from a cooler into a purifying column, the screws were not used to obtain counter-current flow. An early process in which screws were used specifically for this purpose was developed by Scott & Joscelyne (20) although the liquid phase in this process was a solvent rather than a melt. Molten naphthalene was introduced into one end of a horizontal column containing cold methanol and the crystals formed were carried to the other end of the column by a screw conveyor. The crystals were augered out of the trough by another screw and further purification was obtained by cascading methanol over the crystals. This process is not thought to have gone into commercial operation.

A vertical column system utilizing screws was devised by Frevel (21, 22). The cylindrical column was attached, at a point below its top, to a larger diameter vessel and protruded some way into the latter. The feed was supplied to the middle of the column which was cooled at the base. The crystals formed there were augured up the column and out of the liquid by two intermeshing screws, which rotated on their axes, while the latter also precessed about the axis of the column. As the crystals reached warmer zones they commenced to melt and the liquid flowed back down the screws over the ascending crystals to effect washing and/or recrystallization. Crystals reaching the top of the column were discharged over the edge into the wider column and were melted in the annular section between the two columns. Part of the melt was removed, and the remainder flowed over the top of the inner column onto the screws.

Various forms of screw were used by Austin in attempts to devise a purification system for naphthalene and other materials. A ribbon-type helical conveyor was used in the cooling section of a vertical column, and a simple bladed stirrer in the purifying section. In further experiments with a horizontal trough, the conveyor was formed by attaching to a central shaft a series of discs that had been split and twisted so as to provide a screwing action. Neither of these systems proved to be efficient (23).

C. The Schildknecht Rotary Helix System

In 1962 Schildknecht and Vetter described experiments with a small column crystallizer in which a rotating helix was used to effect transport of crystals counter-current to the melt (24). It appeared from the results described then, and in a series of subsequent publications (25-34), that the apparatus operated as a true multi-stage fractional crystallizer which was widely applicable and possibly capable of scale-up for commercial applications.

The basic principle of this apparatus is illustrated in Figure 2.

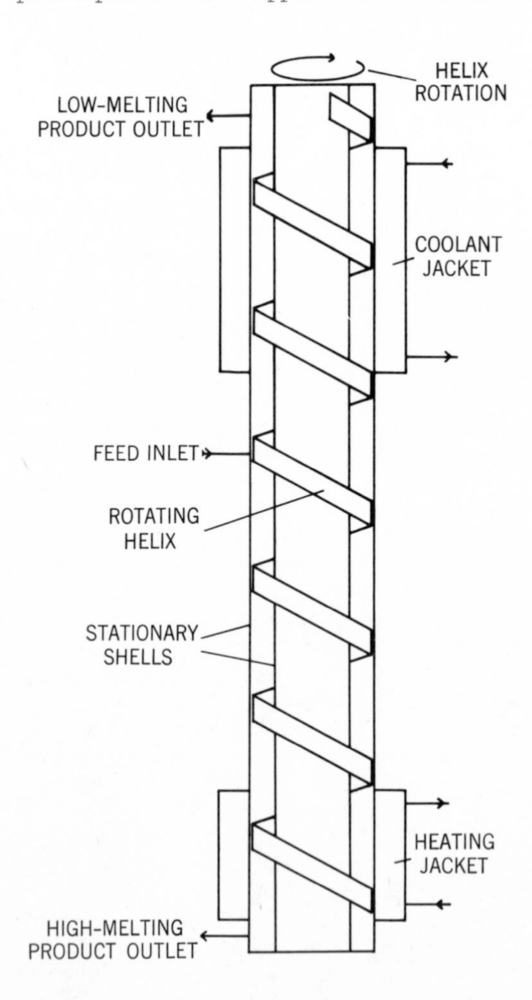

Figure 2. Principle of the Schildknecht column crystallizer.

The material to be crystallized occupies the annular space between two concentric vertical tubes. The upper end of this column is cooled and the lower end warmed, while the intermediate section is adiabatic. A close-fitting helical coil is rotated in the annulus between the two stationary tubes, the direction of rotation being such that crystals formed in the cooling section are driven through the adiabatic purification section to the warm end of the column. Sufficient heat is supplied to the latter to melt all the arriving crystals and displacement of the resulting melt by the arrival of further crystals produces a reflux stream of liquid, counter-current to the crystal movement. The unit may be operated under total reflux i.e. as a batch unit, or as shown in the diagram, feed can be supplied intermittently or continuously to the central section of the column, with the corresponding withdrawal of fractions enriched in the respective components from the top and bottom of the column. Although downward crystal transport is more usual, columns have also been operated with upward transport. The action of the rotary helix in addition to transporting crystals, is thought to promote inter-phase mass-transfer in the purification section and also provides the advantage of scraped-surface heat-exchange in the cooling section.

It is evident that the apparatus was originally conceived as a small laboratory tool for use particularly as an adjunct to zone-melting. Many of the examples quoted by Schildknecht et al refer to operation under total reflux, although in some instances semi-continuous or continuous feed and withdrawal systems were used. The publications refer to several different columns and the dimensions are not always clearly stated, but the largest column had an outer tube of 25mm i.d. and an inner tube of 10mm o.d. while the helix was o.8m long and had a 10mm pitch. A summary of some of the results obtained by Schildknecht et al is given in Table 1, and some relevant physical properties for these and other crystallizable materials is given in Table 2.

The reasons for the success of the Schildknecht type column compared with other devices undoubtedly stem from the differences in design, which include the following:-

(i) The column is flooded with liquid, and the crystal phase is dispersed, so that purification can occur over the whole length of the adiabatic section. Interphase mass-transfer is assisted by the action of the helix although this also promotes axial mixing. In columns using pulsed crystal transport systems a compact crystal bed is obtained and purification is confined to a narrow zone (16). In some screw conveyor systems the crystals are lifted out of the liquid on the screw and are washed only by a stream of melt trickling down the screw (20-22).

(ii) Unlike some other screw systems (20, 23) the column is vertical. As shown in Table 2, the crystals obtained from many materials are denser than the melt so that they tend to fall through it, and in most of the examples quoted by Schildknecht crystal transport would be assisted by gravity. However, this is not an essential feature of the process as separation of materials producing less-dense crystals has been effected with

TABLE 1 PURIFICATION OF VARIOUS MATERIALS BY

Compound	Crystallizing point °C Initial	Final Top	Bottom	Bottom product yield %
Stearyl alcohol	55.0	39.4	57.4	-
Cetyl alcohol	49.0	47.3	49.5	-
" "	>49.4	46.0	50.1	-
" "	>49.4	49.0	53.0	-
2,4 Dinitrotoluene	-	54.5	70.8	-
Aniline	-	-	-	-
Benzene ('crude')	x	-	x+1.60	30
" ('pure')	y	-	y+0.65	-
Myristic acid	53.6	53.1	54.2	-
Benzophenone	47.7	46.5	48.4	-
Azobenzene/stilbene	63-67.5	54.0	100.0	-
2-Naphthol	122.0	121.5	123.5	-
Dibenzyl (98%)	-	53.0	119	-
Phenol	-	35.5	41.5	-
Fluorene	-	103.0	116.5	-
Anthracene	-	-	-	-
Naphthalene	-	-	-	-
Stearyl alcohol	55.0	45.5	57.6	89
" "	"	46.5	57.2	67
" "	"	47.0	57.5	80
" "	"	53.6	57.2	67
" "	"	53.0	57.8	67
" "	"	52.7	56.7	80
" "	"	47.6	55.9	80
Phenanthrene	99-101	95.0	116.0	50
Diphenyl (90%)	65	59-60	68.5-69	50
"	-	63.5	68.5-69	50
"	-	58.0	68-68.5	50
"	-	61-2	68-68.5	50
"	-	62-3	67.5-68	50
Terphenyls	68.5	-	90.185	-
	-	-	195-209	-

COLUMN CRYSTALLIZATION (24-34)

Through-put g/h	Column length mm	Coil speed rev/min	Remarks	Reference
-	150	120-150	T.R.	24
-	150	80	T.R.	24
-	150	130	T.R. Recycle from previous run	24
-	150	150	T.R. " " " "	24
-	-	-	T.R.	24
-	600	-	T.R.	24
-	600	-	4 passes	27,33
30	600	-	'Free' of impurities; 2 passes	33,34
-	170	90	T.R. for 1.5h	29
-	130	-	T.R. for 1.5h	29,33
-	800	35	T.R. pulsed column	29,33
-	120	60	T.R. for 2h	26,27,33
-	50	60	T.R. for 2.5h	26,29,33
-	520	75	T.R. for 2.5h, 70% of column colourless	29,33
-	140	120	T.R. for 2.5h	29,33
-	-	-	T.R.	27,33
-	150	-	T.R. for 2h	26
1.35	-	100	S.C. } column capacity 15g	29
0.9	-	100	S.C. } column capacity 15g	29
0.8	-	100	S.C. } column capacity 15g	29
1.4	-	80	S.C. } column capacity 60g	29
1.8	-	80	S.C. } column capacity 60g	29
7.5	-	80	S.C. } column capacity 60g	29
7.5	-	35	S.C. } column capacity 60g	29
14	-	65	S.C.	27
50	800	60	S.C.	25,33
50	800	60	S.C.	25,33
50	800	60	S.C.	25,33
100	800	60	S.C.	25,33
200	800	60	S.C.	25,33
-	-	-	S.C. 1st pass	27,33
-	-	-	S.C. 2nd pass	27,33

T.R. = Total Reflux S.C. = Semi-Continuous

TABLE 2

Compound	m.p.°C	Density decrease on fusion g/cc	Latent heat of fusion cal/g	Latent heat of fusion Kcal/mole
Cetyl alcohol	49.3		33.8	8.19
Myristic acid	58.0		47.5	10.85
Benzophenone	49.0(α)		23.5	4.28
Azobenzene	68.0		~30.0	~5.5
2,4-Dinitrobenzene	70.0	0.16	26.4	4.81
Dibenzyl	52.5	0.012	31.0	5.64
Fluorene	115.0	0.096	28.9	4.80
Anthracene	216.04	0.139	38.7	6.90
Phenanthrene	99.15	0.050	25.0	4.33
Diphenyl	69.2	0.141	28.9	4.47
Diphenylene oxide	82.1	0.134	25.0	4.2
Fluorene	115.0	0.096	28.9	4.8
2-Naphthol	122.2	≠	31.1	4.48
Naphthalene	80.29	0.154	35.6	4.54
Thionaphthene	31.32			
Indene	- 1.8		19.9	2.31
Benzene	5.53	0.120	30.1	2.35
Thiophene	-38.2		14.1	1.19
Quinoline	-14.94		19.9	2.56
Isoquinoline	26.48		13.8	1.78
Phenol	40.90	0.057	30.8	2.31
o-Cresol	30.99	0.093	30.1	3.08
m-Cresol	12.22	0.114	20.0	2.16
p-Cresol	34.69	0.116	19.0	2.06
2,4-Xylenol	24.54		25.0	3.06
2,5-Xylenol	74.85	0.16	27.1	3.31

≠ 0.13 for 1-Naphthol

TO CRYSTALLIZATION

Viscosity cP	Surface tension dyne/cm	Max. crystallization velocity, mm/min. at given super-cooling
13.4(50°C)		∞
4.8(55°C)	41.8(50°C)	$\sim$60(27.7°) 600(30°)
1.21(114°C)		
0.69(222°C)	17.6(246°C)	
1.93(99°C)	37.2(100.5°C)	
1.35(75°C)	32.3(100°C)	7500(16°)
1.21(114°C)		
	36.5(130°C)	
0.90(85°C)	32.0(81°C)	∞
2.42(40°C)	42.6(35°C)	
1.05(50°C)	37.4(28.5°C)	
0.91(0.15°C)	28.8(20°C)	>2300
0.84(1°C)	33.9(20°C)	
4.81(10°C)	45.6(20°C)	
3.57(25°C)		
4.74(40°C)	39.2(23°C)	∞
4.75(35°C)	40.3(20°C)	236(30.3°)
34.6 (10°C)	37.6(20°C)	$\nless$10
8.06(35°C)	39.2(20°C)	$\nless$230
1.61(80°C)		300

downward acting spirals (35,37), and upward transport of crystals denser than the melt has also been used successfully (39-45).

(iii) The Schildknecht apparatus utilizes a detached helix rotating freely in the annular space between two stationary tubes, the helix being a close fit to both tubes. In previous designs an Archimedean screw, ribbon conveyor, or other device was used, the helical flights being attached to a central shaft and the whole assembly rotated inside an outer tube with an unspecified clearance between the rotor and the wall.

(iv) The geometric proportions of the rotary helix differ from those of the flights in other screw systems; in particular the width of the free helix is proportionately much smaller. In Schildknecht's apparatus the ratio of flight width to outer tube diameter was 0.3:1 and the ratio of flight width to pitch was 0.75:1 but in Austin's apparatus (23) the corresponding ratios were 0.47:1 and 1.25:1 respectively. Experiments with flights of different widths have shown that the apparatus becomes less effective above certain flight width:diameter ratios (35,37) while on the other hand, provided a suitably low ratio is used, separation can also be achieved with Archimedean type screws (37).

IV. DEVELOPMENT OF CONTINUOUS CRYSTALLIZERS

A. Small Columns Using a Rotary Helix

Pouyet used a continuously operated column of the Schildknecht type for the purification of aromatic amines (46). The apparatus consisted of an outer tube of 35mm i.d., an inner tube of 15 mm o.d., and a helical coil of unspecified pitch 60mm long; the ratio of flight width:diameter was 0.28:1. The results obtained with a rotation speed of 70 rev/min and a throughput of 24 c.c./h are given in Table 3. Pouyet considered the results to be inferior to those obtained by zone melting but pointed out that the column could be operated continuously.

Powers and a series of co-workers have studied the performance of a Schildknecht column operating both as a batch unit (39-43), and as a continuous system (43-45), by examining the removal of cyclohexane from benzene. The column was 0.7m long, with an outer tube of 32mm diameter and an 11mm o.d. inner tube. The helix, which had a lenticular cross-section and a pitch of 10mm, was rotated at a speed of 59 rev/min and, as in some other experiments (33), was also subjected to axial oscillations at a frequency of 290 cycles/min. The helix transported crystals upwards, the cooling section being at the base of the column and the melter at the top. The feed material contained several impurities, including up to 28000 ppm wt of cyclohexane, and very substantial removal of the latter compound was effected, e.g. the cyclohexane content of a feed containing 1500 ppm wt was reduced to only 1 ppm with a throughput of 210 g/h, although in this experiment the yield of pure product was only 9% of the feed.

TABLE 3 FRACTIONAL CRYSTALLIZATION OF AMINES (46)

Amine	Crystallizing Point °C		
	Feed	Product from one pass	Product from three passes
Aniline	-6.10 -6.4	-6.00 -6.10	-6.00
o-Toluidine	-23.80 -24.70	-23.70 -23.85	-23.70
m-Toluidine	-30.10 -33.10	-30.05 -30.20	-30.05
2,3-Dimethylaniline	+3.40 -1.20	+ 3.80 + 3.80	+ 3.80
2,4-Dimethylaniline	-15.10 -15.70	-14.90 -15.10	-14.90
2,5-Dimethylaniline	15.30 13.20	15.70 15.70	15.70
2,6-Dimethylaniline	10.80 8.70	11.25 11.10	11.25

The investigations undertaken by the present authors had two main objectives. The first was to examine the applicability of the spiral crystallizer to the separation or purification of materials of industrial importance, particularly in the coal tar and benzole industries respectively, and to assess the feasibility of scale-up for commercial applications. The second was to optimize the design, and by establishing the effects of design on performance to provide a theoretical background for the process. Both the Coal Tar Research Association (C.T.R.A.) and Benzole Producers Ltd. constructed Schildknecht type columns with a diameter of 25mm and, in the light of the successful results obtained, columns of 50mm diameter were built.

The general arrangement of the C.T.R.A. (35) columns is shown in Figure 3. The 25mm column consisted of a glass outer tube

0.72m long in conjunction with alternative steel inner tubes of 18.5, 15.0, and 11.8mm diameter respectively, and helical coils of various cross-section shapes and pitches were used.

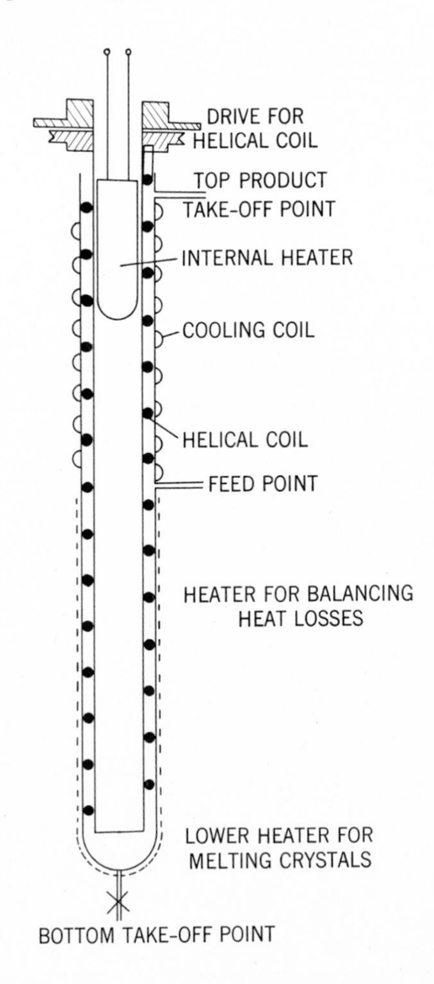

Figure 3. Basic arrangement of the C.T.R.A. continuous column crystallizer.

The 50mm column was constructed similarly and was virtually geometrically proportional to the smaller column, all dimensions having been scaled-up by a factor of 2. Some examples of the feedstocks, operating conditions, and the purification obtained, are given in Tables 4 and 5, while in Figure 4 the crystallizing points of the higher melting products have been plotted against crystal flux, i.e. $g/h/cm^2$ of annular area. The effects of reflux ratio on the purification of crude benzole containing thiophen, and

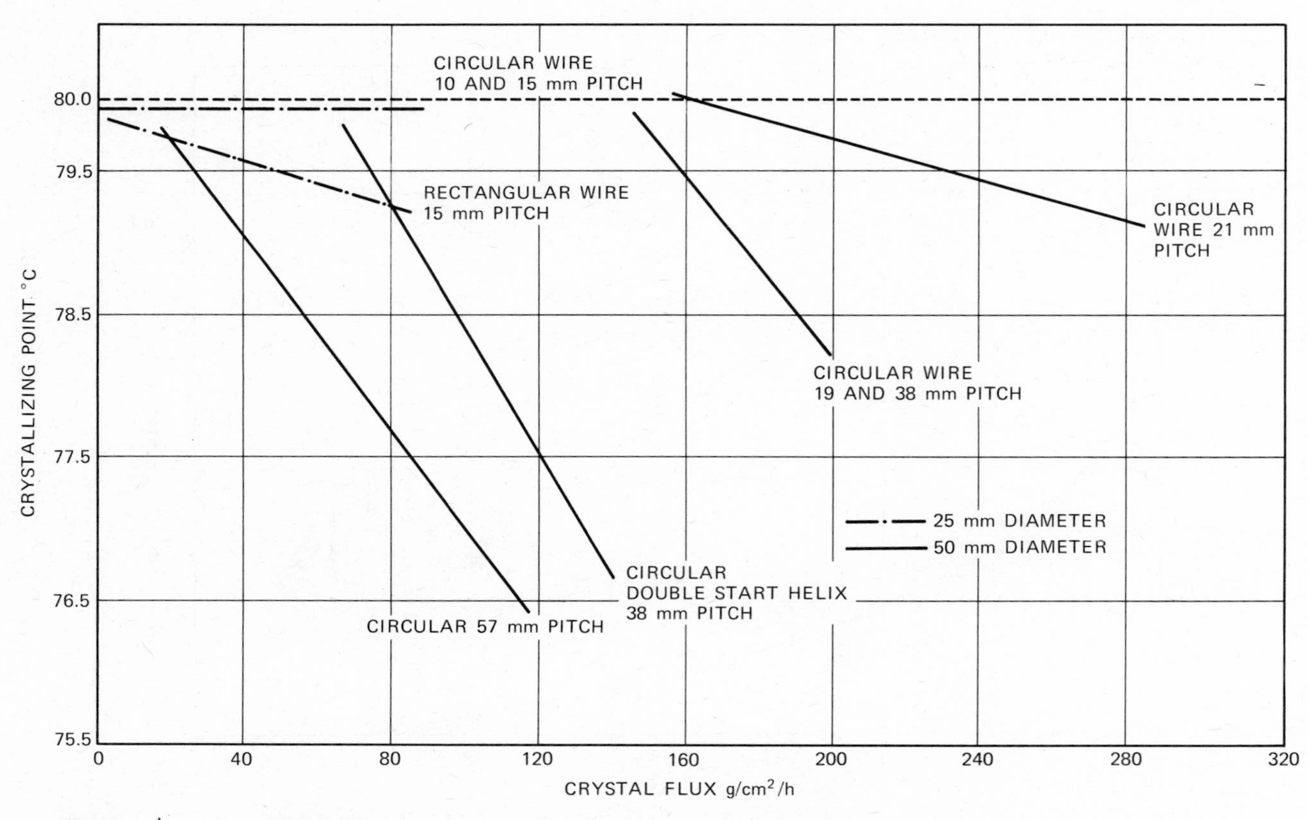

Figure 4. Treatment of napthalene Oils in 25 and 50mm diameter crystallizers.

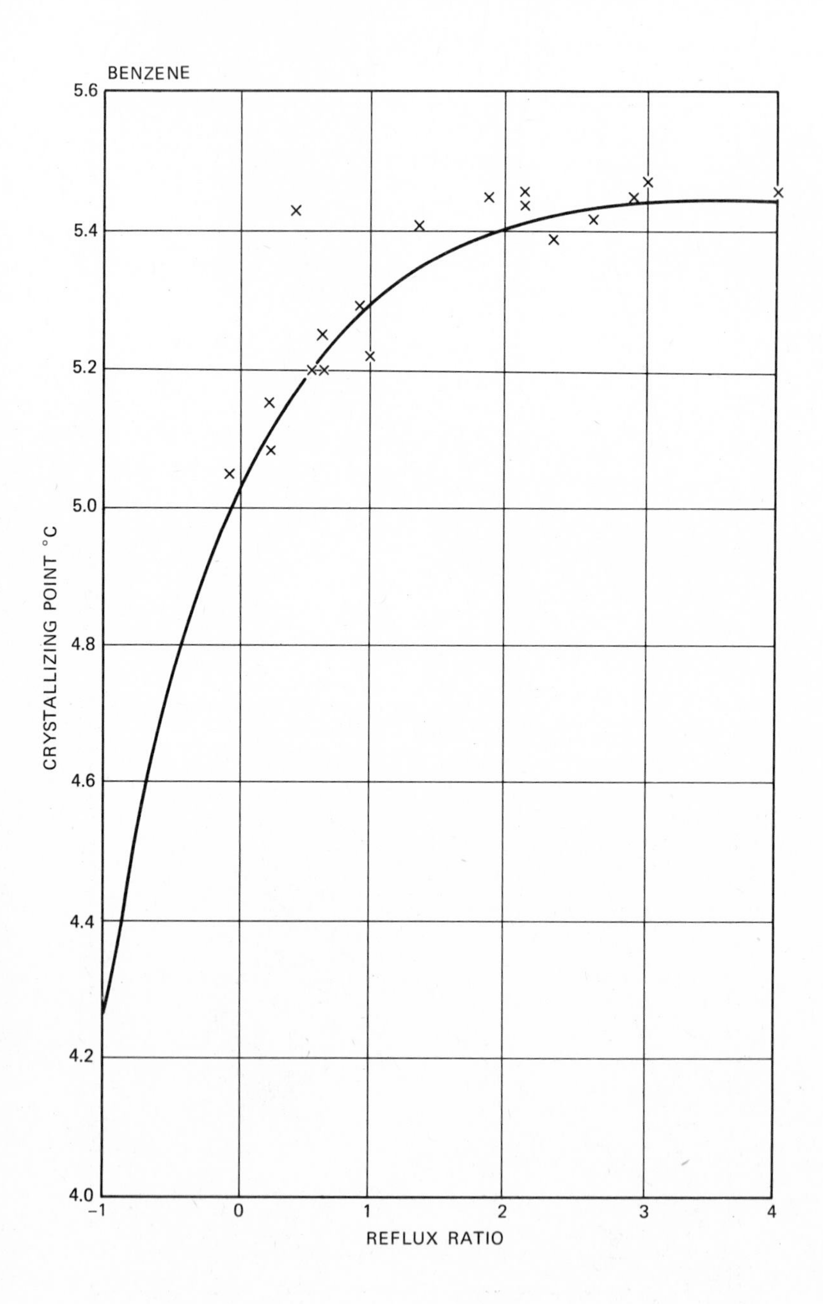
BENZENE
5.6
5.4
5.2
5.0
4.8
4.6
4.4
4.2
4.0
CRYSTALLIZING POINT °C
–1
0
1
2
3
4
REFLUX RATIO

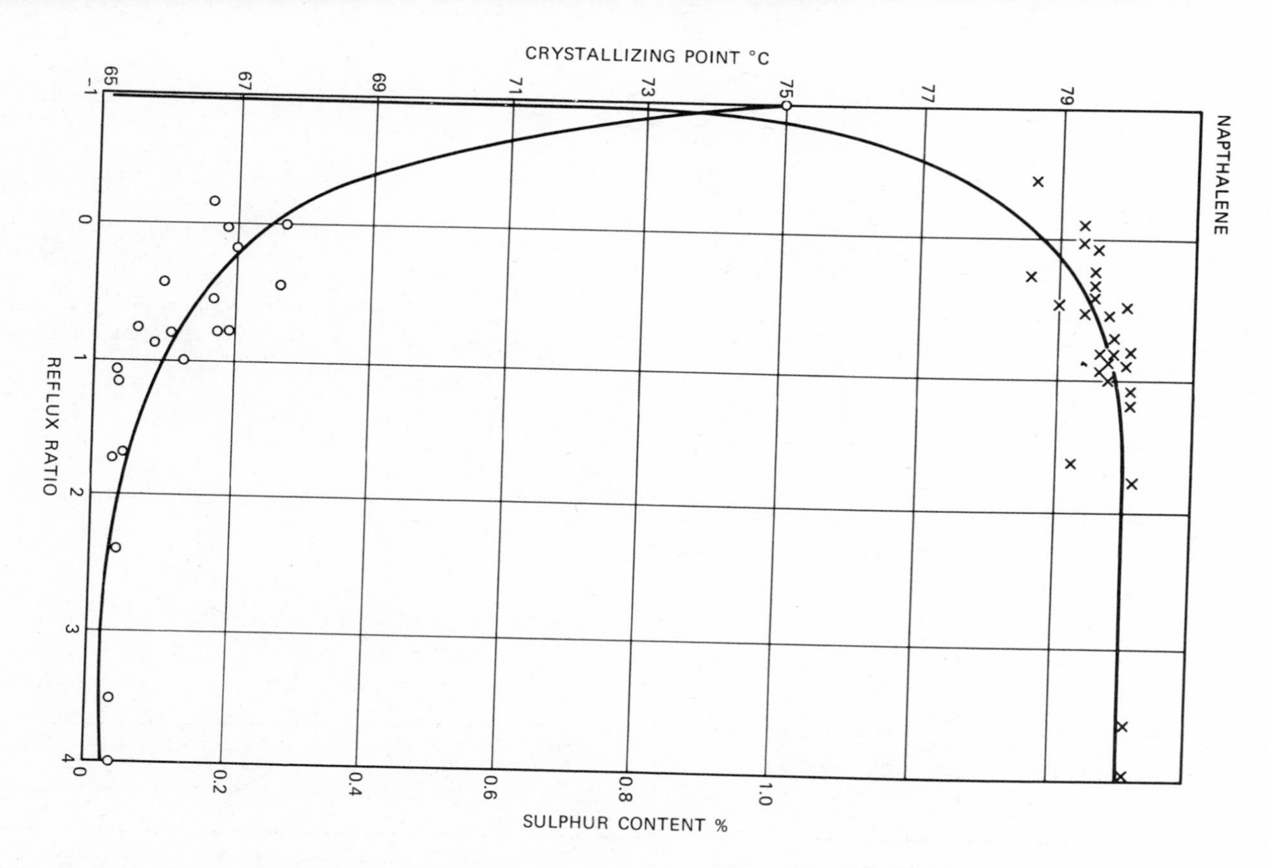

Figure 5. Effect of reflux on purification.

TABLE 4 RESULTS OBTAINED ON 25mm DIAMETER COLUMN CRYSTALLIZER (35)

Feedstock	Helix*		Feed	Top product	Base	Product	No. of
	Pitch mm	Speed, rev/min	Rate g/h	Characteristics	Recovery, % Wt	Characteristics	theoretical stages
Reagent grade benzene + thiophen c.p. 3.1°C, 1.88% S	9.7	100	38	c.p. 1.58°C	42.0	c.p. 5.12°C	5.1
Crude coke-oven benzole 62% B, 13% T, c.p.-40°C, 1.13% S	9.7	28	64	1.23% S	45.5	c.p. 2.00°C 4% T, 0.41%S	
Phthalic grade naphthalene c.p. 78.7°C, 0.66% S	15	100	98	c.p. 77.15°C 0.52% S	59.8	c.p. 79.9°C 0.09% S	
Firelighter crude naphthalene c.p. 77.4°C, 0.44% S	15	312	93	c.p. 73.9°C 0.80% S	45.7	c.p. 80.1°C 0.05% S	
40's Anthracene cake 43% A, 11% P, 17% Carb c.p. 180°C	15	70	55	29% A, 20% P Trace Carb	50.0	48% A, 44% Carb Trace P	
Anthracene paste 19% A, 39% P, 17% Carb c.p. 191°C	15	12	24	1-2% A, 52% P	26.0	38.4% A, 18-22% P	
50.3% Fluorene, 49.7% DO c.p. 94.5°C	10	100	412	c.p. 88.1°C	43.1	c.p. 101.6°C 38.0% DO	4.1
Crystal phenol, c.p. 39.8°C	15	56	373	0.084% S	60.1	c.p. 40.59°C 0.008% S	

(contd.)

TABLE 4 (continued)

Crystal phenol c.p. 39.08°C + 8% water	15	70	89	37.1	75.9	c.p. 40.74°C	
80% Phenol, 20% o-cresol	10	106	73.5	65.5% Phenol 34.5% o-cresol	56.0	90.8% Phenol 92% o-cresol	6.3
90% Phthalic anhydride c.p.128°C, 10% tetralin	15	70	167		82.9	c.p. 130.6°C	
80% Isoquinoline, 20% quinoline c.p. 18°C	15	46	60	60.5% IQ	33.5	85.3% IQ	
Commercial p-cineole c.p. 1°C	15	70	170		65.0	c.p. 1.36°C *ca.* 99.7%	
Bottom product from above	15	70	135		65.0	c.p. 1.39°C *ca.* 99.8%	

Footnotes: S = sulphur; B = benzene; T = toluene; A = anthracene; P = phenanthrene; Carb = carbazole; DO = diphenylene oxide; IQ = isoquinoline; c.p. = crystallizing point; x 3.2mm diameter, circular

TABLE 5 RESULTS OBTAINED ON 50 mm DIAMETER COLUMN CRYSTALLIZER (35)

Feedstock	Centre Tube mm	Helix			Feed Rate g/h	Top Product Characteristics	Recovery % Wt	Base Product Characteristics	Reflux Ratio
		Cross-section + width mm	Pitch mm	Speed rev/min					
Reagent grade benzene + crude benzole c.p. 4.25°C, 0.044% S	36.8	Sq6.4	19	70	897		95.3	c.p. 5.45°C 0.01% S	1.8
Unwashed coke-oven benzole c.p. 4.23°C, 0.6% S	31.2	R.8.4	19	70	816-1077	2.55% S	79-86	c.p. 5.30°C 0.18% S	
Naphthalene oil (coke-oven) c.p. 65.11°C, 0.98% S	36.8	C.6.4	21	70	up to 1500	c.p. 28.5-38.5°C	ca.50	c.p. 80, 10°C 0.05% S	0.8-4.0
Napthalene oil (coke-oven) c.p. 69.1°C, 0.98% S	36.8	C.6.4	21	70	1500	c.p. 44.7°C	51.5	c.p. 80.10°C 0.04% S	1.2
Phthalic grade naphthalene c.p. 77.93°C, 0.54% S	36.8	C.6.4	25.4	130	2004	c.p. 67.25°C	96.8	c.p. 80.05°C 0.086% S	
Anthracene paste 39% P, 17% Carb	36.8	C.6.4	21	70	449	c.p. 56.5°C	22.1	47% A, 8% P, c.p. 185°C	
Acenaphthene + 22% Fluorene	36.8	C.6.4	21	70	458	c.p. 65.6°C	48.3	c.p. 93.1°C 96% Acenaphthene	
95% p-Cresol,0.2% Ph,0.9%o-C 3.3% m-C, others 0.4%	36.8	C.6.4	21	70	1510	1.1% Ph,2.3% o-C, 17.9% m-C	50.3	0.3% m-C	

(Contd....)

TABLE 5 (Continued)

90% γ-Picoline, 10% β-Picoline	36.8	C.6.4	21	70	653		68.9	0.5% β-Picoline
92.1% 2,6-Lutidine, 7.9% β-Picoline	36.8	C.6.4	21	70	273	27.2% β-Pic	78.0	0.5% β-Picoline
96% -Picoline, 3.8% 2,6-Lutidine	36.8	C.6.4	21	70	521	15.9% 2,6-L 1.4% EPy, 2.1% β-Pic	77.5	No trace of impurities by GLC
95% p-Nitrotoluene, 5% o-Nitrotoluene	36.8	C.6.4	21	70	915	43.2% O-, 1.5% m-isomer	91.4	Trace o-nitrotoluene
23% o-xylene, 17% p-xylene, c.p. 3°C	36.8	C.6.4	21	70	383		48.3	c.p. 13.08°C, 99.5% p-xylene
32% m-Xylene, 68% p-xylene c.p. -2°C	36.8	C.6.4	21	70	246		48.8	c.p. 13.0°C 99.5% p-xylene
10% Aqueous sodium chloride	36.8	C.6.4	21	70	549	22.1% NaCl	52.7	0.75% NaCl
7% Aqueous ammonium chloride	36.8	C.6.4	21	70	734	18.2% NH_4Cl	65.9	1.56% NH_4Cl
5% Aqueous ethanol	36.8	C.6.4	21	70	569	14.7% Ethanol	73.1	1.7% Ethanol
Beer, 3.30% Ethanol	36.8	C.6.4	21	70	622	7.68% Ethanol	62.2	0.2% Ethanol, 0.34% solids

Footnotes

S = sulphur	o-C = o-Cresol	EPy = Ethyl pyridine
A = anthracene	m-C = m-Cresol	C = circular
P = phenanthrene	p-C = p-Cresol	Sq = square
Carb = Carbazole	2,6L = 2,6-Lutidine	R = rectangular
Ph = Phenol	β Pic = β-Picoline	c.p. = crystallizing point

naphthalene are shown in Figure 5, and indicate that a reflux ratio of about 1 is adequate for both systems.

The results illustrate the possibility of continuously purifying various materials including both eutectogenic and solid-solution systems. The degree of separation achieved in some of the latter systems can be used to calculate the number of theoretical or equilibrium stages from a knowledge of the appropriate phase diagram, using a graphical method analogous to the McCabe-Thiele method for distillation, and the results obtained for several such systems are included in Table 4. The ease of separation of a particular system is dependent on the distance between the liquidus and solidus on the appropriate phase diagram so that in a given column a greater degree of separation is achieved, e.g. between naphthalene and thionaphthene than between benzene and thiophen. The separation of eutectogenic components appears to be unselective; the degree of removal of cyclohexane, methylcyclohexane and toluene from benzene was the same for each component whether present singly or together and this represents an additional advantage for crystallization over distillation methods. It is clear that scaling-up the apparatus from 25 to 50mm did not impair the separation power, but comparison of the results did not show how throughput was related to size.

The 25mm diameter column constructed by Benzole Producers Ltd. (37) is shown in Figure 6. It had an overall length of about 0.4m and the outer shell was in three sections: an upper metal section fitted with an annular jacket for coolant, a section of glass tube, and a metal base section incorporating an electrical heater. The central stationary tube was fixed at the base but free at the top. The helix was wound from circular section wire of 3mm diameter on a 12mm pitch and was driven by an electric motor through both fixed and variable gears. Cooling was obtained by circulating refrigerated brine through the annular jacket. The feed was supplied by a metering pump and the product offtake stream was controlled by a valve, while the impurities overflowed through the offtake pipe at the top of the column.

Experiments were made with benzene fractions containing 1 to 10% of impurities, mainly C_6 and C_7 hydrocarbons. The feed had been treated previously for removal of thiophen by chemical means since it was thought that any physical method would be unlikely to achieve the 20,000 to 1 reduction in concentration of this compound that is necessitated by current commercial specifications. The helix was rotated at speeds between 100 and 200 rev/min and while the speed did not greatly affect the degree of separation obtained, the higher speeds prevented crystal plugging at the higher throughput rates. Once adequate purification of a particular feed had been obtained the throughput was increased until deterioration in product quality could not be counteracted by increased coolant flow. The bottom product offtake rate was controlled close to the maximum value for the particular feedstock, as calculated from a mass-balance on the basis that the mother liquor should have a minimum c.p. of -15°C, which is the limiting value for this to be saleable as motor spirit.

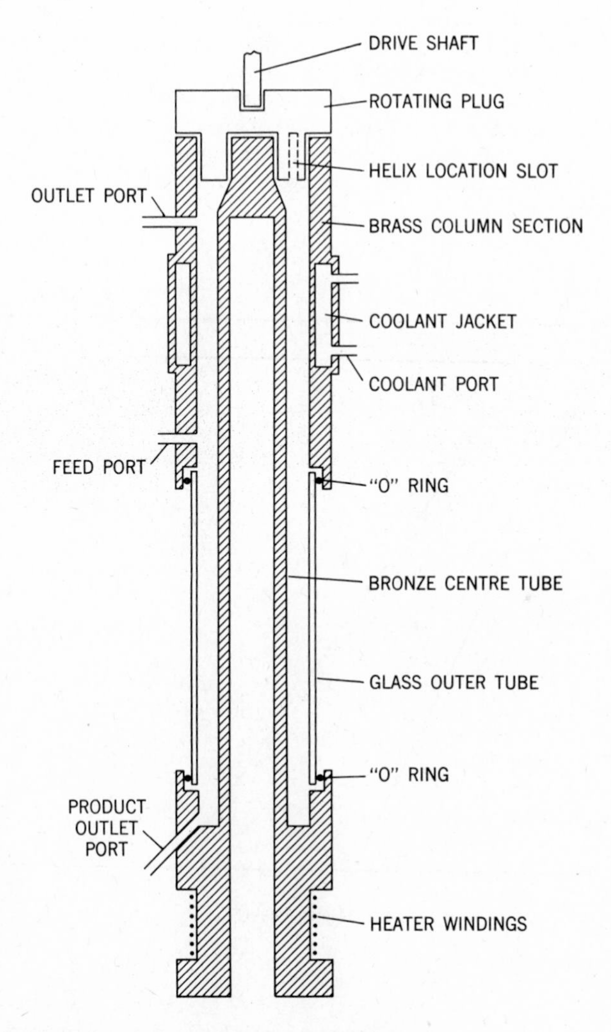

Figure 6. 25mm diameter column for Benzene.

With a feedstock having a c.p. of 5.0°C, i.e. containing 1% impurities, the maximum yield of benzene with a minimum c.p. of 5.50°C is about 97% but to give a slight operational margin a nominal 95% yield was taken. Figure 7 shows plots of product quality against throughput of various feeds. A slight reduction in yield allowed the throughput to be increased from 1.25 to 2 l/h.

The point at which the sharp fall in product quality occurred was found to correspond to that at which the crystal transport rate approached the product offtake rate, i.e. the flow of reflux liquid approached zero. Estimates of the crystal rate and hence the amount of reflux may be obtained both from the heat extracted by the coolant and from the heat supplied to the melter. A graph of product purity against reflux rate showed that with the 1% impurity feed, commercial quality product (99.9% purity) could be obtained when about 15% of the total crystal flow was returned as reflux, while material containing only 10 ppm total impurities was obtained with 40% reflux.

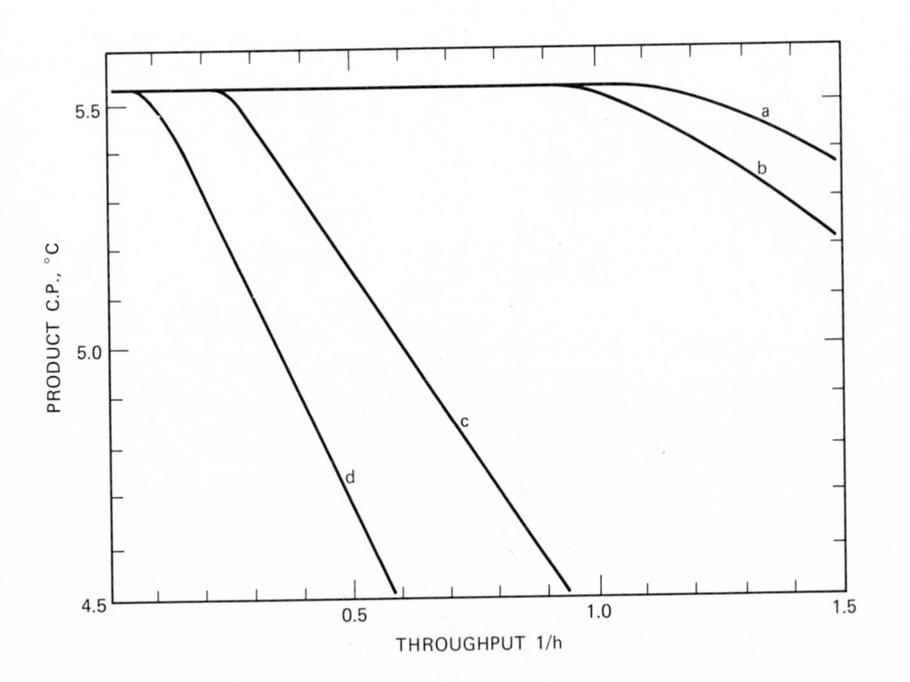

Figure 7. Variation of product quality with benzene throughput.

The 50mm column is shown in Figure 8 and had the same basic layout as the 25mm column and utilized similar equipment for heating, cooling, rotor drive, and material transfer systems. The outer column, however, was constructed entirely of metal and the purification section was provided with several ports which could be used for sampling the column contents, or as alternative outlets for the bottom product so that the effective length of the column could be varied. The throughput was expected to increase with the square of column diameter and to obtain a corresponding increase in heat-transfer area in the cooling section the length had to be doubled compared with the 25mm column.

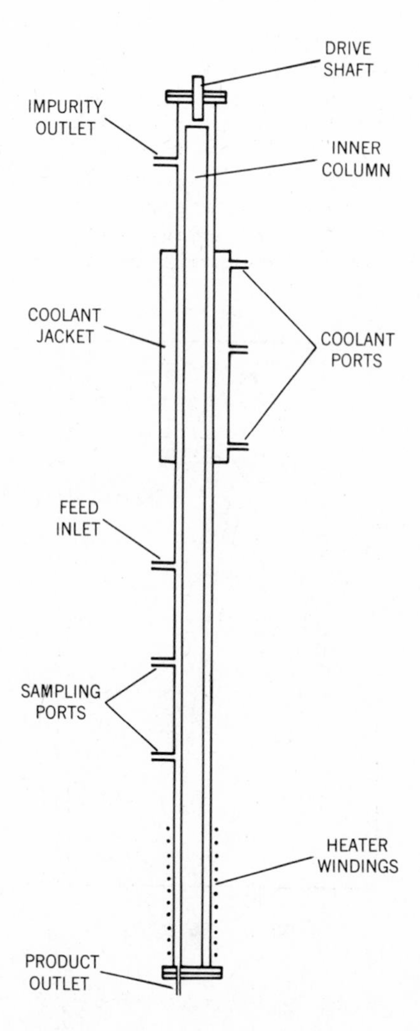

Figure 8. 50mm column for Benzene. (Helix not shown; not to scale).

As it was thought that the factor limiting the throughput of the 25mm column was heat-transfer rather than crystal flux, the length of the cooling jacket was increased by a further 100%, but provision was made for utilizing only part of this if required. The length of the purifying section was also increased and the column had an overall length of 1.25m. The cross-section width and the pitch of the helix were obtained initially by geometric scale-up but subsequently tests were made with several variations of cross-section shape, width, and pitch, some of which gave an improved performance.

Experiments were made with benzene feedstocks containing up to 10% impurities and the column operated successfully on all these, but the majority of experiments were made with a feed containing 1% impurity (c.p. 5.0°C). Taking as criterion the production of benzene with a minimum c.p. of 5.50°C in 95% yield, throughputs of over 5 l/h and crystal fluxes of 550 g/h/cm^2 were obtained and these compare favourably with those reported by other workers. The maximum throughput varied with the design of the helix and was very dependent on helix rotation speed as illustrated in Figure 9.

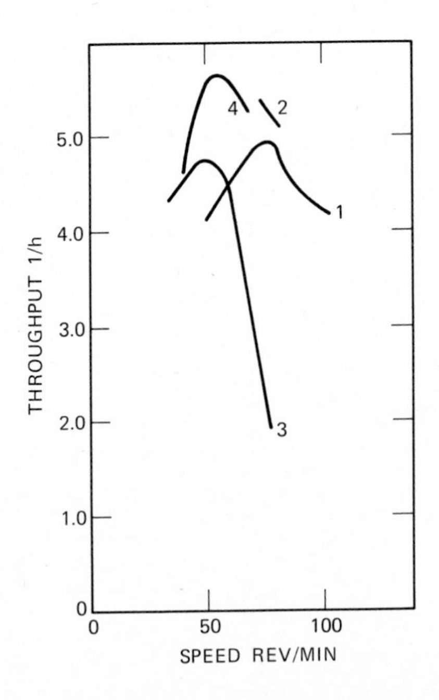

Figure 9. Effect of helix rotation speed on throughput.

Comparison of the results with those from the smaller column supports the premise that throughput is proportional to the square of column diameter and optimum speed inversely proportional to diameter. No further increase in throughput was obtained as a result of the out-of-scale heat transfer surface. It appears that throughput is dependent on both heat-transfer capacity and crystal flux, and that both these factors are dependent on the geometry and speed of the helix.

Figure 10 shows impurity concentration profiles obtained by analysing samples taken from the ports in the purifying section during operation. It appeared from these that the column could be shortened while still producing material of adequate commercial quality and this was confirmed by raising the position of the external melter and taking the pure product through one of the intermediate ports.

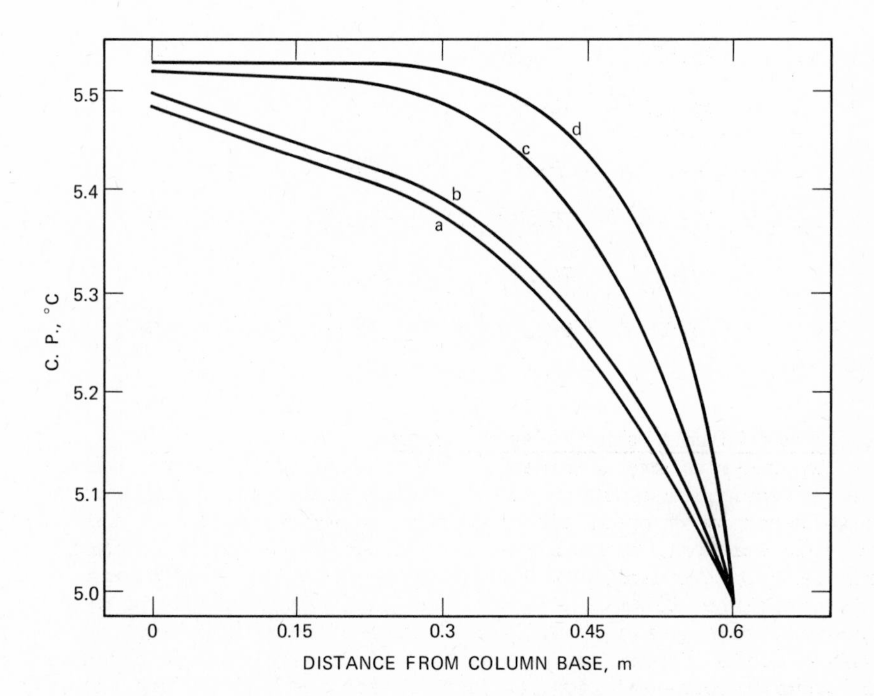

Figure 10. Impurity profiles at various rotor speeds.

Contrary to expectation, it proved more difficult to obtain reliable estimates of the amount of reflux with the 50mm column than with the 25mm column as considerable discrepancies were found between the heat extracted and the heat supplied to the column, but on the basis of the latter it appears that, as with the smaller column, 10 to 20% reflux is adequate for most purposes. Estimates of the power from the rotor drive system dissipated by the helix have been obtained using a dynamometer and it appears that this energy may contribute significantly to the heat balance.

The continuous operation of Schildknecht type crystallizers has also been reported by Middleton (47). Benzene and naphthalene were treated in a column of glass and metal construction, 50mm diameter by 0.75mm long. The results, which are given in Table 6, show a high degree of purification and good yields, but the throughput is not recorded.

TABLE 6 CONTINUOUS CRYSTALLIZATION OF BENZENE AND NAPHTHALENE (47)

Feed		Crystallizing point °C		Product Yield %
		Feed	Product	
Benzene	99.5%	5.03	5.53	92
"	99.2%	4.77	5.53	95
"	98.7	4.20	5.53	92
Naphthalene	90.45	75.4	80.03	91
"	89.45	74.9	79.73	90
"	65.0	60.3	78.2	90

B. Large Columns Using a Rotary Helix

Experiments have been made at the C.T.R.A. (36) with Schildknecht type columns of 75, 100 and 143mm diameter. The 75 and 100mm units were constructed, like the 50mm column described in the previous section, from precision-bore tubing, and were limited initially by the availability of tubing to a length of 0.72m; subsequently a 1.14m length of 100mm tube was obtained and used to construct a longer column. These crystallizers were operated with helical coils of various pitches made from wire of both circular and square cross-sections, 9.5mm wide for the 75mm column and 12.5mm wide for the 100mm column. The center tubes of both units were of metal, and that for the 100mm column was equipped with an internal cooling tank in the upper part and an internal heater in the melting section.

Experiments on the 75mm column were begun using a mild steel helix but no crystal transport was obtained with a naphthalene oil of 65°C c.p. (i.e. 74% naphthalene). Transport was obtained with a 75°C c.p. naphthalene under total reflux, but attempts to operate at finite throughput were unsuccessful as the flights became plugged with crystals and it appeared that these were adhering to the surface of the helix. The use of a polished chromium-plated helix overcame these difficulties and the results obtained are summarized in Figure 11, in which crystal flux is plotted against the c.p. of the higher-melting product. A flux of 200g/h/cm^2 was obtained before there was any significant fall in the c.p. of the product. Similar results, also shown in Figure 11, were obtained in the 100mm column with a geometrically similar helix. Comparison of these results with those for the 50mm column given in Figure 4

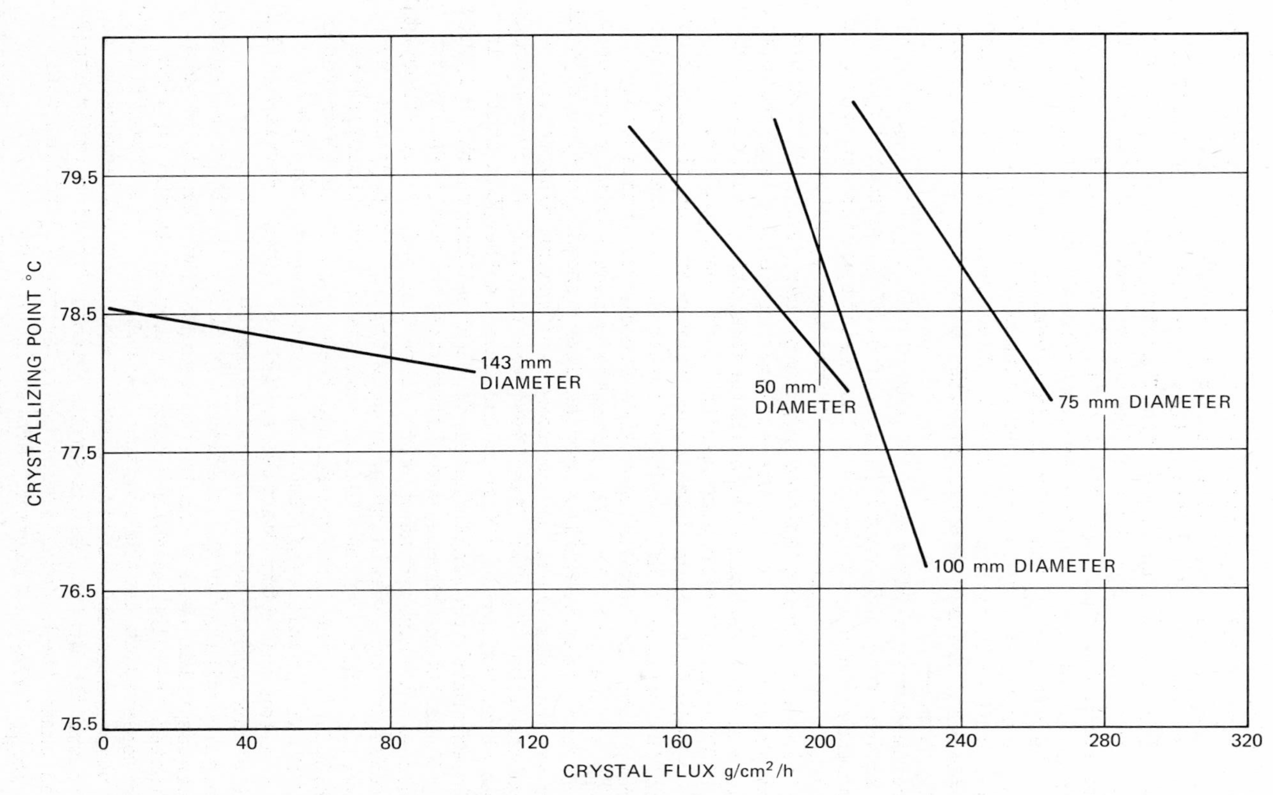

Figure 11. Treatment of napthalene oils in 75, 100 and 143 mm diameter crystallizers.

shows that the limiting flux-density is similar for all three crystallizers.

The 143mm column was constructed entirely from mild steel and was 1.22m long. Operation with naphthalene using a helix constructed from thin flat strip, 48mm wide with a 133mm pitch, resulted in repeated plugging of the flights. Crystal transport was obtained with a circular section helix 14mm wide and continuous feed operation was possible with crystal flux densities up to $80g/h/cm^2$, but attempts to increase this resulted in seizure of the helix. The failure of the wide helix stressed the need to maintain geometric similarity to the smaller columns and the seizure of the 48mm wide coil illustrated the difficulties anticipated in scaling up the detached helix, particularly where the flights are proportionally thinner.

Experiments were also made in each of the crystallizers with a crude benzene fraction having a c.p. of 4.25°C and containing .044% sulphur (mainly thiophen). The results, which are given in Figure 12, show the effectiveness of the various sizes of crystallizer but the ultimate throughputs that could be obtained were not determined. It is interesting to note that appreciable crystal transport was achieved with benzene using the helix 48mm wide by 133mm pitch which was unsuitable for naphthalene.

The conclusions drawn from these experiments, on the basis of the limiting crystal flux densities obtained, are that geometric similarity with the 50mm column is desirable for satisfactory scale-up, at least to the 100mm diameter scale, and that throughput for geometrically similar crystallizers should be proportional to annular area and hence to the square of column diameter. Although the limiting flux densities were similar, a lower product c.p. was obtained in the 100mm column than in the 50mm unit. In order to estimate the scale factor for column length, the H.E.T.S. of the 50 and 100mm columns were determined on test mixtures of naphthalene and thionaphthene at total reflux and finite throughput. The effects of some variations in the geometric proportions and cross-section shape of the helical coils were also investigated. The H.E.T.S. for the 25mm column had been found to be 220mm under total reflux; results for the 50mm column ranged from 310 to 670mm, and for the 100mm column from 530 to 910mm. Results obtained at finite throughput showed a similar scatter and this was attributed partially to the tendency of the system to form crystal agglomerates which were transported rapidly down the column and contaminated the high-melting product. Another factor causing scatter appeared to be the coolant temperature; the lower H.E.T.S. values were generally associated with lower coolant temperatures and this was thought to be due to the higher fraction of crystals obtained under these conditions more effectively limiting axial dispersion. Owing to the wide scatter it was not possible to distinguish clearly the effect of speed of rotation, but the effect of pitch was distinguishable for the 50mm diameter column, 25mm pitch giving rise to a higher H.E.T.S. than did the 19mm pitch,

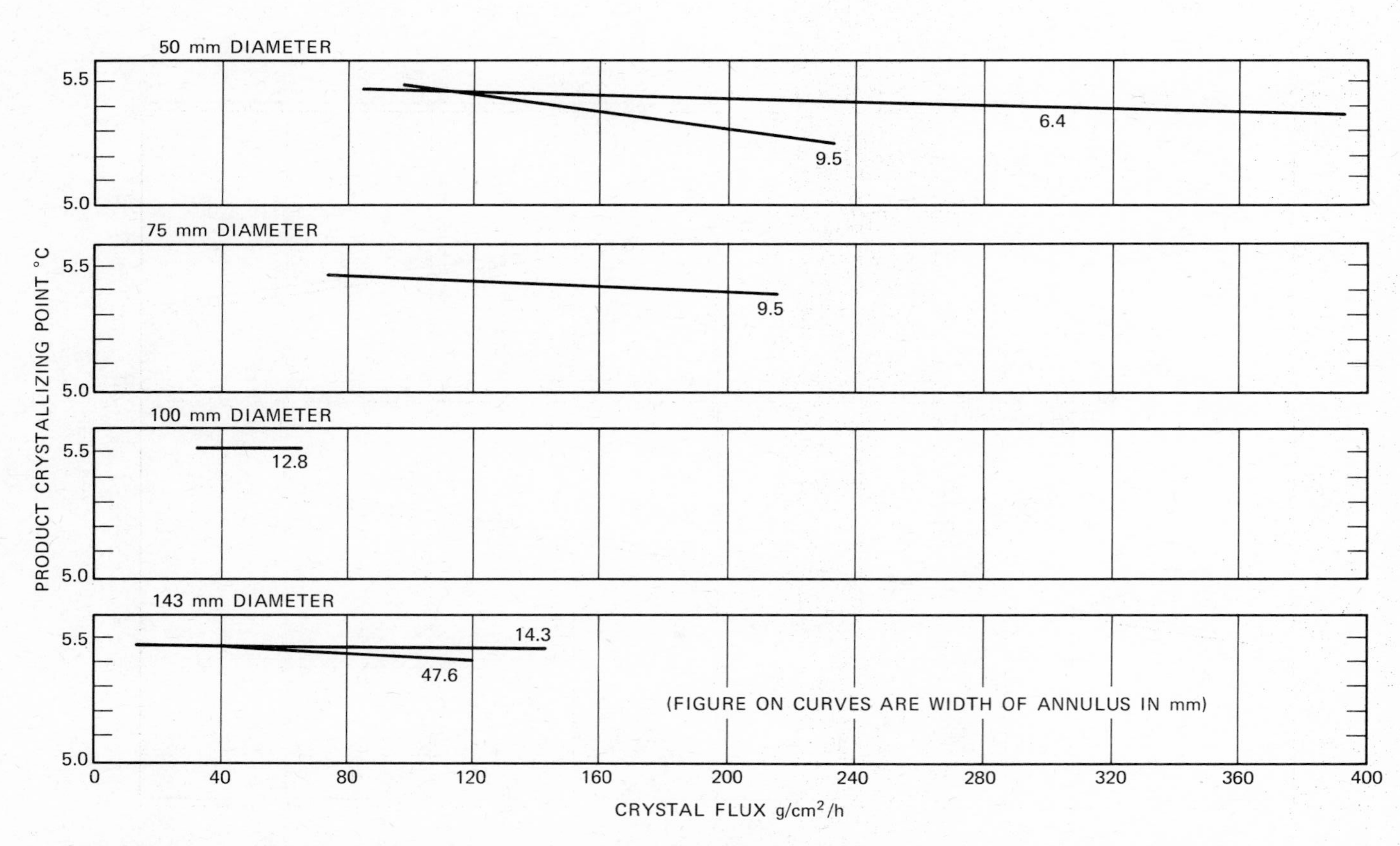

Figure 12. Treatment of benzene on columns of various sizes.

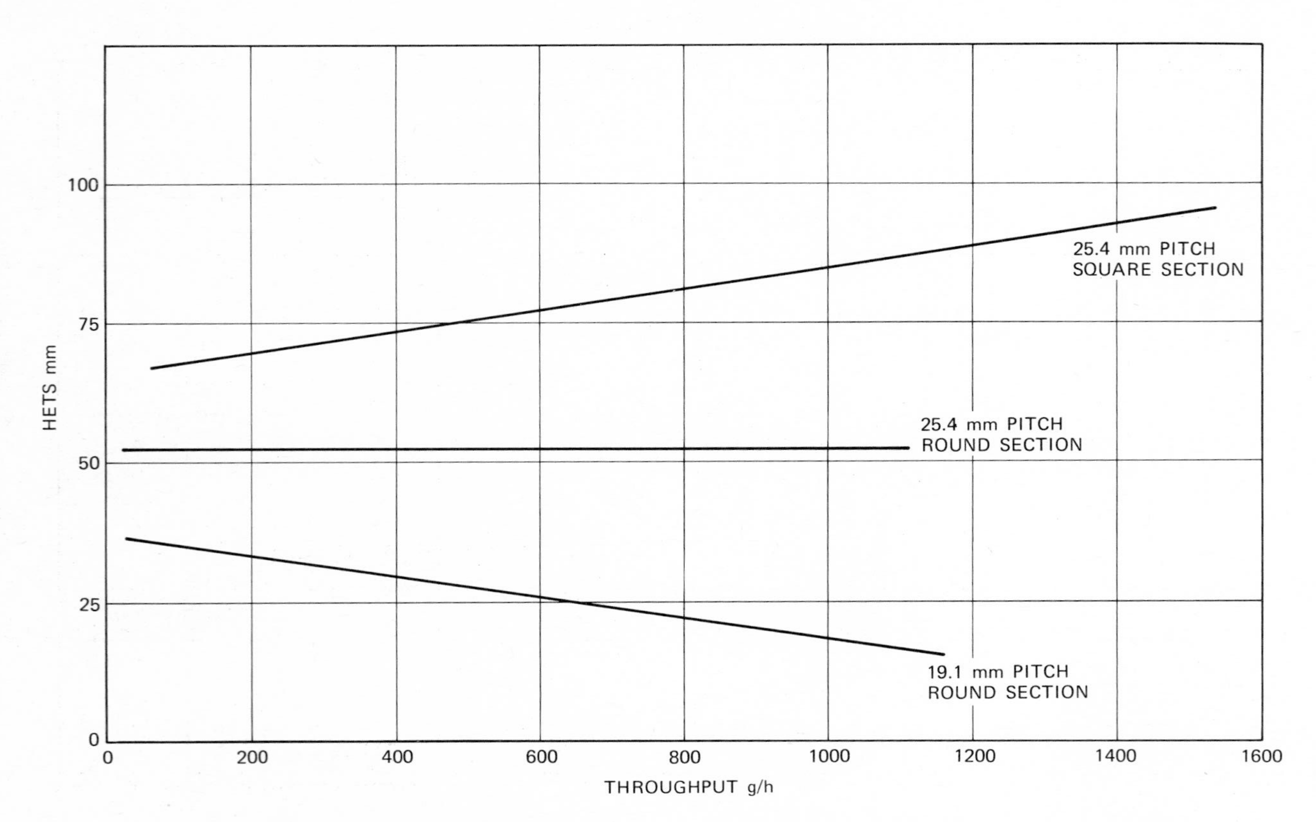

Figure 13. 50mm diameter crystallizer HET_s vs. throughput, napthalene-thionapthene system.

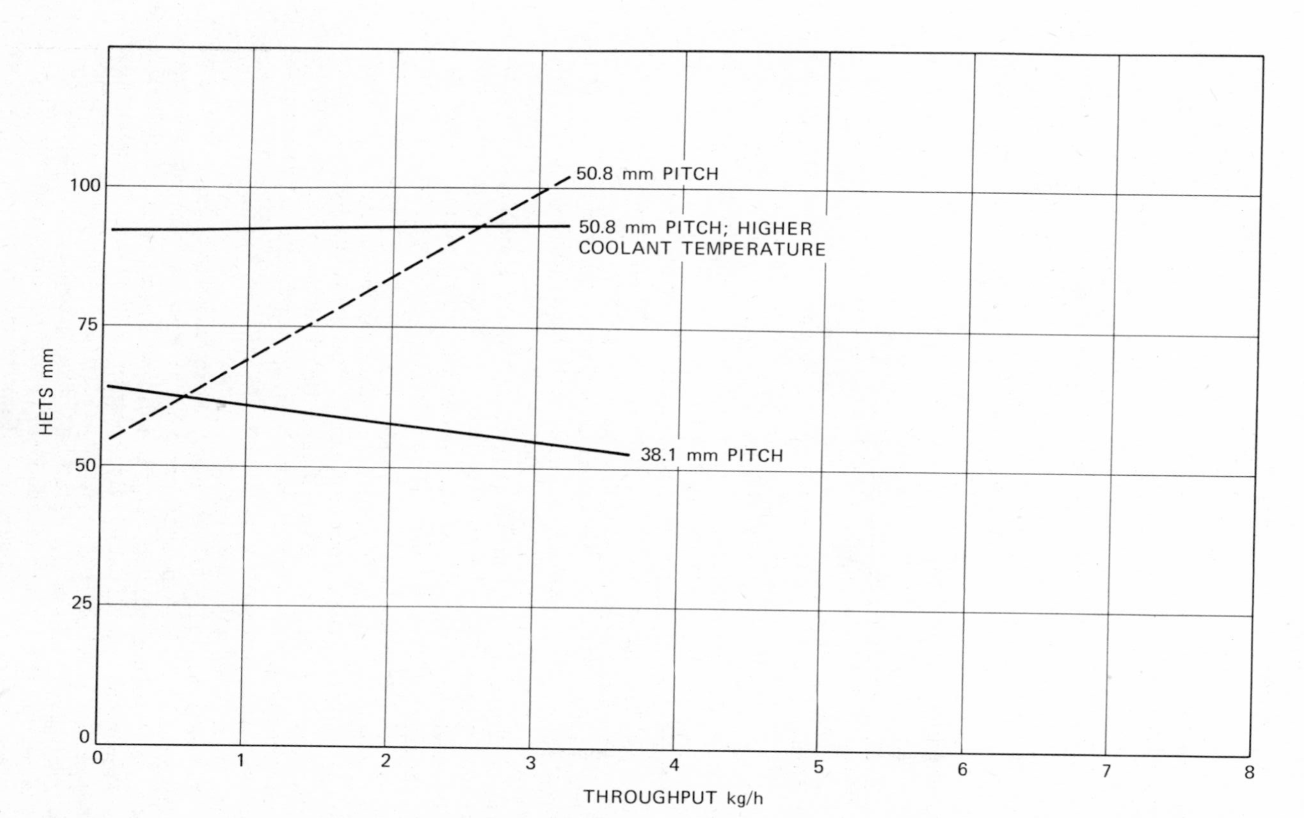

Figure 14. 100mm diameter crystallizer HET_s vs. throughput, napthalene thionapthene system.

while square section coil appeared to be less efficient than round section. The results for the 50 and 100mm columns are shown in Figures 13 and 14 respectively; the only conclusion that can be drawn from comparison of these is that H.E.T.S. increases with column diameter.

A crystallizer of 200mm diameter, 1.8m long, with a nominal throughput of 25 gal/h has been erected by Newton Chambers Ltd. It has been operated with benzene and is reported to have given encouraging results but some unexpected problems were encountered requiring additional laboratory work, for which purpose a glass and metal unit of 100mm diameter and 1.2m long was constructed. No details of the results have been published and it is not known if satisfactory operation on naphthalene has yet been achieved (47).

C. Columns using Archimedean Type Rotors

Because of the mechanical difficulties of scale-up inherent in the free helix design of Schildknecht, it was thought advisable to investigate other forms of rotor. Although it had been reported (24, 35) that Archimedean screws, in which the helix is attached to a central shaft and rotates with it, did not give effective separation it was thought that this should be proved before developing more elaborate designs (37).

Rotors for the 50mm column (Figure 8) were prepared by attaching flights of various dimensions to close-fitting central shafts and mounting these in bearings at the top and bottom of the column. These rotors were tested against the criterion previously used, i.e. a 95% yield of benzene with a minimum c.p. of 5.50° from a feed with a c.p. of 5.0°C. The performance varied with the pitch and cross-section dimensions of the screw flight, probably to a greater extent than with the helical coils. The throughput varied markedly with rotation speed but the optimum speed bands were higher and wider than those for the helical coils. By using the correct combination of flight dimensions, pitch, and rotation speed, a throughput of over 5 l/h was obtained, equal to that of the best Schildknecht type column. Figure 15 shows the variation in throughput with rotation speed for various rotors. These include one of rectangular section which is of particular interest as it suggests that the rotor could be fabricated by methods similar to those used for screw conveyors.

The diagram also illustrates that, as with the Schildknecht rotary helix (35), the performance of some of the rotors could be modified considerably by polishing and chromium-plating the surface. The throughput obtainable with rotors of optimal dimensions was not affected by plating but with other rotors the throughput was considerably increased. With all the rotors, plating caused a reduction in the optimum speed of rotation and resulted in smoother operation of crystallizer with less tendency for the development of pressure within the column.

Some further variations on rotor design have been examined but with little success. A short Archimedean type rotor was used

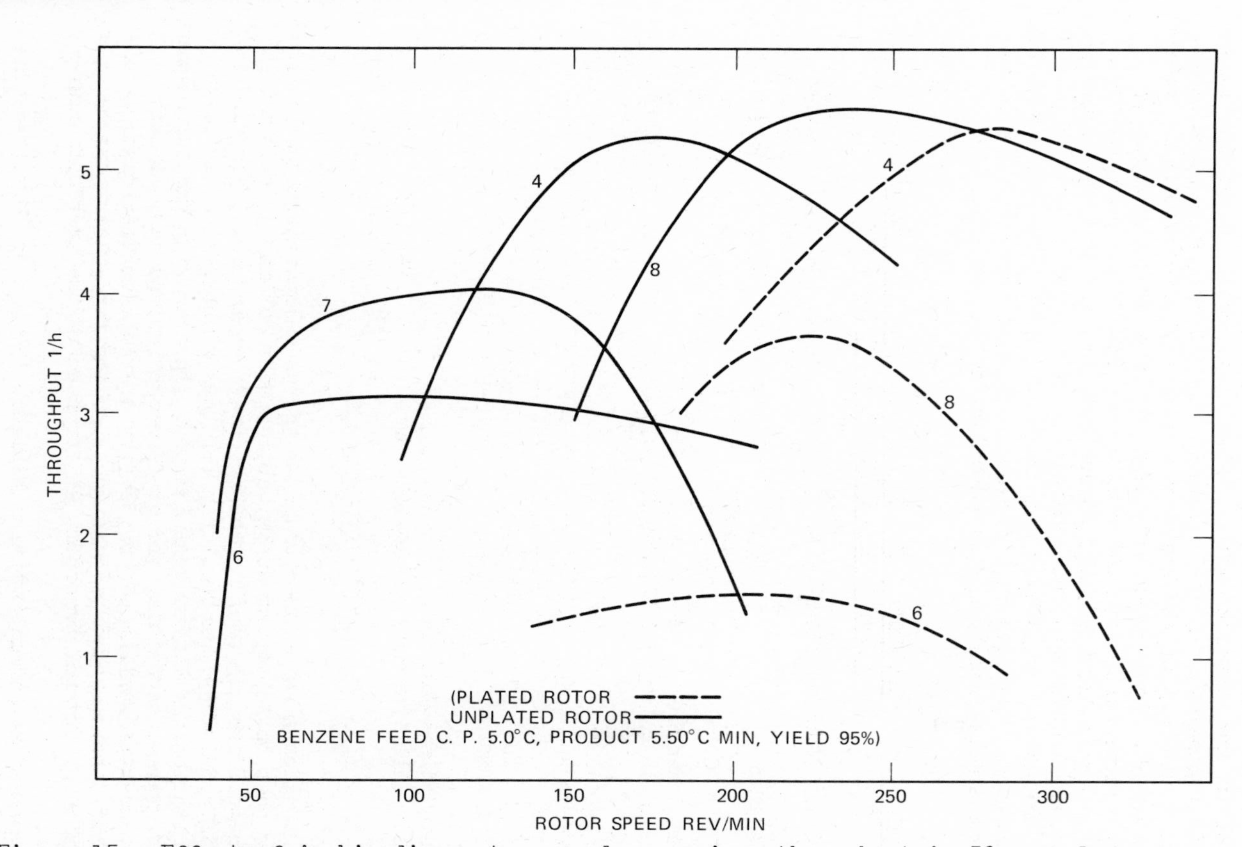

Figure 15. Effect of Archimedian rotor speed on maximum throughput in 50 mm column.

extending only over the top half of the column and while adequate purification of the 1% impurity benzene was obtained, the throughput was limited to 1.5 l/h. The pressure developed in the column was similar to that obtained with a full length rotor suggesting that it is mainly generated in the cooling zone, and torque measurements suggest that most of the power dissipation of the rotor also occurs in this zone. Replacement of the whole of the screw flight by sprung scraper blades in the cooling section only, gave very unsatisfactory results and these were not improved when the screw flight was restored in the remainder of the column.

A column with an Archimedean type rotor has also been used with systems other than benzene, including xylenes and aqueous solutions, and some of the results obtained are given in Table 7.

TABLE 7 TREATMENT OF VARIOUS FEEDSTOCKS IN 50 mm COLUMN WITH ARCHIMEDEAN ROTOR (38)

Composition % W/W			Feed Rate l/h	Bottom Product Yield, %
Feed	Top Product	Bottom Product		
Aqueous Ethanol				
Ethanol 5	12.7-9.0	0.1-0.4	0.6-1.1	70-80
Mixed Xylenes:				
p-Xylene 90	52-57	99.5-98.2	0.5-1.8	80
" 85	52-60	98.7-97.3	0.6-1.1	70
" 80	49-53	97.5-94.0	0.6-1.0	60
" 80	51-60	95.2-93.5	0.3-1.1	60
Aqueous Sodium Chloride:				
Sodium Chloride 3.5	10.9-8.5	0.1-0.6	0.4-1.8	65-75

The throughputs were lower than for the usual benzene feedstock, but better than those reported for Schildknecht columns of this size, and the performance could probably have been improved by optimization of conditions, particularly for the xylenes as the available refrigeration facilities were limited. In comparing the throughputs with different materials, the differences in feed composition, latent heat of fusion and the density of the crystals relative to the liquid should also be considered. With the aqueous systems the direction of crystal transport was opposed to the gravitational effect.

Renewed experiments with naphthalene (36) have shown that this material also can be purified with an Archimedean type rotor with throughputs at least as good as with the detached helix, provided (i) the screw has the same ratio of swept wall area (per turn of flight) to annular area as an effective detached helix, (ii) the surface finish is adequately smooth and resistant to corrosion, (iii) heat conduction along the screw to the cooling section is minimized and (iv) the speed of rotation is at least doubled compared with the detached helix. Some results are given in Table 8 for the conversion of phthalic grade naphthalene to pure naphthalene in 50mm (Figure 3) and 100mm crystallizers fitted with Archimedean screws.

TABLE 8 TREATMENT OF NAPHTHALENE IN COLUMNS WITH ARCHIMEDEAN ROTORS

Feed 97.07% naphthalene c.p. 78.76°C
Sulphur content: 0.37% (1.55% as thionaphthalene).

Feedrate kg/h	% Take off at bottom	Bottom product c.p. °C	Bottom product Sulphur content % w/w	Purity % w/w
50mm diameter crystallizer:-				
1.3	88.9	80.07	0.07	99.64
2.2	91.25	80.06	0.09	99.62
3.2	89.0	79.81	0.17	99.17
3.2	89.8	79.71	0.24	98.97
4.5	91.2	79.83	0.19	99.21
5.2	89.0	79.50	-	98.55
100mm diameter crystallizer:-				
3.8	91.3	79.94	-	99.41
4.4	91.6	80.13	0.04	99.76
4.9	91.8	80.09	0.08	99.68
7.0	90.2	79.93	-	99.40
7.2	94.0	79.93	0.14	99.40
8.0	93.7	79.96	0.13	99.44
8.2	92.5	79.73	0.20	99.01
11.1	91.6	79.83	0.19	99.21

The separation of various materials with a column, apparently of the Archimedean type, has also been reported by Gel'perin et al (48). A rotor having a helical flight pierced by numerous holes or fabricated from wire mesh has been described, the column contents also being subjected to axial pulsations. This system was operated with caprolactam and phenol (49).

V. CRYSTALLIZER CONSTRUCTION & DESIGN

A. Rotors

It is evident from the experimental results given earlier that the performance of a Schildknecht column is highly dependent on the geometric proportions of the helix. In order to obtain effective transport of crystals, the flight width and pitch must not be too large relative to the diameter of the column. In treating materials in which transport is relatively difficult, e.g. naphthalene, flight width:diameter ratios (W/D) greater than 0.2, and flight width:pitch ratios (W/P) in excess of 0.4, are unsatisfactory. The effect of coil geometry on performance is shown in Figure 16.

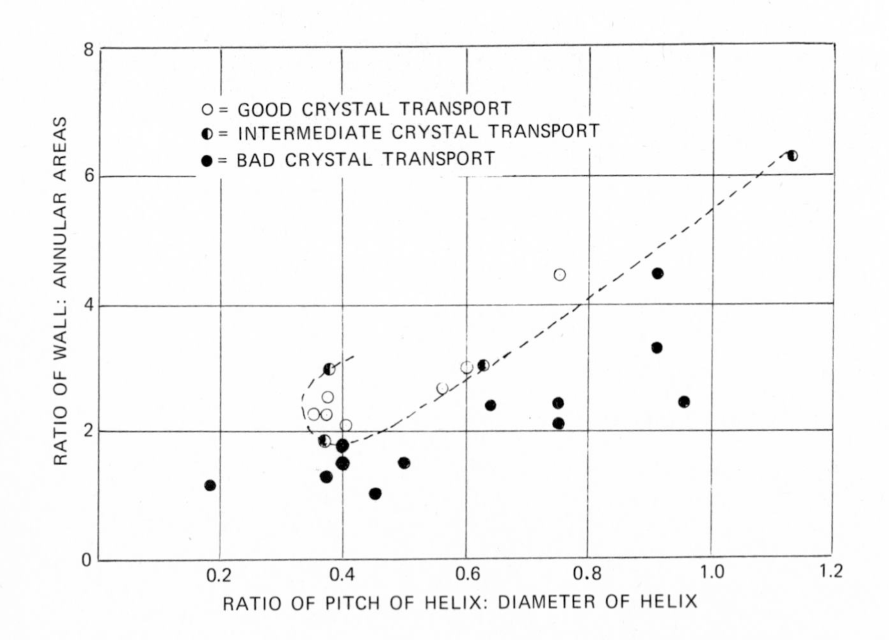

Figure 16. Effect of geometrical proportion of crystallizer on crystal transport.

These values were exceeded in the apparatus used by Austin, although the rotor was of different form, and this may well account for its lack of success in operation. Other feedstocks are not as sensitive as naphthalene to these ratios; the helical coils used by Schildknecht (25,27), Pouyet (46), and by Powers et al (39-45), evidently gave effective transport with numerous materials although the W/D values were in the range 0.2 to 0.33. However, even with systems in which crystal transport can be effected fairly easily, e.g. benzene, the throughput that can be obtained, for a given degree of purification, varies with the coil geometry. This is particularly evident with Archimedean rotors, and while crystal transport has been obtained with rotors of W/D from 0.12 to 0.19, and W/P between 0.125 and 0.25, the throughput varies widely and optimum performance is obtained only from a rotor having a narrowly defined combination of dimensions (37). Further consideration of the factors governing crystal transport will be given in a later section.

The lower limits for W/D and W/P will be determined by mechanical factors. During operation a helix will be subject to various forces tending to cause distortion, including a torque arising from friction at the walls between the helix and the crystals adhering to the cooling surfaces, and axial loadings caused by ploughing off crystals from the walls and propelling these through the liquid.

The axial deflection f for a helical coil of circular cross-section is given by the expression:-

$$f = Ra = \frac{64\ P\ell R^2}{Ed^4}$$

This can be manipulated (36) to give the change in diameter δD when a torque T is applied at the driven end:-

$$\delta D = \frac{12\Pi\ D^2\ \mathrm{Sec}\phi}{d^4}\ \frac{T}{E}$$

where a = angular deflection
d = diameter of wire
D = diameter of coil
E = Youngs' Modulus
f = axial deflection
ℓ = length of wire in coil
R = radius of coil
T = Torque
ϕ = pitch angle of coil

With a mild steel coil of 50mm o.d., made from 6mm o.d. circular wire with a pitch of 18mm, a contraction of 1.7mm in the driven end diameter would be expected for a power input of 50 watts at 70 rev/min. In practice a power consumption of this order has sometimes been encountered during operation with naphthalene (36).

The power input might be expected to scale, for square section wire, according to the expression

$$\text{Power} \propto b\ \ell\ \mathrm{Cosec}\ \phi$$

where b is the thickness of the flight in contact with the wall. If $b \propto D$, $\ell \propto D$, then for constant ωD, power $\propto D^2$. This means that $-\delta D \propto D$, i.e. the deformation of the coil would be proportional to diameter in geometrically similar crystallizers.

It has generally been found satisfactory to provide a clearance of 0.25mm between the helix and both inner and outer walls of the column, this being consistent with a small amount of by-passing of the helix by crystals and liquid while allowing rotation without seizure. The clearance is obviously a crucial factor affecting the difficulty of construction of larger columns. In order to maintain the same degree of scraping in the cooling section it would be necessary to keep the clearance constant during scale-up, and this may mean that a smaller proportion of the total crystal flux will be susceptible to slip past the edge of the flight in larger columns.

The experiments also indicate the importance of the surface finish of the helix. A smooth surface, obtained e.g. by polishing and chromium-plating, promotes crystal transport by reducing the tendency for crystals to adhere to the surface of the coil. Some helical coils that were previously ineffective in treating naphthalene have been used successfully after plating. The throughput of benzene obtainable from non-optimal rotor designs was considerably improved by plating, and with optimal designs plating caused a reduction in the optimum speed of rotation and resulted in the development of lower pressures in the column. It appears from a consideration of the forces involved in crystal transport that this would be aided by a smooth surfaced helix and by rough-surfaced column walls. The properties of surface finishes other than chrome plating are currently under examination.

According to Schildknecht, the most effective helical coils were fabricated from wire of lenticular cross-section (25,27) and a helix of this form was used by Powers _et al_ (39-45). It appears, however, that the improved performance claimed for this shape can also be obtained with circular or rectangular cross-sections if these have appropriate dimensions, so that the inherent weakness of the lenticular wire can be avoided. Circular cross-section wire has been most frequently used in the laboratory work because of the relative ease of winding coils from this material. The deformation resulting from torsion may be reduced by driving both ends of the coil but this introduces further constructional complications.

An Archimedean rotor is not subject to radial deformation in the same manner as a detached helical coil but the flights must be sufficiently rigid to prevent distortion. The integral center shaft facilitates the provision of bearings and setting up of the rotor for true-running within the column, but the provision and maintenance of a close-clearance between the rotor and the outer wall is still a prime consideration in the fabrication and operation of the column.

It has proved most difficult to obtain crystal transport in systems of high viscosity; good transport was obtained relatively easily in benzene with a viscosity of 1cP whereas in phenolic systems, with viscosities greater than 5cP, it was much more difficult. It was observed in the high viscosity systems that the crystals produced were extremely small and in operation some of these slipped past the rotating helix. If this is the cause of the poor transport then remedial measures, such as increasing the crystal size or reducing the clearance between the helix and the walls, might lead to a considerable widening of the scope of application of this type of crystallizer. It is proposed (36) to incorporate a wiping blade of plastic material into a rotor to give zero clearance between it and the walls, and to investigate factors leading to growth of larger crystals by (i) use of purer starting materials (ii) pre-crystallization and control of nucleation rate (iii) use of solvents or additives.

B. Rotor Drive.

Powers et al operated their crystallizers with an upward crystal movement, i.e. with the crystallizing section at the bottom and melting section at the top of the column, with the result that for the benzene/cyclohexane system crystal transport was unaided by gravity. Most workers, however, have preferred downward transport and the present authors have found this convenient and probably more effective as greater crystal flux densities were obtained. In treating aqueous systems where the crystals are buoyant, upward transport might be advantageous.

The experimental results demonstrate the important effects of rotation speed on column performance; each size and design of rotor has an optimum speed of rotation and it appears that this may vary with the feedstock. It is thought that at lower speeds crystal transport becomes sluggish, whilst at higher speeds purification is hindered by back-mixing of the column contents. McPhee (38) has suggested that the curves in Figure 15 relating throughput, F, to rotation speed, S, can be represented by an equation of the type:-

$$F = AS - Be^{S} + C$$

where A is a factor related to crystal transport determined by column geometry and the properties of the feedstock; and B is a factor related to phenomena opposing separation, such as back mixing, which will also be determined by the column and feed. On scaling-up it seems reasonable to maintain constant peripheral velocity for the rotor so that rotational speed will be inversely proportional to column diameter, and this premise is generally borne out by experience. However, one result of this approach is that the heat-transfer surface is scraped less frequently and this may result in a reduced heat-transfer coefficient as the column size increases, or even allow sufficient crystal build-up on the surface to interfere with the rotation.

In some of Schildknecht's experiments, (33) and in those of

Powers (39-45), small vertical oscillations were imposed on the helix, in addition to the rotation, in order to obtain good dispersion of the crystals in the liquid. In the present authors' opinion this procedure is an unnecessary complication that can be avoided by using the appropriate combination of rotor speed and design.

The power required to drive the rotor is of considerable interest, particularly with larger units. Dissipation of power by the rotor in the warm end of the column would reduce the heat loading of the melter but if, as seems more likely, the major part of the dissipation occurs in the cooled section then the heat extraction duty will be correspondingly increased. The effect of scale on power requirements, with particular reference to that consumed in detaching crystals from the wall, has been briefly considered in an earlier section. Additional power will be required to transport the crystals and to rotate the helix against the viscous drag of the column contents. In experiments with naphthalene (36), the power consumption was determined from the electrical input to the motor driving the helix. At a throughput of 2 kg/h the power for detachment of crystals was about 20 W in the 50mm crystallizer and 40 W in the 100mm column. The power for transporting crystals has been calculated as only 0.01 Wh/kg, and that consumed in turning the rotor in the absence of crystals was about 180 and 210 W respectively, in the two columns. Assuming that the power for crystal removal is proportional to throughput, then for a throughput of 16 kg/h in the 100mm crystallizer the requirement would be 320 W, i.e. 20 Wh/kg. In a geometrically similar column of 305mm diameter, the power consumption would be at the rate of 2.9 kW, i.e. about 4 hp, plus the losses in the transmission system. In experiments with benzene (38) the power consumption was measured by a mechanical dynamometer attached to the rotor drive shaft. The power consumption in the absence of crystals was only a few watts, but in operation at throughputs of about 3 kg/h was at the rate of 60 to 160 W, i.e. 20 to 50 Wh/kg, which is of the same order as that found for naphthalene.

C. Heat Transfer

Whilst the majority of crystallizers have been constructed with outer tubes of glass, which allows visual observation of crystal behavior, the 25mm column for benzene had metal heating and cooling sections to facilitate heat transfer, and the 50mm benzene column (37) and some of the larger columns (36, 47) were constructed wholly of metal. There is little doubt, from the heat transfer data in Table 9, of the advantage of using metal rather than glass for the outer tube. The differences in heat-transfer rates for the benzene and naphthalene feedstocks may be related, at least in part, to the large difference in purity of these materials. A theoretical expression for scraped-surface heat exchangers due to Kool (55) can be simplified over a reasonable range of conditions to:-

$$\text{Film coefficient} = 1.2(kc_p\ \rho/\gamma)^{\frac{1}{2}}$$

where c_p = specific heat

k = conductivity

ρ = density

γ = scraping interval

which gives a film coefficient for benzene of about 95 Btu/h/ft^2/°F.

TABLE 9 HEAT TRANSFER COEFFICIENTS FOR COOLING SECTION

Feed	Feedrate kg/h	Column Diameter mm	Tube Wall	Coefficient Btu/h/ft^2/°F	Reference
Benzene	0.7		glass	35-55	36
"	0		steel	145	36
"	1.0	25	brass	70	38
"	1-5	50	brass	35-55	38
Naphthalene Oil	0.4		glass	20	36
"	0.9		glass	12	36
"	0		steel	24	36
"	0.95		steel	31	36

If, however, the surface is not properly scraped, additional resistance to heat transfer occurs in the residual film.

When geometric similarity is used as the basis for scaling up crystallizer length, the heat transfer area will increase with the square of the diameter; since the throughput increases similarly, there should be no difficulty in removing heat unless the transfer coefficient deteriorates on scale-up, and there is so far no evidence of this effect. However, in order to minimize the overall length of the larger crystallizers, it would be advantageous to reduce the length of the cooling section. With high-melting feedstocks it may be possible to reduce the heat-transfer area by increasing the temperature differential between the coolant and the column contents, although care will be necessary to prevent plugging of the column due to the rapid onset of crystallization in the lower part of the cooling section. With feedstocks which necessitate refrigerated coolants, lowering the temperature would entail higher capital and running costs for the refrigeration system. In

Schildknecht type columns additional heat transfer area can be obtained by utilizing the inner column for this purpose, but with Archimedean columns this would be likely to cause adherence of crystals to the rotor, and indeed even slight warming of the rotor has been suggested to prevent this (48). The heat-transfer capability could be usefully augmented by using a precrystallizer to convert the feed to a slurry before entering the main column, although the extent of this technique is limited by the need to maintain a mobile slurry, and because all the additional cooling required to increase the crystal flux to equal the product off-take rate, plus that required to produce reflux, must be effected within the main column. It has been found possible to cool a 65°C c.p. naphthalene oil to 60°C, forming a slurry containing 38% solids, i.e., 50% of the naphthalene content as crystals, and to feed this to the centre of the crystallizer column. No effect on performance was observed other than a corresponding reduction in the heat removed by the cooling section (36). Benzene slurries remain mobile with up to about 50% crystals, and a slurry of this composition was generated by passing the feed through a pre-crystallizing column, of similar construction to a 50mm Archimedean column, and introduced into the main column. The purification was not impaired and while the use of the precrystallizer did not increase the total throughput obtainable from the column it allowed this to be maintained when the length of the cooling zone was reduced by half (37).

Melting in the small columns was effected by electrical windings on the exterior of the column or attached to the inner column. Alternative means available, particularly for larger columns, include electrical immersion heaters within the annulus, external jackets and internal coils for circulation of suitable heat-transfer fluids, and various forms of external heater through which part of the product stream is recirculated to the base of the column. In general, the melting section can be considerably shorter than the cooling section and will consequently make only a small contribution to the total column length. Heat transfer coefficients are likely to be higher than in the cooling section, and in working with low-melting products it will be possible to use a relatively large temperature differential between the product and the heating medium. Where it is necessary to restrict the heat flux to avoid local overheating of materials that are higher-melting, more viscous, or heat sensitive, the use of the recirculatory system may well be attractive. In some circumstances it may be possible to effect an economy in the overall energy requirement by utilizing the heat output from the refrigeration system to supply at least part of that required for melting.

VI. THEORY

A. Analysis of Column Performance

Theoretical analyses of the separation obtained in column crystallizers have been made by Anikin (17,50,51), Yagi et al (17,

52) and Powers (17,53). The last author approached the problem from the concept of differential counter-current contacting, rather than that of the ideal stage, assuming the existence of two distinct phases whose compositions change continuously in the direction of flow. Based on this model, a differential equation may be set up expressing the net rate of transport of a component through the column in terms of mass-transfer between phases and simple mass-balance considerations. Two distinct cases arise, for systems showing solid solubility and eutectogenic systems, respectively.

For systems exhibiting solid solubility Powers derived the expression:-

$$\frac{Z}{HTU} = \int_{x_o}^{x_z} \frac{dx}{y^I - x}$$

where

$$HTU = \frac{C}{KaA\rho} + \frac{\rho DA\phi}{C}$$

(for list of symbols see end of section)

This expression assumes (i) steady-state operation under total reflux so that there is no net transport of either component through the column, (ii) that mass flow rates and HTU are independent of position in the column.

The expression was used by Powers to calculate the HTU for the system azobenzene/stilbene in a column of unspecified dimensions and gave a value of 33mm. Similar calculations have been made by Freeman (36) for other systems with the results given in Table 10.

TABLE 10 TERMS IN POWERS' HTU EXPRESSION (25mm column (36))

System	Coil Rotation speed rev/min	HTU cm	$\frac{\phi D}{C}$ cm	$\frac{C}{Ka}$ cm	$\frac{\phi D}{Ka}$ cm^2	HTU min cm
Azobenzene-stilbene (Powers)	-	3.3	0.13	3.2	0.4	1.2
Naphthalene-thionaphthene	60	17.6	10.0	7.6	76	17.4
Naphthalene-2-naphthol	50	29.4	16.1	13.3	214	29.2
Naphthalene-2--naphthol	105	21.2	12.6	8.6	108	20.8
o-Cresol-phenol	31	47.8	23.7	24.1	571	47.8
Diphenylene oxide-fluorene	74	5.5	2.1	3.4	7.1	5.3
Diphenylene oxide-fluorene	74	5.4	1.9	3.5	6.7	5.2
Diphenylene oxide-fluorene	49	4.8	1.4	3.4	4.8	4.4

Freeman observed that for HTU to be a minimum, regarding C as a variable, then:-

$$C/A = Ka\ \phi\ \rho^2 D)^{\frac{1}{2}}$$

and

$$HTU_{min} = \frac{2(\phi D)^{\frac{1}{2}}}{Ka}$$

He has been able to show that the HTU values in Table 10 are close to HTU_{min} for the conditions used, but that Powers' value for azobenzene/stilbene is about 3 times larger and would be lower at reduced crystal flows.

In dealing with eutectic systems, Powers introduced the simplifying assumption that crystals produced in the cooling section consisted of one pure component, contaminated only by an adhering film of mother-liquor. Combining the rate of mass-transfer of impurity from the adhering liquid to the free liquid phase with the mass balance relationships, and specifying batch steady-state conditions, then:-

$$\frac{x_Z}{x_o} = \exp \frac{-Z}{HTU}$$

where

$$HTU = \frac{(\ell + \ell^1/C)\ \ell^1}{KaA\rho} + \frac{\rho DA\phi}{C}$$

Assuming the impurities in recrystallized azobenzene to be eutectogenic and the relationship between composition and temperature in the column to be linear, a value of 123mm for the HTU was obtained.

Powers extended his treatment to cover steady-state operation with continuous supply of feed and removal of products, and obtained the expression:-

$$\frac{x_E}{x_s} = \frac{(E/C - 1)}{(S/c + 1)}\ \frac{S/c + eaP[\ -\ (1 + \delta/c)\ Z_S}{E/c + eaP[(1 - E/c)\ Z_E}\ \begin{vmatrix} HTU_s] \\ HTU_E] \end{vmatrix}$$

Powers original treatment for batch operation with a eutectogenic system predicted that the axial composition profile for the free liquid would be exponential if the crystal phase were pure. Working with benzene/cyclohexane, Albertins found that the profile was not exponential in the melter end of the purifying section and explained this by the presence of a constant level of impurity in the crystal phase. He concluded that the separation achieved was limited by this constant impurity and by axial dispersion. (39,40). Gates pointed out that this treatment neglected washing of the adhering liquid and he developed improved models for both eutectic and solid solution systems, which he tested against experimental results. In eutectic systems he found that when mass-transfer is considered together with the effects of constant crystal phase impurity, both axial dispersion and mass-transfer between phases are significant effects but that the former is the greater. (41,42). The model was further developed by Henry for a continuously operating crystallizer to give satisfactory predictions for the effects of several variables on the composition of the product streams and the axial concentration profile. The slope of the latter is determined by the crystal flow rate, the withdrawal rate of the pure product, mass-transfer, and axial dispersion. The dominant factor limiting separation in continuous operation is axial dispersion, its effect being more pronounced under these conditions than under total reflux because of an additional dependence of the mass-transfer factor on the pure product flow rate. Mechanical factors, such as plugging and poor dispersion of the solid phase, cause reduced separation and deviations from the theoretical predictions (43,45).

Powers' treatment has been criticized by Arkenbout & Smit (54), particularly in its use of an unusually defined HTU concept. These authors have developed a general column equation applicable, under total reflux conditions, to both mixed crystal and eutectic systems. When purification is mainly by recrystallization and the impurity concentration is low, the general equation can be simplified to:-

$$Z\,\frac{\ell\,(Q-1)}{P} = Q \ln \frac{y_z}{y_o}$$

Where the purification is mainly effected by washing, as in eutectic systems, the equation reduces to:

$$Z\,\frac{\ell}{P} = \ln \frac{y_z}{y_o}$$

The above treatments consider only the enrichment occurring in the purification section, the crystallizing and melting sections being regarded as well-mixed entities. Wiegandt and Lafay, however, consider the separation occurring in the cooling zone with a eutectic system and neglect the effect of mother liquor adhering to the crystals. They have developed a graphical method for determining the number of theoretical stages in the column, taking into account both material and enthalpy balances, the temperature-concentration equilibrium and operating lines. (9). Application of this method to the results obtained in a 25mm column with benzene gives a result between 1 and 2 theoretical stages. (38). A graphical analysis similar to the Ponchon-Savarit method for distillation and including both mass and enthalpy balances has been proposed by Matz. (10).

Symbols:

a = interfacial area for mass transfer, per unit volume of column

A = area of column cross section

C = crystal flow rate

D = effective coefficient of diffusion in the liquid

E = product flow rate from enriching (cold) end of column

E = (subscript) enriching section

K = mass transfer coefficient

ℓ = free liquid flow rate

ℓ^1 = adhering liquid flow rate

P = column parameter

Q = 'effective simple process factor'

S = product flow rate from stripping (warm) end of column

s = (subscript) stripping section

x = mole fraction of component in crystals

y = ' ' ' ' in liquid

y^1 = ' ' ' ' ' ' at equilibrium

z = position in column, measured from freezing section

Z = total length of purifying section

ϕ = volume fraction of liquid in slurry

ρ = density of liquid

B. Crystal Transport Mechanism

It is evident from the work of Powers et al (39-45) in which benzene crystals were propelled upwards against gravity, and the present authors' experiments in which ice crystals were transported downwards against the natural buoyancy effect, that the dominant factor producing crystal transport is not gravity but the action of the rotor. The mechanism of transport in Schildknecht type columns has been studied by Freeman (36) in experiments with plastic particles in aqueous media. The use of this system obviated the complications of crystal melting, recrystallization, and mass-transfer between phases; it also permitted the concentration of solid phase to be controlled, and enabled variations to be made in the density difference between the phases and in the shape and size of the particles.

The experiments were made in a column simulator as shown in Figure 17.

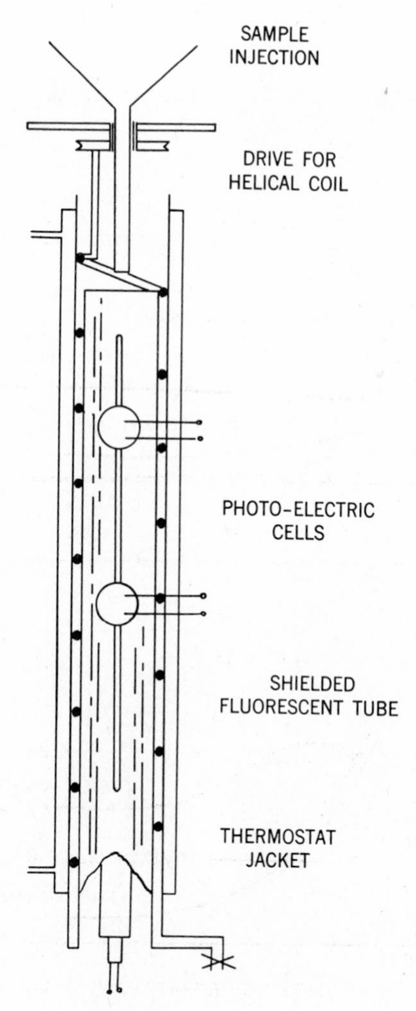

Figure 17. Crystal transport simulator.

This was essentially a column crystallizer in which the central tube had been replaced by a cylindrical fluorescent light source, so that the passage down the column of a batch of particles added to the top could be followed with the aid of photo-electric cells located outside the outer glass tube. Three types of particles were employed: 1.59mm squares, cut from 0.79 and 0.10mm gauge polythene sheet ('thin' and 'thick' squares), and 2 x 2mm polythene cylinders.

The effects of particle shape, size, and density difference on the transport velocity down the column are shown in Figure 18, the velocity being expressed in a dimensionless form as a percentage of the apparent downward velocity of the helix.

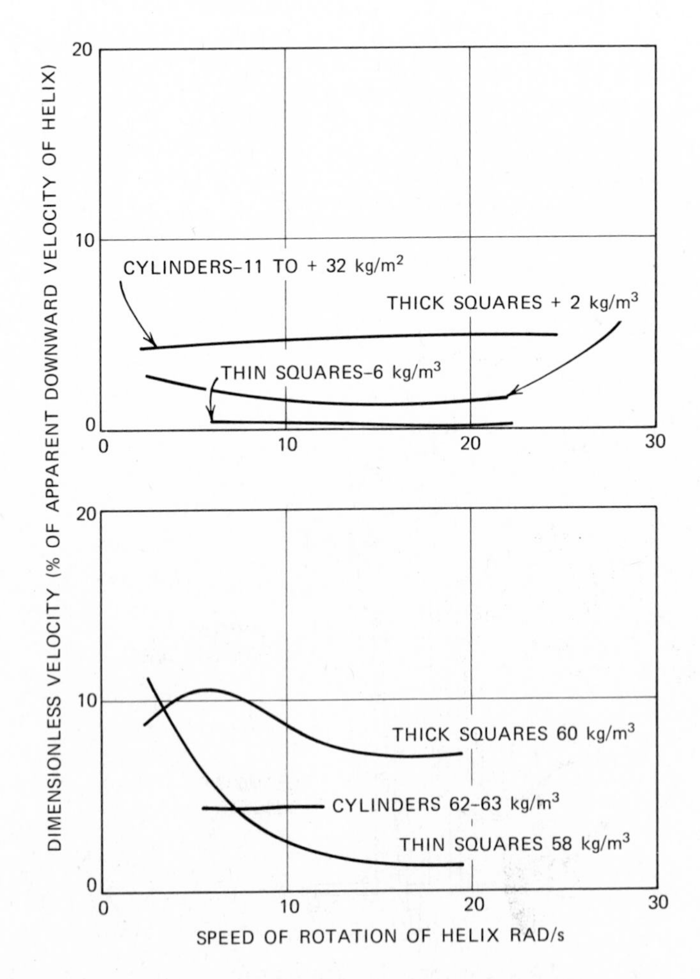

Figure 18. Effect of particle shape, size and excess density difference on transport of plastic particles in simulators.

Figure 19 shows the effect of liquid phase viscosity;

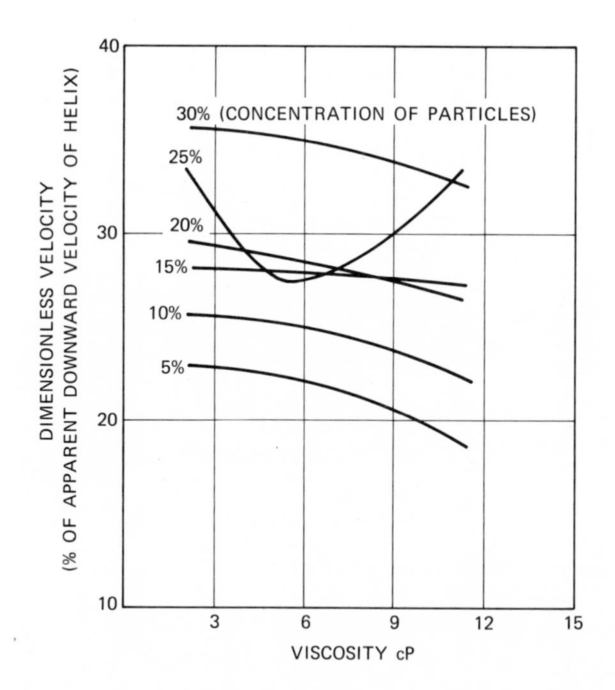

Figure 19. Effects of viscosity on transport of plastic particles in simulator.

Figure 20 shows the effect of particle concentration, for thick squares and cylinders, at different helix rotation speeds and with helical coils of various pitches. The coil, of 19mm pitch, was of optimal design for the transport of naphthalene crystals; no transport was obtained with coils wound from wire of larger diameter.

The results obtained can be classified according to the size of the particles. With particles that were too large to slip between the helix and the walls, three regimes were observed: (i) Particles of small positive or negative buoyancy were uniformly dispersed in the liquid, and transport appeared to be by a combination of viscous interactions between the helix and the walls. The speed of transport was proportional to the rotational speed of the helix, the proportionality coefficient being concentration dependent, but varied only slightly with viscosity over the range investigated. (ii) Particles of moderate to large excess densities tended to settle on the helix at moderate speeds of rotation and travelled mainly as a loosely packed bed, sliding down

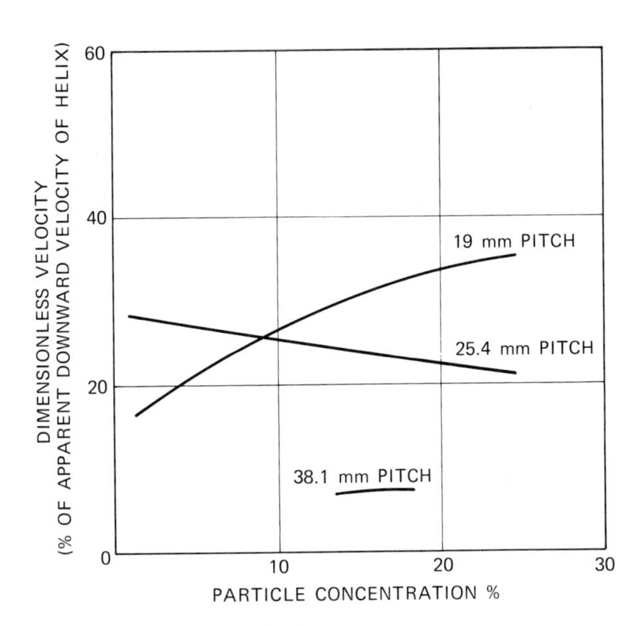

(a) with rotors of various pitch.

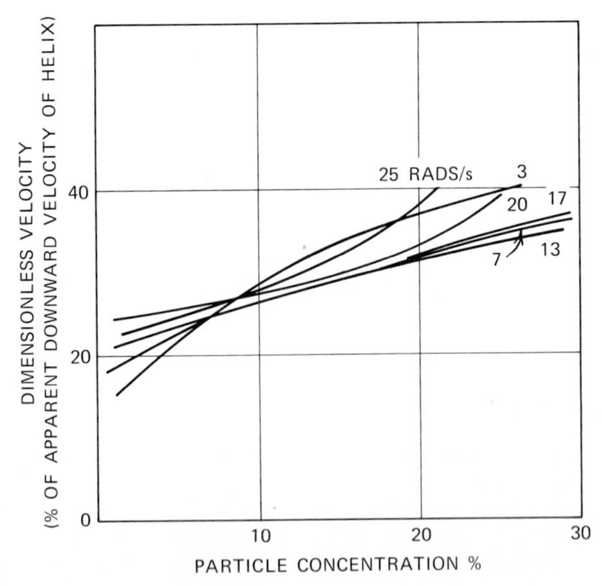

(b) at various rotation speeds.

Figure 20. Effects of particle concentration on transport in simulator.

the coil. Although the evidence is not given here, the speed of transport was again largely independent of viscosity, and the velocities attained were substantially less than the calculated Stokes free fall velocities. (iii) Particles of the same type as in (ii) became uniformly dispersed at higher rotation speeds and transport appeared to be by a combination of the mechanism in (i) and gravity.

With particles that could slip past the helix: (i) Particles with small negative or positive buoyancy underwent little transport at low concentrations other than by dispersion. (ii) Particles with moderate to large excess densities tended to slip past the helis at low concentrations, or to settle on the coil and slip down it.

Freeman has suggested that, where gravity is not the dominant force producing crystal transport, the rotation tends to produce a swirling motion of the crystal slurry due to frictional forces at the surface of the helix and, in order to obtain axial transport of crystals in the apparent direction of motion of the helix, the swirling motion must be restrained by a frictional force originating at the column walls. The crystals are also subject to forces arising from the viscous drag of the liquid, gravity, and the axial pressure gradient in the liquid. The liquid is subjected to forces similar in origin, but differing in magnitude, to those acting on the crystals.

An element of the helix can be represented by a rectangular slab of dimensions t x a x Δ ℓ, and its pitch is then given by a+t/cosΘ.

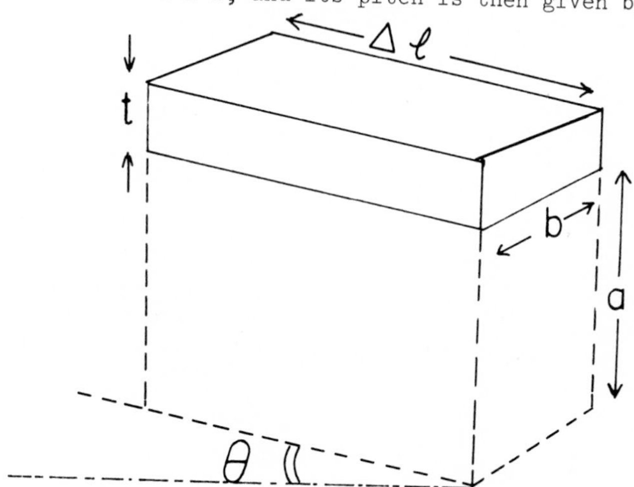

The velocities of the solid, V, and liquid, U, can be resolved into components parallel and perpendicular to the surface of the helix:-

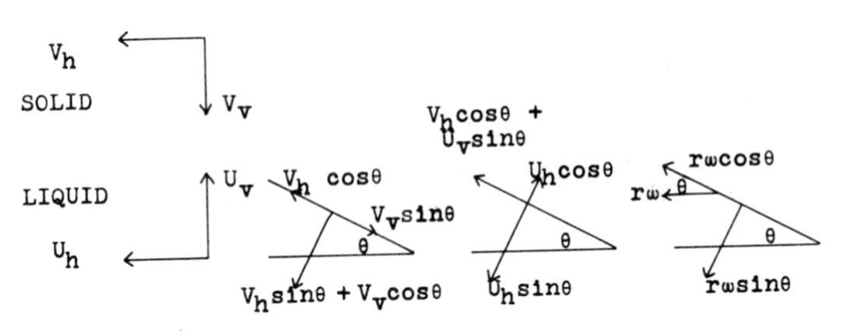

The forces acting parallel to the surface of the helix are:

Solid:

Gravity: $\phi s \Delta\rho g \, \text{Sin}\Theta \, a \, b \, \Delta l$;

Wall friction: $2a \, \Delta l \, (r\omega \, \text{Cos} \, \Theta - V_v \, \text{Cosec} \, \Theta) \, K_w n$;

Helical coil friction: $2b\Delta l \, (-V_v \, \text{Cosec} \, \Theta) \, K_c n$;

Viscous drag: $ab\Delta l \, \dfrac{9z\phi s}{2R^2} \, \left(-V_v \dfrac{\text{Cosec} \, \Theta}{\phi_L}\right) f_s \, (\phi_s)$;

Pressure drop: $P \, a \, b \, \Delta \, l \, \phi_s$;

Sum of forces zero: $\Sigma = 0$

Liquid:

Wall friction: $2a\Delta l - \dfrac{6z}{b} \left(- r\omega \text{Cos} \, \Theta - V_v \dfrac{\phi_s}{\phi_L} \text{Cosec}\Theta\right) f_L \, (\phi_s)$;

Helical coil friction: $2b\Delta l - \dfrac{6z}{a} \left(-V_v \dfrac{\phi_s}{\phi_L} \text{Cosec} \, \Theta\right) f_L \, (\phi_s)$;

Viscous drag (due to particles): $-ab\Delta l \, \dfrac{9z\phi_s}{2R^2} \, \left(-V_v \dfrac{\text{Cosec} \, \Theta}{\phi_L}\right) f_s \, (\phi_s)$;

Pressure drop: $P \, ab\Delta l \phi_L$;

Sum of forces zero: $\Sigma = 0$

Taking the sum of these forces as zero, and assuming that there is no relative motion perpendicular to the surface of the helix, the following expression is obtained for the axial velocity of the crystals:-

$$Vv = \frac{\left[\dfrac{Kw}{\phi_s} - \dfrac{Cw}{n\phi_L}\right] \omega r \, \text{Cos} \, \Theta \, \text{Sin} \, \Theta}{\left[\dfrac{Kw}{\phi_s} + \dfrac{K_c}{\phi_s} + \dfrac{D}{\phi_s \phi_L{}^2} + \dfrac{(Cw + C_c)}{n} \dfrac{\phi_s}{\phi_L{}^2}\right]}$$

Key:

V	= velocity of solids
U	= velocity of liquid
v (subs)	= vertical
h (subs)	= horizontal
ω	= angular velocity of coil
r	= radius of coil
Θ	= pitch angle of helix
a	= perpendicular distance between flights
b	= width of helix
t	= thickness of helix
Δl	= a length of helix
$\Delta\rho$	= density difference between solid and liquid
ϕ	= fraction
s (subs)	= solid
L (subs)	= liquid
f	= function
z	= viscosity
R	= radius of solid particles
n	= number of particles per unit volume
K	= solid frictional coefficient per particle
w (subs)	= wall
c (subs)	= helical coil
ρ	= pressure gradient

$$f_L(\phi_s) = \frac{Cwb^2}{12z} = \frac{C_c a^2}{12z} \; ; \; f_s(\phi s) = \frac{D.2R^2}{\phi_s 9z}$$

C	= liquid frictional coefficient
D	= viscous drag coefficient

Other forms of this expression can be obtained; e.g. insertion of the expressions for C_c and C_w for spherical particles, D and n:

$$C_w = \frac{12z}{b^2}\phi_L^{\;1-\alpha} \; ; \; C_c = \frac{12z}{a^2}\phi_L^{\;1-\alpha} \; ;$$

$$D = 6\Pi Rz\phi_L^{\;1-\alpha} \qquad n = \frac{\phi_s}{\frac{4}{3}\Pi R^3}$$

gives an expression containing terms which include the dimensions of the crystals and of the helix:-

$$V^* = \frac{\left[K_W - \dfrac{16\,\Pi\,R^3 z}{\phi_L^{\alpha}\,\phi_s\,b^2}\right]\cos^2\Theta}{K_W + K_c + \dfrac{6\,\Pi\,Rz}{\phi_L^{1+\alpha}} + 16\,R^3 z\,\Pi\,\dfrac{\phi_s}{\phi_L^{1+\alpha}}\left[\dfrac{1}{a^2} + \dfrac{1}{b^2}\right]}$$

The velocity of the crystals can be expressed in a dimensionless form $V^* = V_V/r\,\omega\,\tan\Theta$. It was found experimentally that the dimensionless velocity was independent of rotation speed, and that this is in accord with theory may be seen by comparing the transport equation for V_V and the above definition of V^*. The velocity was also almost independent of viscosity and, in order to account for this, the solid frictional factors must be proportional to viscosity. Available values of D & C, together with values for K estimated from sedimentation data for particles in tubes (56), can be inserted in the transport equation and this then becomes consistent with the observed results.

It can also be shown from the transport equation that $V^* = 0$ if the particle radius is zero, that large values of K_c are unfavourable, and that K_W should preferably be large. The expression for V_V shows a maximum at a pitch angle of 45°, but this is not consistent with the experimental results, and this probably arises from approximations in the derivation, especially those ignoring the curvature of the system. By substitution of the expressions for K_W and K_c (56)

$$K_W = 12\frac{a}{b}\,RZ\phi_L^{-\alpha}\,(\phi_L^{-40R/b} - 1)$$

$$K_c = 12\frac{b}{a}\,RZ\phi_L^{-\alpha}\,(\phi_L^{-40R/a} - 1)$$

it can be demonstrated that increasing the width of the annulus should reduce V_V. In the experiments with plastic particles transport ceased abruptly when the annulus was increased by 50% over the optimum value.

It may be concluded from these considerations that the forces responsible for crystal transport are probably as postulated, transport depending on the frictional forces acting on the crystals. Ideally, the friction should be large at the wall, and small at the rotor, implying rough surfaced walls and a smooth surfaced rotor. The viscosity of the liquid phase does not have a large direct effect on transport but affects it indirectly due to its influence on crystal growth, and it has been shown that the crystals should preferably be large. Crystal transport should be encouraged by using a small annular width relative to column diameter.

VII. CONCLUSION

A continuous method of fractional crystallization has been developed from the prototype devised by Schildknecht and is capable of separating mixtures to give high-purity materials in good yields. The high throughputs obtainable from small units, e.g. over 5 l/h from a 50mm column, make these immediately attractive for the preparation of special batches of material for research purposes or the regular production of fine chemicals. On the larger scale, the low energy requirements of the process become an important advantage. Columns of up to 200mm diameter have been operated successfully; the experiments indicating geometric similarity as the basis for scale-up. A 2 times increase in diameter produces a 4 times increase in throughput, and assuming that this scale factor can be extrapolated, throughputs of 1000 to 3000 tons p.a. would be obtained with columns of 300 to 500mm diameter and 4 to 9m long, although for any given throughput of this order it would probably be preferable to use two or more columns rather than a single larger one in order to facilitate mechanical maintenance. Power requirements for driving the rotor are not likely to be excessive.

As the column size is increased, the detached helical coil used in the original crystallizer design becomes progressively more difficult to fabricate and operate with the necessary small clearance between the helix and the walls. It has been found possible, however, to replace the helical coil by an Archimedean screw provided this has the appropriate, closely defined, geometric proportions and that it is operated at a suitably higher speed than the detached helix. Using this system the standard of engineering required is within the scope of current practice.

On the basis of geometric similarity for the cooling section, scale-up should not be limited by heat-transfer capability but, if desirable, the heat-transfer surface within the column may be partly replaced or augmented by using a pre-crystallizer to convert the feed to a slurry before it enters the main column.

The process is applicable to both mixed-crystal and eutectogenic systems provided the composition is sufficiently removed from that of any eutectic, intermediate compound or minimum-melting point mixture that may occur. Various analyses of column performance have been proposed but these have not so far gained full acceptance, nor have they been fully tested against actual column behavior. These analytical treatments, whilst providing information on the required length for a column of particular design and a specific separation, do not give any guide to rotor design which remains largely empirical. Crystal transport simulation studies suggest that the application of the process is limited to systems giving rise to crystals larger than the clearance between the rotor and the shell.

With the columns used so far, this criterion restricts applicability to materials with a viscosity less than 10 cP, or preferably less than 5 cP, but it is hoped that work on the promotion of crystal growth or on methods of reducing the clearance, will remove this limitation.

In order to test the validity of the scale-up predictions, and to examine those factors not readily investigated in the smaller crystallizers, a 230mm diameter column with an Archimedean rotor is under construction by Benzole Producers Ltd. It has been designed primarily to treat benzene and has a nominal throughput of 115 l/h (37). The C.T.R.A. is designing a 300mm column to treat naphthalene and hopes to erect this during 1970.

VIII. REFERENCES

(1) G.M. Wolten & W.R. Wilcox, in 'Fractional Solidification', Vol. 1, M. Zief & W. R. Wilcox Eds., Edward Arnold, London, 1967, p. 21.
(2) Farbwerke Hoechst A.G., Neth. Pat. Appls. 6,505,427, (1965), 6,611,646, (1967).
(3) Battelle Development Corp., Neth. Pat. Appl, 6,505,171 (1965).
(4) H.F. Wiegandt, U.S. Pat. 3,251,193, (1966).
(5) A.C.J. Kuivenhoven, Bull. Inst. Int. Froid. Annexe No. 3, 85, (1966).
(6) M. Landau & A. Martindale, Proceedings European Symposium on Fresh Water from the Sea, Athens, 1967.
(7) Institut Francais du Petrole, des Carburants et Lubrifiants Neth. Pat. Appl,, 6,614,087, (1967).
(8) R. Lafay, S. African Pat. 68,03,832 (1968).
(9) H.F. Wiegandt & M.R. Lafay, 7th World Petrol Congr. 1967, Proceedings, 4, 47.
(10) G. Matz, Symposium uber Zonenschmelzen und Kolonnen-Krystallisieren 1963; Kernforschungszentrum, Karlsruhe, p. 345.
(11) J. Lindley (I.C.I.) Brit. Pat. 1,142,864, (1966).
(12) E. S. Grimmett (Phillips Petroleum Co.) U.S. Pat. 3,392,539, (1968).
(13) J. Sladky, S. Kulla, & V. Kalab, Czech Pat. 126,518, (1966).
(14) K. H. Hachmuth (Phillips Petroleum), U.S. Pat. 2,593,300, (1952).
(15) P. M. Arnold (Phillips Petroleum), U.S. Pat. 2,540,977, (1951).
(16) D. L. McKay in 'Fractional Solidification', Vol. 1, M. Zief & W. R. Wilcox, Eds, Edward Arnold, London 1967, p. 427.
(17) R. Albertins, W. C. Gates & J. E. Powers, ibid., p. 343.
(18) Imperial Chemical Industries, Neth. Pat. Appl., 6,610,208 (1967)
(19) Imperial Chemical Industries, Neth. Pat. Appl., 6,515,674, (1966).
(20) R. Scott & E. H. Joscelyne, Brit. Pat. 630,387.
(21) L. K. Frevel, U.S. Pat. 2,659,761, (1953).
(22) Dow Chemical Co., Brit. Pat. 669,868.

(23) R. V. Austin, Ph.D. Thesis, University of Leeds, 1951.
(24) H. Schildknecht & H. Vetter, Angew, Chem. (1961), 73, 612.
(25) H. Schildknecht & K. Maas, Warme (1963), 67, 121.
(26) H. Schildknecht, Z. Analyt. Chem. (1961), 181, 254.
(27) H. Schildknecht, K. Maas, & W. Kraus, Chem.-Ing. Tech., (1962), 34(10), 697.
(28) H. Schildknecht, Chimia (Aarau), 1963, 17, 145.
(29) H. Schildknecht, S. Rossler & K. Maas, Glas-Instr. Tech., (1963), 7, 281.
(30) K. Maas & H. Schildknecht, in 'Symposium uber Zonenschmelzen und Kolonnenkristallisieren', H. Schildknecht, Ed; Kernforschungszentrum, Karlsruhe, 1963, p. 373.
(31) F. Schegelmilch & H. Schildknecht, ibid, p. 437.
(32) H. Schildknecht, presented at Symposium on Crystallization, 56th Natl. Meeting A.I. Chem. E., San Francisco, 1965.
(33) H. Schildknecht & J. E. Powers, Chemiker-Ztg/Chem. App. (1966), 90(5), 135.
(34) H. Schildknecht, Anal. Chim. Acta, (1967), 38, 261.
(35) W. D. Betts, J. W. Freeman and D. McNeil, J. Appl. Chem. (1968), 17, 180.
(36) W. D. Betts & J. W. Freeman, unpublished work.

ENDLESS BELT ELECTROPHORESIS*

Alexander Kolin and Stephen J. Luner
Department of Biophysics
School of Medicine
University of California, Los Angeles

*This work was supported by the Office of Naval Research, U.S.P.H. Grants GRSG 1 SOL FR 05354 and 5 TI-MH-6415, and Cancer Detection Services, Inc.

I. DEVIATION ELECTROPHORESIS

The earliest successful attempts to achieve continuous flow separation by deviation electrophoresis utilized a porous matrix curtain to inhibit thermal convection in the electrophoretic buffer column (1). A buffer solution derived from a reservoir descended through a vertical rectangular buffer-soaked curtain of sand or glass beads and left the curtain drop-wise. The mixture of molecular solutes to be separated entered at the top of the curtain as a streak. Two lateral vertical metal strips lining the left and right edges of the porous bed imposed a horizontal electric field upon the ions in the wet curtain causing the buffer ions as well as the ions in the streak to migrate and the latter to separate according to differences in electrophoretic mobilities.

The application of this configuration to the separation of particle suspensions containing cells or subcellular particles required elimination of the porous bed or filter paper sheet later substituted for it so as to avoid adsorption and immobilization of the particulate ingredients by the solid matrix. Thus, Barrolier, Watzke and Gibian (2) replaced the filter paper by a rectangular liquid sheath sandwiched (Figure 1) between two glass plates which were inclined slightly against the horizontal plane.

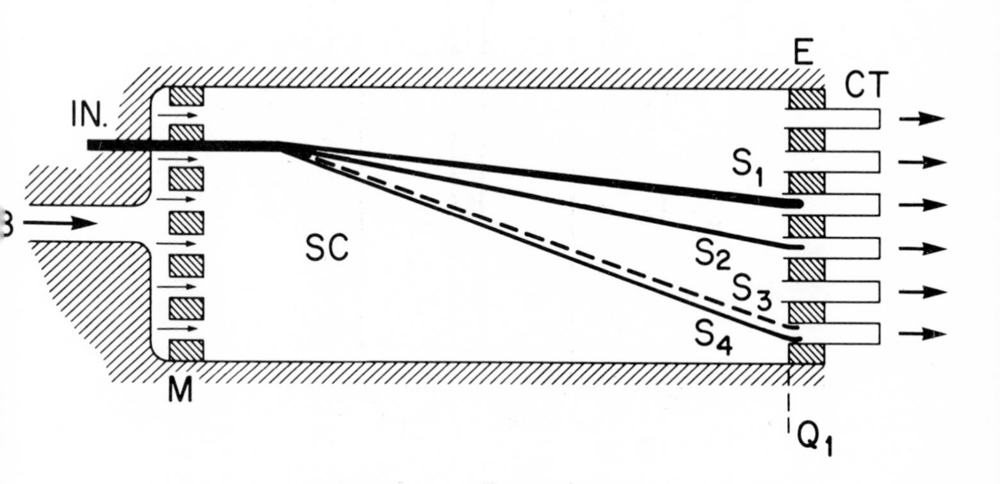

Fig. 1. Scheme of flowing curtain electrophoresis (top view). Continuous flow separation of four components S_1-S_4 entering from injector IN into the separation cell SC and leaving the cell through collector tubes CT. M = manifold. B = tube conveying the buffer solution. O_1 = section at the location of the end plate E. (From A. Kolin, J. Chromatog., 17, (1965) 532-537).

The use of an inclined plane has the advantage over a vertical plane of reducing thermal convection. One pays, however, for this improvement by a drawback which is aggravated as the horizontal position is approached. The streak which is normally denser than the surrounding buffer tends to sink toward the lower plate and denser particles suspended in it tend to sediment upon the bottom plate. Hannig preferred to avoid the sedimentation by vertical placement of the plates (3) and by using an elaborate system for a very close temperature control within the buffer column to reduce thermal convection (3).

II. THE PROBLEM OF THERMAL CONVECTION

Thermal convection is the main experimental difficulty encountered in carrying out electrophoresis in vertical fluid sheaths or columns. Figure 2 shows the cross-section of a rectangular fluid column which we can imagine to be carrying a vertical electric current.

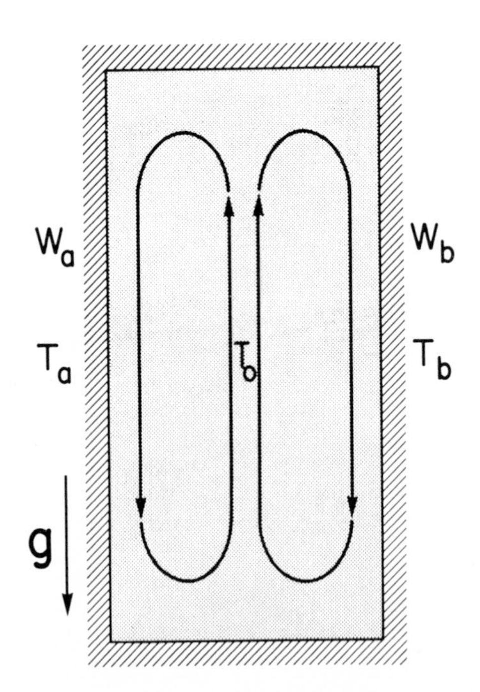

Fig. 2. Type of convection engendered when fluid of temperature T_0 is introduced between walls W_a, W_b maintained at temperatures T_a and T_b ($T_a < T_0 > T_b$). (From A. Kolin, (4)).

If the column is of negligible width as compared to its length and height, the heat can escape mainly through its side walls so that a horizontal temperature gradient will develop in the liquid column.

This temperature gradient will create a concomitant horizontal density gradient with the highest density near the cool walls and the lowest density at the central temperature maximum (provided the coolest point is above 4°C). Thus a circulation pattern will result in the column (4) as shown in Figure 2.

An analogous convection pattern will result in a horizontal annular fluid column traversed by an axis-parallel current. Such a convection pattern is shown in Figure 3 (4).

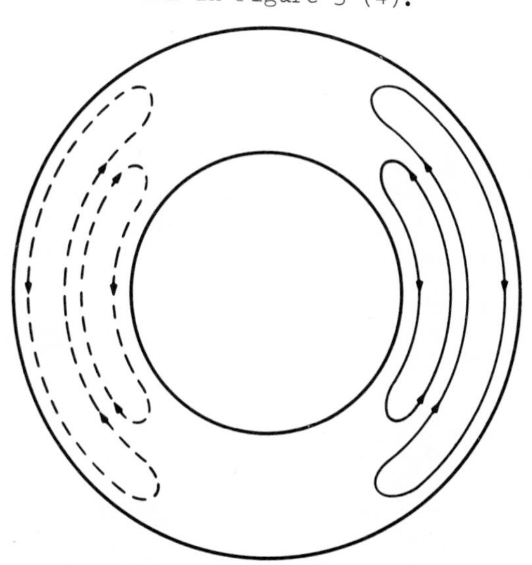

Fig. 3. Drawing of convection pattern observed by means of dye streaks when water at room temperature is introduced between two horizontal cylinders kept at 0°C. The convection pattern is analogous to the pattern between walls shown in Fig. 2. (From A. Kolin, (4)).

It is obvious that such convection patterns will tend to cause remixing in separation patterns achieved in fluid columns heated by electrophoretic currents. It is the objective of this paper to consider a process which allows suppression of such convection by appropriate circulation of the fluid in the electrophoretic column and to show how high-resolution deviation electrophoresis has been made possible in liquid sheaths thus freed from thermal convection.

A. Suppression of Thermal Convection by Inversion of Vortices

Figure 4a shows a vortex drawn as a solid line on the right-hand side of the fluid-filled annular space between two horizontal concentric cylinders. We can imagine this vortex to have been formed by maintaining the inner cylinder surface at a higher temperature than the outer cylinder surface.

Let us imagine now that the fluid in the annular space has rotated 180° so that the vortex now occupies the position shown on the left as a dashed line. We see that the points of the vortex loop which moved downward before the rotation are now moving upward after the rotation and *vice versa*. The vortex motions in the new directions are, however, now inhibited by the forces of buoyancy and gravity. The motion is now downward near the warm inner wall where the predominant force of buoyancy acts upward on the descending warm fluid of low density, whereas the upward motion of the fluid in the vortex near the cool outer wall is retarded by the predominant force of gravity acting on the cool dense fluid. The end result is a retardation of the circulation in the vortex which eventually comes to a standstill (4). The thermally engendered vortices can thus be extinguished simply by circulation of the fluid (as shown in Figure 4a). This circulatory motion does not allow sufficient time for accumulation of disturbing amounts of rotational kinetic energy in incipient vortices and consistently applies a braking torque to nascent vortices which inhibits their development. The circulation does not have to follow a circular path as is shown in the self-explanatory Figure 4b (4).

B. Suppression of Thermally Engendered Vortices by Electromagnetic Circulation

We can now ask the question as to how to generate the fluid circulation which, as was shown above, inhibits formation of thermal convection. It turns out that there is a simple way of accomplishing it by electromagnetic pumping. The pumping action is based on exertion of an electrodynamic force upon a current-carrying fluid by a magnetic field perpendicular to the current. The force is at right angles to the current and to the magnetic field.

Figure 5 shows a configuration similar to that shown in Figure 4a. The inner cylinder is labeled C_1 and the outer one C_2. The fluid between them is traversed by an electric current flowing from electrode E_1 to electrode E_2. The electrodes are in compartments EC_1 and EC_2 which communicate through the annular space between the cylinders C_1 and C_2. To impart a rotational motion to the fluid in the annulus, a tangential force was used: Since there is an axis-parallel current flowing through the annular space, a radial magnetic field was utilized to create such a force through interaction with the axial current.

Figure 6 shows a configuration of bar magnets which generates a substantially radial magnetic field over an extended distance by sandwiching a soft iron bar in between two like poles of identical bar magnets. This configuration of magnets is used in Figure 5 inside the cylinder C_1. The buffer solution in the annular space is, thus, set in rotational motion of great constancy after establishment of a steady state subsequent to turning on the current. The circulation of the fluid through the annular space accomplishes the desire inversion of vortices and, thus, the desired stabilization against thermal convection.

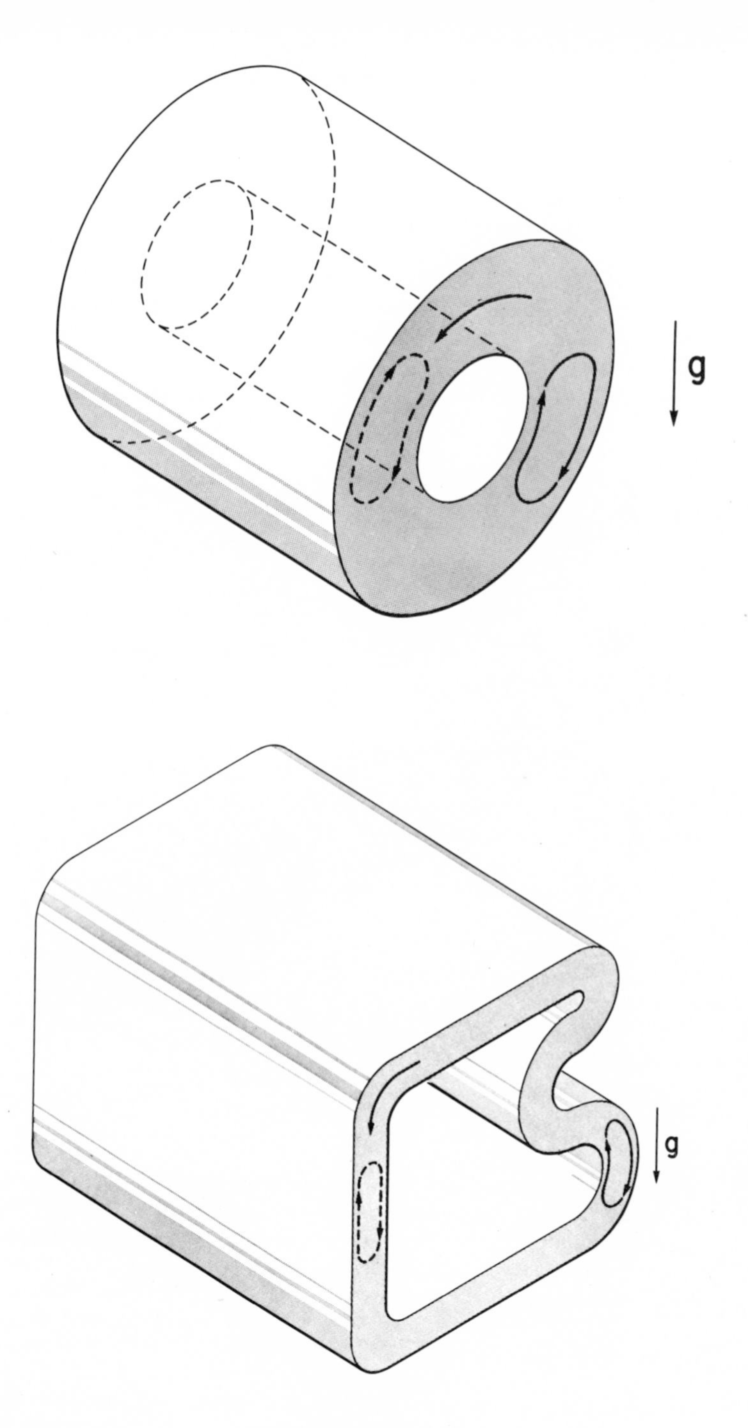

Figure 4, a, b.

Fig. 4. (a) Fluid rotating in the annular space between two horizontal concentric cylinders. g: gravitational field vector. The convection pattern shown as a solid loop on the right is inverted relative to the g vector when it is transported to the left side, as shown by the dashed loop. (b) Fluid circulating through a noncircular channel similarly to the motion shown in (a). The convection pattern is inverted relative to g by transfer from the right to the left channel section as in the case shown in (a). (From A. Kolin, (4)).

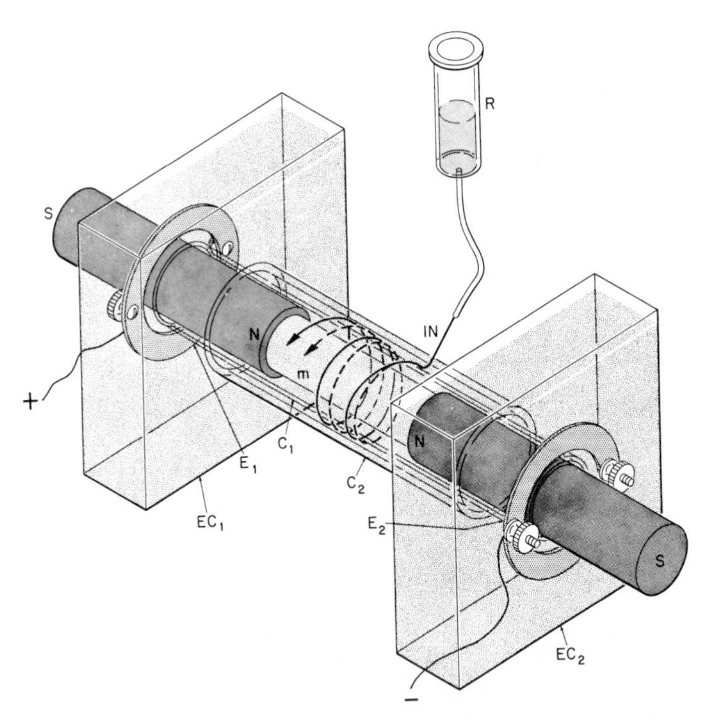

Fig. 5. Scheme of helical path electrophoresis. N-S = Bar magnets; m = soft-iron cylinder E_1, E_2 = electrodes; EC_1, EC_2 = electrode compartments; C_1 = inner lucite cylinder; C_2 = outer lucite cylinder; IN = injector; R = reservoir for mixture to be analyzed. (From A. Kolin, (6)).

III. HELICAL PATH DEVIATION ELECTROPHORESIS

A non-conducting injector IN (for instance, a fine L-shaped glass capillary) terminates midway between the cylinders C_1 and C_2 (Figure 5). It is connected to a reservoir R from which a fluid containing ions or charged suspended particles enters the annular space as a fine streak (about 0.2 mm in diameter). If the injected particles are neutral, they go into a circular orbit in which they accumulate. If they are electrically charged, their motion is a superposition of linear translation along electrical field lines with a revolution about the central axis which is imposed upon them by the electromagnetic rotation of the fluid. The resultant streak formed by the injected particles is thus a right- or left-handed helix depending on the sign of the charge. If the streak contains components differing in electrophoretic mobilities they will form streaks of different helical pitch as illustrated by the solid and dashed helix in Figure 5. The pitch of the helix is a measure of the electrophoretic mobility.

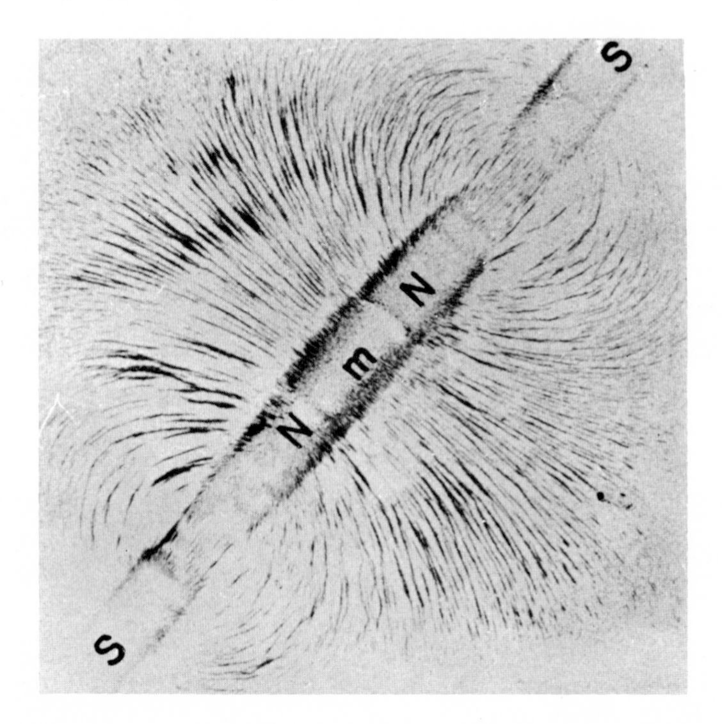

Fig. 6. Magnetic field surrounding the configuration of two bar magnets NS with an intermediate soft-iron cylinder m. (From A. Kolin, (6)).

The actual apparatus has two lucite walls and a bottom which are cemented to the compartments EC_1, EC_2, thus forming a tub around the cylinder bridge. This tub is filled with an ice-water mixture for the purpose of cooling the electrophoretic column. This generates a very steep temperature gradient which, for a temperature of, say, 40°C at the center of a 1 mm wide annulus, would be on the order of 150° C /cm. In spite of this extremely steep temperature gradient, the streaks show no evidence of thermal convection, as can be seen from figure 7 (5). In a typical cell of about 3 cm mean annular diameter and 1.5 mm annulus width, the normal current density would

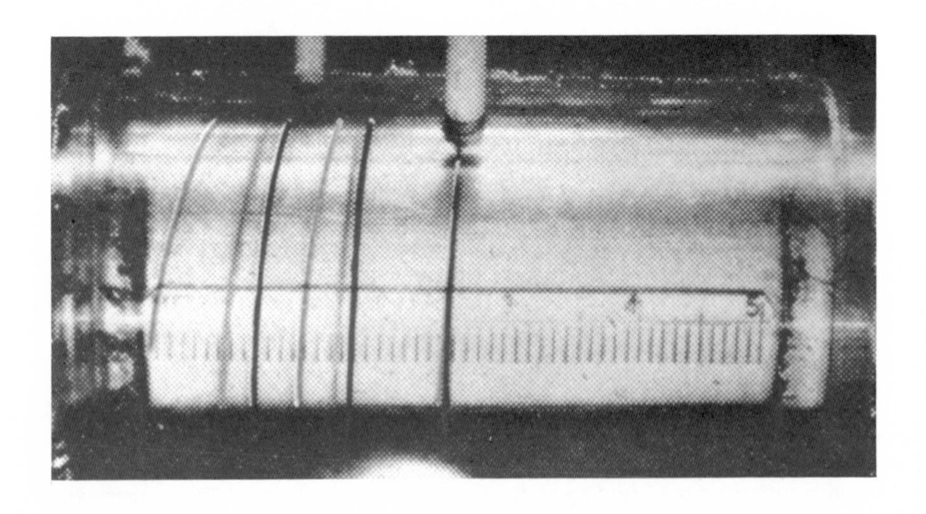

Figure 7

Fig. 7. Separation of a mixture of India ink, rose bengal, and Evans blue. The sequence of lines seen from right to left is: (1) initial streak in beginning stage of separation, (2) rose bengal, (3) India ink, (4) Evans blue, (5) rose bengal, (6) India ink, (7) Evans blue. (From A. Kolin, (5)).

be 10^{-2} amp/cm^2, the potential gradient 50 volts/cm, and the intensity of the radial magnetic field 150 gauss. For an ion of 10^{-4} cm/sec (volt/cm) mobility, the helical pitch will be in the order of 1 cm and the revolution time of the fluid in the annular column in the neighborhood of half a minute. Figure 8 shows how rapidly a clearly visible separation of two dyes (rose Bengal and Evans blue) can be obtained.

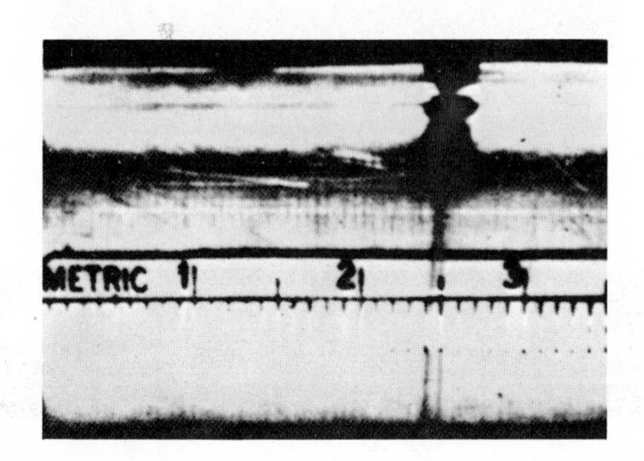

Fig. 8. Illustration of separation speed. Separation of rose bengal from Evans blue achieved within approximately 4 sec at a cell voltage of 700 volts. (From A. Kolin, (5)).

The time required to traverse the distance shown was about 4 seconds. The dyes achieved a distinct separation in a fraction of this time interval.

Figure 9 shows an example of a separation of two microorganisms, the yeasts _Saccharomyces cerevisiae_ and _Rhodotorula_. In this case, the central soft iron cylinder m of Figure 5 has a "window", i.e., a gap which has been milled out. This gap permits transillumination of the electrophoretic column and visualization of particles suspended in the streaks by light scattering.

If a separation of d mm has been achieved after one helical turn, an N-fold separation between two streaks will be obtained after N helical turns. It is possible to insert into the annulus a collector, comprising a number of adjacent fine plastic tubes oriented at their terminal parallel to the tangential velocity of the rotating fluid and to collect the separated components in different tubes, as shown in Figure 10 after 4 helical turns.

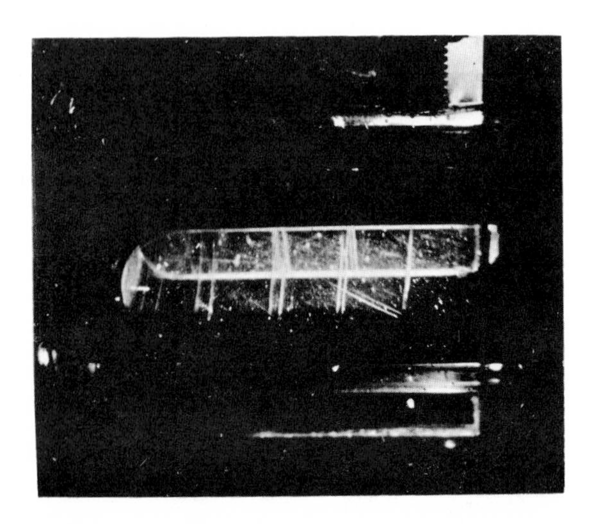

Fig. 9. Separation of two types of fungi: Saccharomyces cerevisiae and Rhodotorula in dilute buffer solution. (From A. Kolin, (6)).

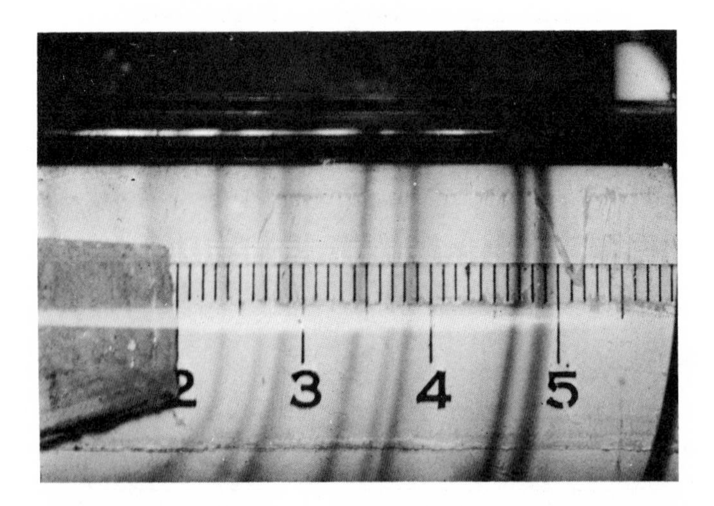

Fig. 10. Pattern of three separated dyes (Evans blue, rose Bengal and "Brush"green recording ink). The streaks are entering the activated collector. (From A. Kolin, (6)).

It is also possible to impose an axial fluid velocity upon the buffer in the annular space by allowing buffer to enter one electrode compartment and/or to flow out of the other. For a steady state, one normally does both at the same time and at the same rate. Figure 11 shows how the helical pitch can thus be modified.

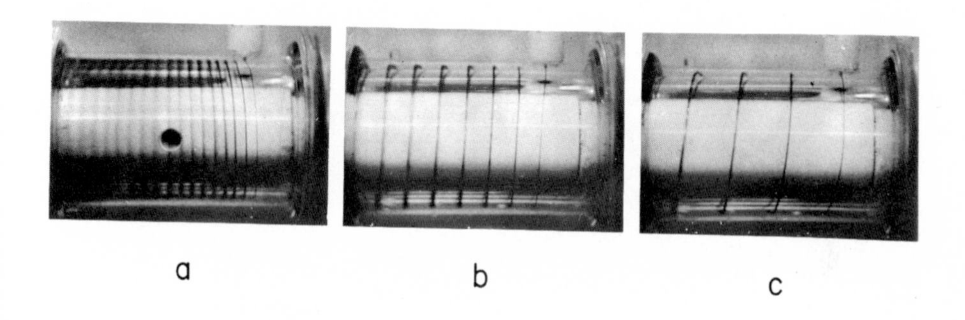

Fig. 11. Effect of axial streaming on helical pitch. The axial flow is toward the left and is increasing as we process from a to c. The injector is seen at the top on the right in each photograph. (From A. Kolin, (6)).

A. Velocity Distribution in the Circulating Belt

In the preceding discussion we disregarded the fact that the fluid velocity of the circulating buffer solution is not uniform throughout the annulus. For an annulus whose thickness, h, is negligible in comparison with its mean radius of curvature R, the rotational flow can be treated like a laminar flow between two closely spaced flat parallel plates (5, 7). The velocity distribution is parabolic and is given by the expression

$$u = \frac{z(z-h)}{2\eta} \; (dp/dx), \qquad [1]$$

where u is the tangential fluid velocity at a radial distance z from the inner cylinder C_1, η the viscosity, and (dp/dx) the force per unit volume directed tangentially (5). Since the materials under investigation are injected at the center of the annulus, we are interested particularly in the velocity u at z = h/2. Remembering the electrodynamic origin of the force density (dp/dx), we get (5):

$$u = \frac{h^2}{80\eta} \{\vec{J} \times \vec{B}\}, \qquad [2]$$

where $\vec{J}$ is the current density and $\vec{B}$ the magnetic field vector. The revolution time τ of a particle in the streak is (5)

$$\tau = 2\pi R/u_o = 160\eta\pi R/(h^2 JB). \qquad [3]$$

The pitch of the helical streak is given by the displacement s_o of an ion migrating electrophoretically, with velocity v in the streak in the course of one revolution (5):

$$s_o = v\tau_o = 160\,\pi\eta RU/(h^2 B\sigma). \qquad [4]$$

The pitch s_o is thus independent of the current density J (σ is the electrical conductivity of the buffer): This paradoxical result is due to the fact that both axial electrophoretic migration velocity of the ion and also the electromagnetically engendered circulation velocity are proportional to J.

B. Electro-osmotic Flow

In the preceding calculations we considered the axial velocity of the ions to be solely due to electrophoresis and to be independent of the distance z from the inner cylinder surface. Actually, a second electrokinetic phenomenon is produced by the applied electric field, electro-osmosis, which superimposes an axial parabolic velocity distribution upon the tangential fluid circulation so that the ions combine their axial electrophoretic migration with the axial electro-osmotic flow which is superimposed upon the tangential electromagnetically engendered circulation.

The phenomenon of electro-osmosis arises from the presence of an ionic atmosphere in the electrolyte adjacent to the cell walls. The charges of the ionic atmosphere are set in motion by the electric field and a streaming motion in the direction of their movement results from the interaction of these ions with the ambient fluid. Since there is no axial transfer of fluid in the steady state, the fluid streaming near the walls is compensated by fluid transfer in the opposite direction in the central region between the walls. It has been shown (8) that the resultant velocity profile is parabolic, as shown in Figure 12 (5). The diagram is to be understood as depicting the interspace of width h between the inner and outer plastic cylinders C_1 and C_2 which confine the annular electrophoretic buffer column. Noteworthy of this velocity distribution is the existence of two zones of zero axial velocity at a distance of 0.21 h from either wall and the fact that the fluid moves in opposite directions between these zones as compared to the motion of fluid beyond them. Another remarkable feature is the non-zero velocity adjacent to the walls.

It is thus clear that the fluid elements of the electrolyte will move in circular orbits only in the Smoluchowski zones of zero

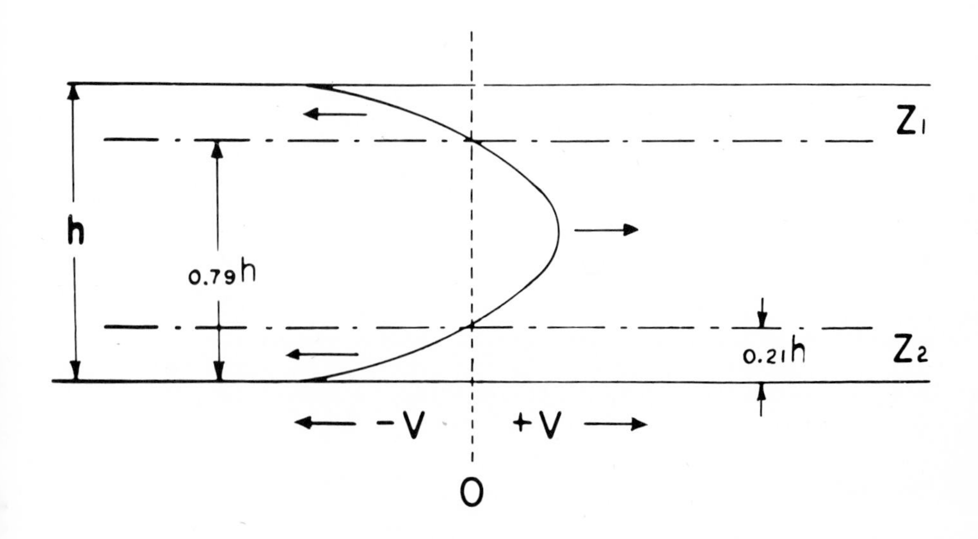

Fig. 12. Velocity distribution between parallel planes (perpendicular to the page) due to electro-osmotic streaming. The applied electric field is parallel to the arrows. The arrows indicate the direction of flow. Z_1, Z_2: zones of zero velocity (parallel to the walls). h: distance between the parallel planes. (From A. Kolin, (5)).

axial velocity. Outside these zones the paths will be helical. In the case of electro-osmosis generating a right-handed helix in between these zones, the helix beyond them on either side will be a left-handed one. The electrophoretic migration of the charged particles in the injected streak will thus be, in general, superimposed upon the axial electro-osmotic streaming which will thus modify the pitch of the resultant helical streak except for the cases where the particles are injected into and confined to the Smoluchowski zones.

In practice, one chooses materials exhibiting a low zeta potential in the chosen buffer or coats the walls with a material which will bring about this condition. Electro-osmosis is thus minimized by choice of suitable hydrophobic materials for the walls of the electrophoretic column or by coating hydrophilic walls which exhibit a high zeta potential (for instance, glass or quartz), with suitable materials such as "methocel" or silicones which reduce the electrokinetic potential of the walls.

In preparative work, electro-osmosis is not disturbing. In determinations of the electrophoretic mobilities of the separated fractions, the result would be, of course, incorrect without making allowance for electro-osmosis. This can be easily accomplished by injecting a reference dye (e.g., Apolon) whose electrophoretic mobility remains zero over a wide pH range (approximately pH = 3 to pH = 10). The Apolon streak provides thus a reference line of zero mobility even in cases where an axial streaming is deliberately imposed upon the electrophoretic column as shown in Figure 11. A second dye (e.g. Brilliant blue) whose electrophoretic mobility remains constant over the same pH range provides a standard of known mobility in terms of which the mobilities of the unknown fractions can be expressed. The mobility can also be determined absolutely from the voltage and period of circulation of buffer. This period is conveniently measured by imposing a sudden lateral disturbance upon the streak and timing its passage across the same horizontal line after one revolution. The ratio of the electrophoretic displacement to the revolution time yields the electrophoretic velocity from which the mobility follows since the voltage drop per centimeter of axial distance in the column can be determined with a vacuum tube volt meter.

C. Resolving Power

Two components which form two distinct streaks separated by a distance d after one revolution will be Nd cm apart after N revolutions. If the width of the individual streaks would not increase with successive turns of the helix, there would be a distinct gain in resolving power in separations and observations performed after many helical turns. This would be expected especially in separations of microscopic particles and macromolecules where diffusion broadening of the streak would be minimal. This expectation is only partly fulfilled since there is a broadening effect which tends to widen the streak as a result of the parabolic tangential velocity distribution in the annulus. We shall refer to this effect illustrated in Figure 13 (11) as "parabolic divergence" (6).

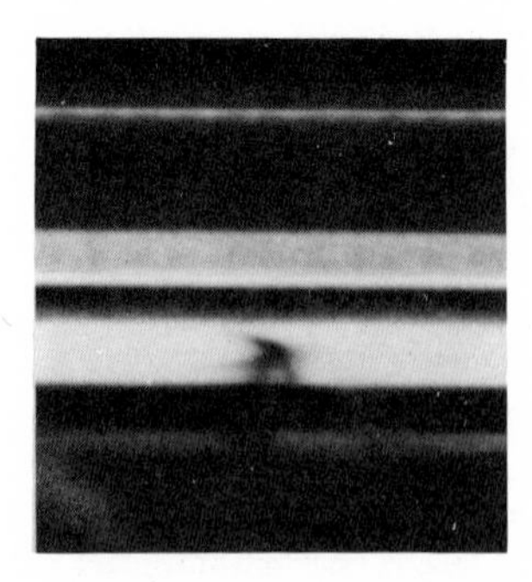

Fig. 13. Parabolic distortion of the cross-section of an originally circular streak. (From S. Luner (11)).

Because of the finite diameter of the streak, its edges which are nearest the walls of the annulus are located in regions of lower tangential velocity than the streak center or the streak edges located in the central region between the annulus walls. As a result, the pitch of the helices described by the particles on those off-center edges will be larger than for the most rapidly revolving particles at the streak center and at the central locations on the streak edges. Due to the angle between the trajectories of the particles occupying different positions in the streak, its boundaries will diverge. This broadening of the streak reduces the resolving power below the ideal value for an infinitely thin perfectly centered streak.

We can define the resolving power R of the electrophoretic separator as the reciprocal of the ratio between the mobility difference ΔU of two electrophoretically distinct components of mobilities U_1 and $U_2 > U_1$, which can just be resolved when the front edge of the lagging streak contacts the rear edge of the leading streak, and the mobility of the faster component (6):

$$R = \frac{U_2}{\Delta U}. \qquad [5]$$

It can be shown (6) that a separator with an annular gap h and an injector of inner diameter d (which determines the width of the entering streak) will have a resolving power given by

$$R = \left(\frac{h}{d}\right)^2. \qquad [6]$$

We see thus that R increases with the square of the annulus width/streak diameter ratio and that it can be made as large as we please by diminishing d. We have less liberty in increasing h because of dangers of turbulence and ensuing thermal convection for large annular gaps. The scale of separation can be increased by increasing d but this gain is obviously achieved at the expense of resolution.

D. Modes of Operation

Separations can be carried out in several different modes (6).

(1) Single-order collection

The apparatus can be adjusted so that the components of the *n*-th helical turn enter the collector in the same order in which they appear in the separation pattern. The decision on after how many helical turns to intercept the streaks by the collector is to be based on the two least clearly resolved components. The separation between two adjacent streaks should be at least equal to the distance between two neighboring collector tubes to minimize the hazard of re-uniting the split components by collecting them in the same collector tube. The optimum condition for collection of two separated components is achieved when component A is located midway between two consecutive turns of component B near the approach to the collector. Any deviation from the central location during the preceding or subsequent helical turns will diminish the distance of the streak of component A from one of the adjacent streaks of component B.

(2) Split-order collection

The terminal points of the helical separation pattern can also be intercepted by the collector so as to collect the desired number of the fastest components in the *n*-th turn while permitting the slower components to complete an additional turn before collection. This may be advantageous in cases where the fastest components are clearly resolved while the slower ones require an additional turn for adequate resolution.

(3) Isoelectric accumulation

Isoelectric accumulation of a chosen component can be achieved in the absence of electro-osmotic streaming at zero axial flow in a buffer corresponding to the isoelectric pH of the component in question. In this case, the pitch of the helical path is zero and the isoelectric material accumulates in a circular orbit while charged components of either sign wander away in both directions in right handed and left handed helical paths. When desired, the accumulated electrically neutral material can be transported to the collector by establishing axial streaming.

(4) Non-isoelectric accumulation

The accumulation method described in (C) is not limited to ampholytes at their isoelectric point. One can compensate the electrophoretic migration of any ionized species by an axial counter-flow so as to accumulate it in a stationary circular orbit. This makes it possible to detect and accumulate materials present in low concentration.

(5) Zonal separation

Zonal separation of very minute amounts of suspended or dissolved material injected in volumes of the order of a few tenths of a microliter can be accomplished by simply injecting a very fine streak about 1 cm long, or shorter, permitting it to revolve about the inner cylinder on a helical path. In the course of this motion, the short streak breaks up into separate component streaks which land in different collector tubes. It is advantageous to adjust the axial flow so as to place the slowest component into a stationary circular orbit so that the full length of the migration path becomes available for the separation pattern.

(6) Multiple separations

The separation scale of continuous fractionation can be increased by mounting several injectors side by side so as to inject the same mixture whose separated components can be collected by one collector. The same arrangement can be used for simultaneous analysis of several different mixtures if the mixtures introduced by the injectors are not identical.

(7) Multi-stage separations

Inadequately purified fractions can be fed directly into the reservoir R of a second separator operated in series with the first one or they can be re-injected after collection into the same separator for reprocessing.

E. The Scheme of the Vertical Endless Fluid Belt

As we can see from equation {6}, it is not advisable to increase the separation capacity by increasing the width of the emerging streak. This can, however, be achieved without altering the geometry of the apparatus by increasing the concentration of the injected solution or suspension. In both instances the density of the streak will be increased with materials normally used in biological work. The consequence of excessive streak density is shown in Figure 14a.

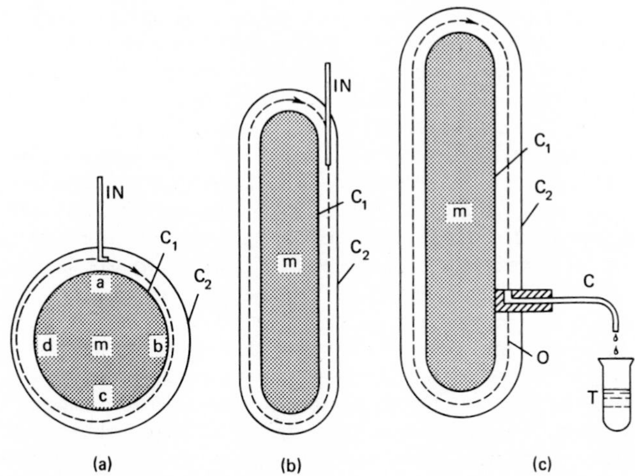

Fig. 14. Principles of stabilization against thermal convection and of suppression of particle sedimentation by meandering and rotational flows. (a) Cross section of an annular horizontal buffer column maintained in rotational motion (indicated by arrow) by interaction between a radial

magnetic field (emanating from iron core m) and an axial electric current (perpendicular to the plane of the paper). IN, injector; C_1, C_2, the inner and outer boundaries, respectively, of the rotating annular buffer column. Dashed line: path of a dense particle emerging from IN and performing a motion which combines sedimentation with rotation about the center of cylinder m. (b) Configuration similar to the scheme of (a) except that m and the shape of the rotating fluid ribbon are no longer circular. IN is the injector, and the dashed line represents again the path of a dense particle. Sedimentation is no longer noticeable because of the shortness of the horizontal component of the particle motion as compared to the total path length. C_1 and C_2, inner and outer boundaries of the noncircular annular horizontal fluid column. (c) Scheme of collection of separated fractions in the method shown in (b). Whereas (b) shows the path of a particle (dashed line) as it emerges from the injector near the buffer compartment B_2, (c) shows it as it approaches and enters the collector C located near the buffer compartment B_1. C_1, C_2, inner and outer boundaries of the rotating buffer belt; t, tube connected to the L-shaped channel in the collector; T, test tube; O, point at which the nylon lace separating the buffer compartment B_1 from the electrophoretic column is provided with an opening to limit low hydraulic resistance communication between these two fluid compartments to this point. (From A. Kolin (9)).

The dense streak deviates from the central circular path and strays from it in oscillatory fashion. This creates the danger of deposition of particles suspended in the streak on the walls of the annular electrophoretic column and, moreover, brings the streak periodically into close proximity of the walls where the velocity gradient is steep and, hence, the parabolic divergence large.

This drawback can be remedied by a simple modification of the scheme. We pointed out by referring to Figure 4b that the shape of the circulation path is of no essence in suppressing thermal convection by inversion of vortices. Figure 14b shows a closed trajectory in which the particles in the streak follow mostly a vertical path. The downward sedimentation within the streak in the descending path slightly accelerating the descent of the particles is balanced by the sedimentation in the ascending path where a compensatory retardation takes place. The slight tendency of the particles to drop below the central path at the bottom of the loop which would bring them closer to the outer boundary of the fluid belt is compensated by the particles' tendency to sediment toward the inner boundary in the passage over the upper curved section of the path. This form of the electrophoretic column, which now resembles in its shape and motion the belt of a belt sander, permits the use of very dense suspensions and concentrated solutions without

objectionable artifacts and serious reduction in resolving power (9).
It turned out that the same principle of electromagnetic propulsion as used in the circular apparatus (5) can be utilized in the endless fluid belt scheme (9). Instead of a circular soft-iron cylinder m, one merely has to use a soft iron bar of rectangular cross-section and instead of 2 bar magnets with a circular cross-section, 4 alnico bar magnets of rectangular cross-section are used. The soft-iron piece can be hollowed out to provide a window similar to the one shown in a circular iron core of Figure 9, without introducing disturbing electromagnetic convection by magnetic field inhomogeneities. It is also possible to introduce a collector into the annular space, as shown schematically in Figures 14c and 15, which has many adjacent separate escape channels located along a line perpendicular to the page through which the separated fractions are guided to a row of collection tubes.

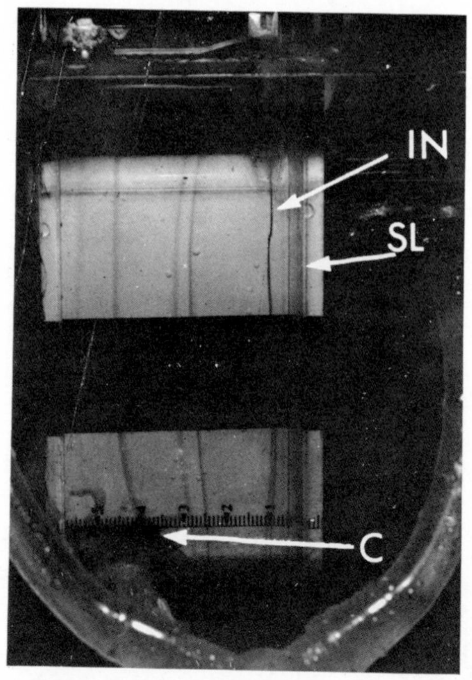

Fig. 15. A streak of Scripps #62 black ink (made unusually thick to improve photographic reproduction) issues from injector IN and splits into two visible components, a red one (of higher mobility) and a blue one. A third component (yellow) directly behind the red one can be seen only after collection in the test tubes. C, collector; SL, nylon lace. (From A. Kolin (9)).

Figure 15 shows the front view of the non-circular helical paths of two dyes separated in an early version of the endless fluid belt apparatus (9). Figure 16 shows the scheme of this apparatus in which the electrode compartments are separated from the buffer compartments which are contiguous with the endless fluid belt by dialyzing membranes. The mode of operation of the apparatus is explained in the extensive legend. A more detailed description will be given in the following section of the improved preparative instrument in which the membranes have been abandoned. All the construction details and essential principles of operation of the instrument as well as a discussion of conditions for effective use of the method will be described in connection with the latter apparatus (10).

Fig. 16. Schematic view of the electrophoretic separator. MB: Mariotte bottle; Res, reservoir delivering the mixture to be analyzed to the separator via the injector IN; BT, tube delivering buffer to the apparatus; SC, stopcocks; P, pump. CT_1, CT_2: tubes conveying coolant flowing through the cooling jacket CC which surrounds the electrophoretic column, which, in turn, surrounds the soft-iron core m. O_1, O_2: overflow tubes draining buffer from electrode compartments E_1, E_2. d_1, d_2: Bundles of thin plastic tubes delivering buffer to the buffer compartments (in actual practice these tubes are submerged into the buffer solution). MF, manifold: E^+, E^-, electrodes; V, air outlet valve; N, S, north and south poles of magnets; m_1, m_2, m_3, m_4, bar magnets of rectangular cross section; C, collector: t, t ....collector tubes; e, e.... collector tube exits. T,T.... test tubes; SL, nylon laces closing the annular space surrounding m on both sides so as to provide a path of increased hydrodynamic resistance between the electrophoretic sheath and the buffer compartments. O, opening in the left nylon lace SL establishing a low hydraulic-resistance combi-communication between the buffer compartment B_1 and the electrophoretic sheath surrounding the soft-iron core m; E_1, E_2, electrode compartments; B_1, B_2, buffer compartments; J_1, M_2, dialysing membranes separating the buffer compartments from the electrode compartments; A, annular space surrounding m which is filled with a buffer solution which rotates in the fashion of an endless belt. The solid and the dashed lines emanating from the injector represent a slow and a fast electrophoretic component which form noncircular helices in the rotating endless fluid belt. Only the front sides of the turns of the helix are shown until the points where the helical streaks enter the collector C. J, Lucite jacket corresponding to the outer boundary C_2 of Figs. 14 b and 14c. (From A. Kolin, (9)).

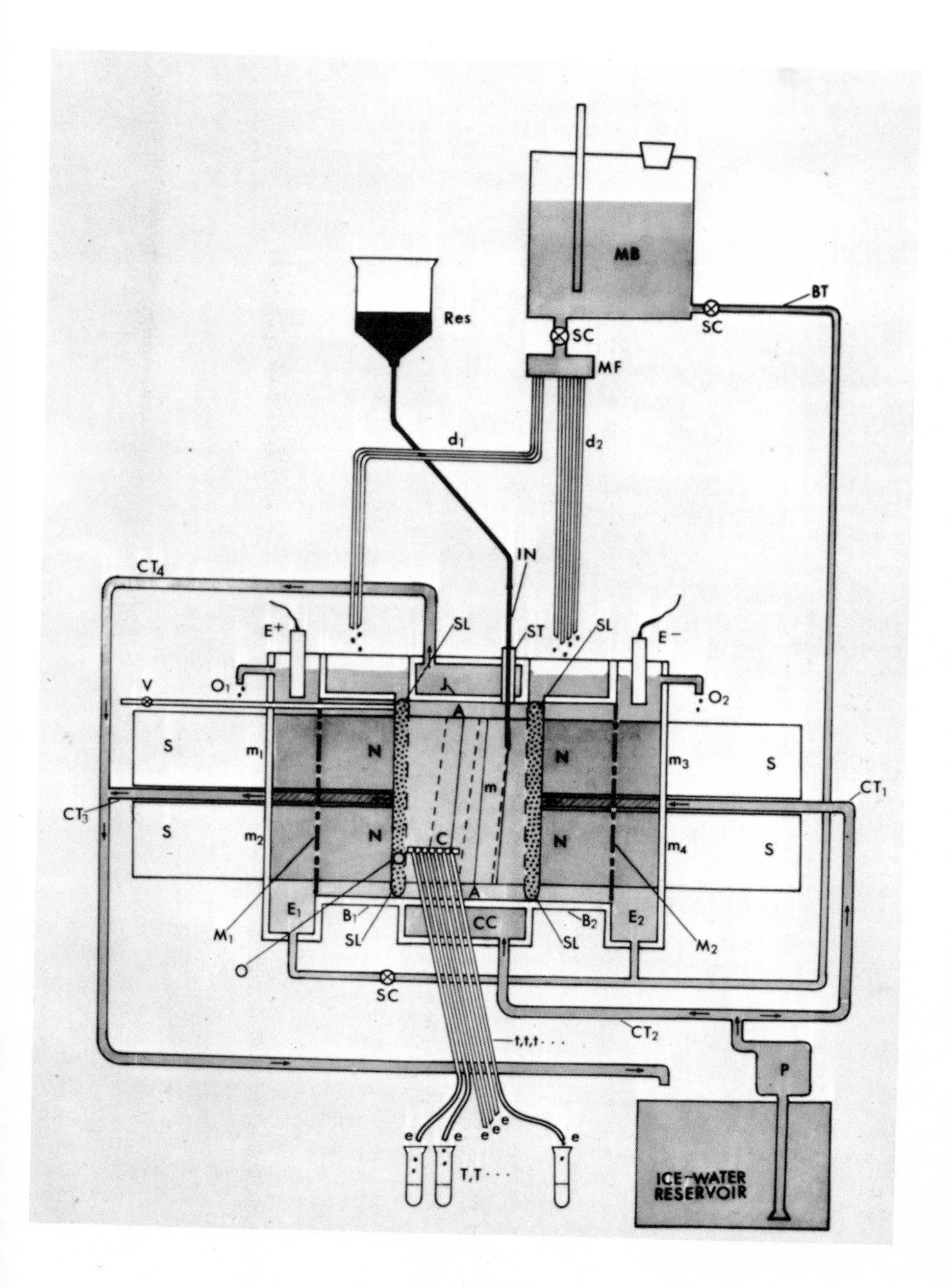
MB
Res
SC
SC
BT
MF
d1
d2
IN
CT4
SL
ST
SL
E+
E−
O1
O2
V
J
A
S
m1
N
m
N
m3
S
CT1
CT3
S
m2
N
C
N
m4
S
A
E1
B1
CC
B2
E2
M1
SL
SL
M2
O
SC
CT2
t,t,t . . .
P
e
e
e
e
e
e
T,T . . .
ICE-WATER
RESERVOIR

Figure 16

IV. THE ENDLESS FLUID BELT ELECTROPHORETIC SEPARATOR

The apparatus currently being employed in a number of investigations has been described by the authors in reference 10. It differs from earlier apparatus in a number of construction details. It can be assembled from several sections sealed by gaskets. A gap in the center of the iron core accommodates a quartz window which permits several methods of visualizing particle streaks. Flow of buffer past the electrodes directed away from the endless fluid belt in the new apparatus permits removal of electrolysis production in the absence of membranes which had previously been used to isolate the electrodes hydraulically from the endless fluid belt.

In order to allow for the introduction of improved newly devised components into the apparatus, as well as to facilitate cleaning it, the apparatus is designed to be disassembled in the manner illustrated in the exploded view of Figure 17. The insulated iron core around which the buffer belt circulates is shown here with the quartz window in its center and the tube through which coolant enters on the near end between the two rubber shock absorbers SA which provide protection against impact due to magnetic attraction as the north poles of two bar magnets are brought into contact with each end of the soft iron core. The two halves of the Lucite mantle which forms the outer boundary of the endless belt come together around the core (see also Figure 18). The two buffer compartments, B_1 and B_2, one on each side of the mantle, then seal onto each end of the core-mantle assembly forming a water-tight unit.

The two halves of the Lucite mantle are sealed together by the linear gaskets G_1 at top and bottom (Figure 18) to form a hollow block with half-inch thick walls. The walls have been hollowed to form a cooling jacket composed of 4 separate compartments, W. Only 0.38 mm of Lucite separate the coolant in these compartments from the buffer belt except at points where a Lucite grid contacts these thin walls to provide rigidity.

Penetrating through the mantle to the fluid belt are holes lined with O-ring gaskets. A round one about 3 mm in diameter permits introduction of the sample injector, IN, and an elongated one 3 mm. high and 18 mm wide harbors the collector, C. A row of four channels drilled beneath the collector permits influx of buffer to compensate for distortion of the helical flow pattern by buffer drainage from the collector by replacing beneath it buffer solution at the same rate at which it leaves the collector. This collector compensator CC is fed by a manifold machined into the mantle. A quartz block 1/2 in. high x 2 in. wide is cemented with silicone rubber (General Electric RTV-112) into the center of each removable half of the mantle as well as into the iron core. The mantle is provided with two vents V for the removal of air trapped in the belt during filling as well as for voltage measurements to determine the electric field in the belt. Twelve 1.5 mm Lucite spacers separate the iron core from the inner wall of the mantle, thus centering the core within it.

The collector with its channel openings facing upward (Figure 14c) passes through a slot in the mantle to abut against the core thereby intercepting the flow in the fluid belt and apportioning

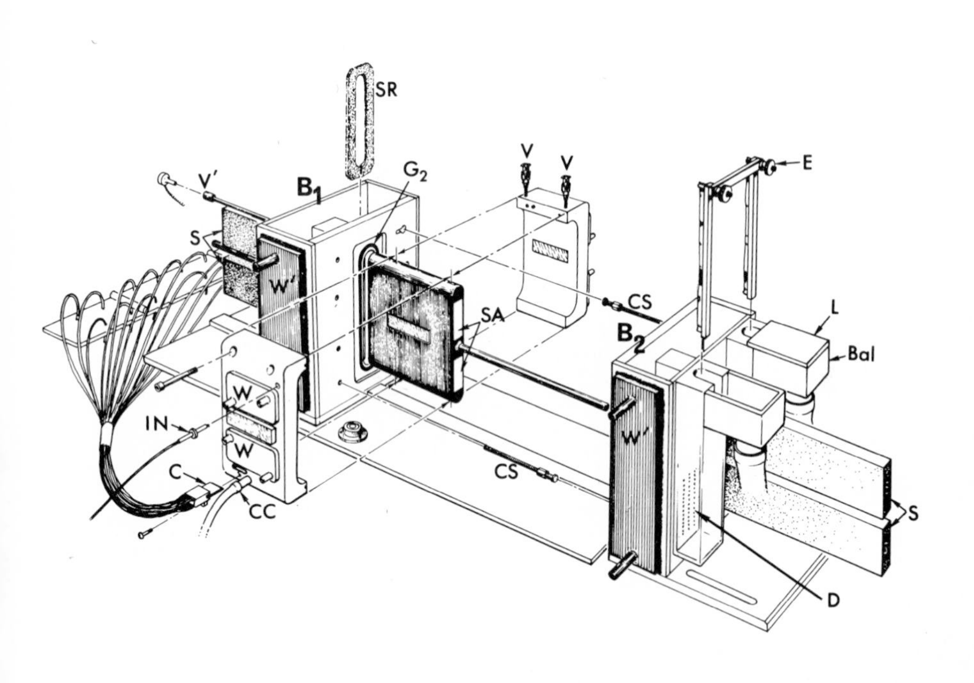

Fig. 17. Exploded view showing assembly of the apparatus. In the front half of the Lucite mantle: IN, sample injector; C, collector; CC, collector compensator; W, cooling water compartments. In the rear half of the Lucite mantle: V, vents leading to the electrolyte belt. On the core: BP, bumper pads of silicon rubber (2 on each end) where north poles of the magnets contact the core; V' air vent. On each side: B_1 and B_2, buffer compartments; S, south poles of the 4 bar magnets; W', cooling water compartments; E, electrode pairs (only cathode pair is shown); D, diaphragm; Bal, balcony; SR, spongy ring of polyurethane foam; G_2, gasket sealing assembled mantle to buffer compartment when CS, compression screws are tightened; L, lid (on all compartments to protect against electric shock). (From Kolin & Luner (10)).

the buffer and sample streaks which enter the collection channels among a row of 15 slots machined at 1 mm intervals into the plug-in collector. The fluid entering each of these slots is channeled into a length of PVC tubing which discharges over a test tube.

The injector consists of a glass capillary tube of 0.2 mm bore (Drummond "microcaps" 2 λ size) running through a 0.6 mm hole in a 3 mm diameter Lucite rod. The portion of the capillary extending into the belt bends downwards through 90°, so that when a stop on

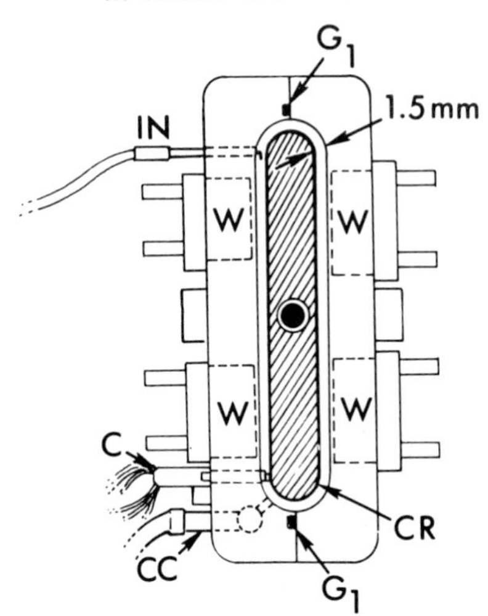

Fig. 18. End view seen from the right, of central section showing the 1.5 mm electrolyte belt between the insulated iron core and the lucite mantle. W, cooling water compartments. IN, injector; C, collector; CC, collector compensator.

the Lucite rod contacts the mantle, the capillary opening is centered between core and mantle and is pointing in the direction of buffer flow. An alternate configuration is to leave the capillary unbent but to introduce it from above so that the capillary still discharges in the same direction and at the same position (comp. Figure 14b).

The core around which the buffer circulates consists of a hollow 10 x 10 x 1 cm slab of 1020 cold rolled steel which consists of two halves which have been hollowed out (Figure 19) for the passage of coolant (water-ethylene glycol, containing 0.1% Na_2CrO_4 as a corrosion inhibitor. The two halves are subsequently joined and soldered together with silver solder. The surface is sandblasted for better adhesion of an insulating varnish coating. Since the iron core insulation must withstand potential differences on the order of a kilovolt while submerged in an electrolyte, it is absolutely essential that the coating be pinhole-free since a miniscule point of insulation failure would be transformed through electrolysis into a large bubbling crater. Seven consecutive coats of electrode-insulating varnish (6001-M Epoxylite Corporation, South El Monte, California) were applied by dipping and baking to build up a total thickness of 150 μ.

In the three years which have elapsed since three such cores were insulated in this manner, there have been only two pinholes observed, each of which was readily sealed with the same varnish. This suggests that with somewhat greater care (dust-free atmosphere, etc.) in the coating procedure a smaller number of coats should be required. Reduction in thickness of the insulation will significantly improve the removal of heat generated in the endless belt.

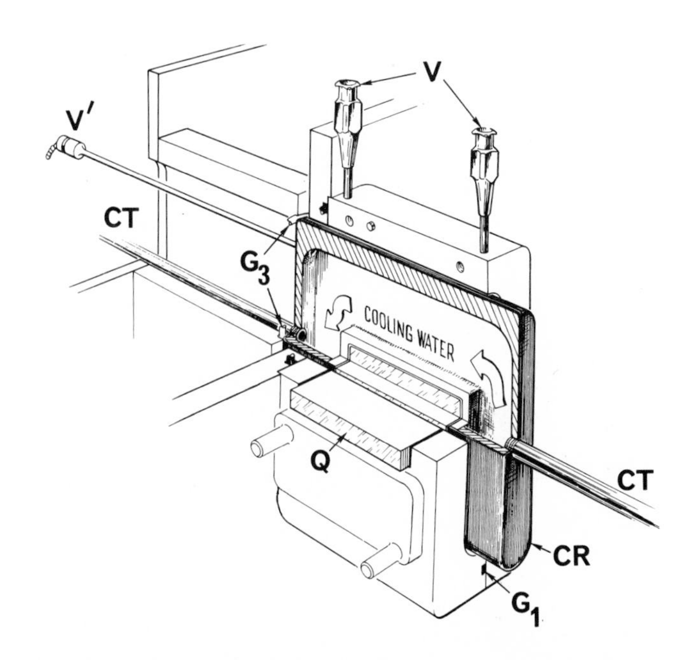

Fig. 19. Cutaway view of core-mantle section. CT, cooling water tube; CR, core; Q, quartz window; V and V', air vents leading to electrolyte belt and interior of the core, respectively; G_1, gaskets sealing two halves of the mantle together; G_3, gaskets sealing each end of the core to magnet tunnels on each side; MT, Lucite tunnel allowing magnets to pass through buffer compartments.

The endless belt of buffer is continuous with the buffer in two 600 ml compartments, one at the left (B_1) and one at the right (B_2) (see Figure 17). Gaskets G_2 seal the mantle to the near wall of each compartment, while gaskets G_3 seal the core to the hollow Lucite blocks extending inward from the far wall and containing the alnico bar magnets whose north poles contact the core. The annular space between these Lucite magnet tunnels and the near walls of the buffer compartments constitute the electrical and hydraulic connection between the main body of the buffer compartments and the endless belt. To prevent convection at these two locations they are loosely packed with a spongy matrix of polyurethane foam, SR in Figure 17. Tight sealing of the gaskets G_2 and G_3 is achieved by means of compression screws CS which draw the two buffer compartments together. Heat exchanger panels W', cool the buffer in B_1 and B_2.

The cathode consists of a pair of vertical platinum strips at the far end of B_2 while the anode (not shown in Figure 17) is a similar pair at the far end of B_1. Flow of buffer upwards and past the electrodes and out over the balconies Bal is relied upon to isolate the endless belt from electrolysis products. The perforated Lucite diaphragms D isolate the electrodes from the inner region of the buffer compartments to prevent disturbance of this flow by gross convection.

Prior to the introduction of these diaphragms, the cellophane dialyzing membranes which were previously employed were found to concentrate the buffer in B_1 while diluting it in B_2 upon the passage of current. This was explained (11) as a result of the transference numbers of the various ions being different in the membrane from their values in the bulk solution. Conductivity drifts of 10% in 20 minutes were observed. These would generally show up as a decrease in conductivity in the endless belt since a right to left buffer flow was generally used to direct the streaks towards the collector.

The replacement of the membranes by perforated plates allows the employment of a scheme of buffer flow which constantly renews the buffer in B_1 and B_2. As in earlier apparatus (6) a Mariotte bottle, MB in Figure 20, is used to deliver buffer at a constant pressure head to a manifold Man. A stainless steel screen (support screen xx 30 025 10 Millipore Filter Corp., Bedford, Mass.) is placed in the path of flow to catch any dust or mold particles in the buffer solution. The manifold feeds the collector compensator, cc, and 25 lengths of 20 gauge vinyl tubing (Alphlex tubing, Alpha Wire, Elizabeth, N.J.) apportioned between B_1 and B_2. Coarse control over the buffer flow from right to left which is used to direct streaks towards the collector is achieved by moving one or more lengths of tubing from B_1 to B_2 or *vice versa*.

The buffer leaves B_1 and B_2 through the perforated diaphragms flowing up past the electrodes and over the balconies, Bal, where bubbles formed on the electrodes rise to the surface. The flow then proceeds down 3/4 in. i.d. tubing leading to 56 cm lengths of 1/16 in. i.d. tubing which discharge at points T_1 and T_2 above a trough.

The rate of buffer flow from the Mariotte bottle into B_1 or B_2 is proportional to the height difference between the bottom of the bubble tube in the Mariotte bottle and the level of buffer in B_1 or B_2. The rate of discharge out of B_1 and B_2 at T_1 and T_2, respectively, is likewise proportional to the hydrostatic head between the level of buffer in B_1 or B_2 and the discharge points T_1 or T_2. In addition, a flow through the belt from right to left (or *vice versa*) can be established by the difference in buffer level between B_2 and B_1. A vernier control rotates a shaft about which the bar connecting T_1 and T_2 pivots. This serves as the fine control of helical pitch. A rotation through a small angle raising T_2 above T_1 will increase the rate of buffer withdrawal from B_1 simultaneously decreasing the flow from B_2 so as to leave the net efflux approximately constant. The result of this will be to lower the level in B_1 while raising that in B_2. This increases, then, the rate of right to left flow through the buffer belt. Flow rates of 3 ml/sec are most commonly used.

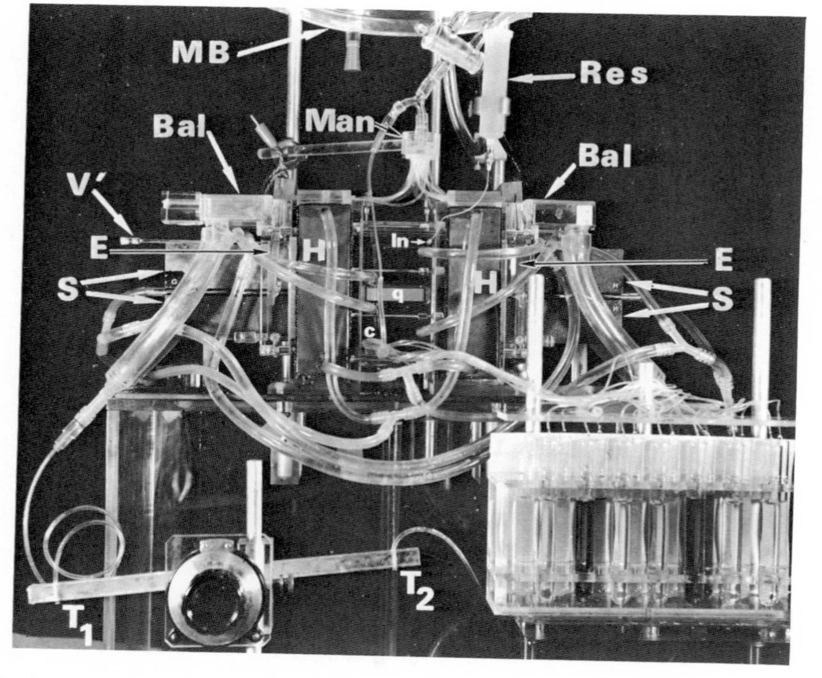

Fig. 20. Photograph of the electrophoresis apparatus. The Mariotte bottle is not shown. It supplies buffer at a constant pressure head to Man, manifold feeding buffer compartments & collector compensator, CC; R, sample reservoir feeding IN, injector; Q, quartz window; C, collector with tubings leading to test tube rack; V', air vent to core interior; H, heat exchangers for cooling buffer compartments; S, south poles of 4 bar magnets; E, platinum electrodes; Bal, balconies on electrode compartments drained by 3/4 in. i.d. tubing leading to 1/16 in. i.d. tubing discharging at T_1 and T_2. The sections of 1/4 in. tubing extending across the apparatus carry cooling water. (From Kolin & Luner (10)).

V. A SIMPLIFIED CELL FOR ANALYTICAL USE

With typical periods of buffer circulation in the 30-40 second range and fields around 100 v/cm, the method of endless belt electrophoresis achieves its best resolution of components in less than five minutes. For this reason, the buffer renewal flow system can be omitted. In a purely analytical instrument the collector-collector compensator dipole is likewise unnecessary. The electrodes still cannot be located at the central region of the buffer compartment, since an unstable pH gradient of up to 0.4 pH units can be established across the belt between B_1 and B_2 in only 3 minutes. If, however, they are removed somewhat from the belt, and the current reversed between runs of a few minutes, the composition of buffer should remain constant (12).

VI. VISUALIZATION OF HELICAL STREAKS

For the purpose of determining the location of streaks in the electrophoresis apparatus, most of the materials which have been studied can be divided into 3 classes. First there are a number of pigments ranging in complexity from the permanganate ion through diazo dyes to the colored proteins such as the hemoglobins and cytochromes. In this case the separation pattern is plainly visible through absorption as a series of colored streaks against a diffusely illuminated white background. Occasionally a filter is necessary to increase contrast in black and white photography of faint streaks such as the yellow streak of the neutral dye Apolon (13) which shows up well through a Kodak Wratten No. 34 purple filter.

The second class of materials comprises substances including nucleic acids and colorless proteins which absorb ultraviolet, but not visible, light. Here again they are conveniently photographed as silhouettes against a diffusely luminous background (14) which is in this case a low-pressure mercury lamp radiating primarily at 2537Å (for instance, the 15-watt G.E. germicidal lamp G15T8). Polaroid 3000 ASA film is used in a camera fitted with a quartz lens of 11.5 cm focal length. Mounted in front of the latter is a bracket to support a glass converging lens (110 cm focal length). This correcting lens is used in conjunction with the quartz lens to focus an image of the endless belt on the plane of the film. To take the photograph in the ultraviolet the glass lens is replaced with a visible absorbing, ultraviolet transmitting filter (Corning C.S. No.7-54). Both the front and back of the belt are in focus simultaneously.

Because the streaks used in this method are only a fraction of a millimeter thick, and because the light absorption at 2537Å is considerably lower for proteins than for viruses or nucleic acids, streaks of protein can be discerned on the photograph only when high protein concentrations are used. For example, when normal human serum is injected into the buffer belt, only the albumin fraction has sufficient concentration to form a discernible streak. To overcome this drawback a light source in the range of 2100-

2200Å where the peptide bond starts to absorb strongly is currently being investigated for use with proteins.

The third class of sample materials is made up of particulate suspensions which are visibly turbid. For this case a source of parallel light is directed through the quartz window so as to miss by a few degrees the eye of the camera lens. Intense bright streaks of scattered light will then be seen or photographed. Indeed, individual mammalian cells or small clumps of cells or tissue will, in an intense enough beam, trace out a line in the course of a several second time exposure.

The inner surfaces of the quartz windows must be kept scrupulously clean to eliminate a disturbing background "noise." Oily films which can attach themselves to the quartz can be minimized by filling the clean apparatus first with a detergent solution which is then displaced or diluted out without allowing the surface to contact the quartz. Figure 21 (10) illustrates the use of all 3 of these methods to visualize streaks.

VII. TYPICAL OPERATING PARAMETERS

The configuration of magnets used in the endless belt apparatus gives a magnetic field close to 410 gauss over the iron core in the endless belt. The thickness of the belt is 1.5 mm. The viscosity is close to one centipoise, since the belt is generally close to room temperature after a steady state has been reached between Joule heating by the current and transfer of heat to the coolant. From these parameters and equation {2} the velocity of buffer circulating in the belt is determined in ratio to the current.

Stabilization against convection is provided both by the shearing of incipient vortices in the parabolic flow profile and by the frequency of their inversion. Both of these effects are enhanced by an increased current. On the other hand, the Joule heating which engenders the convection is proportional to the square of the current divided by the conductivity. Our experience has been that the onset of convection is much less dependent upon current in the range of 100-500 ma than it is upon conductivity. To get as high an electric field as possible in the absence of thermal convection the conductivity measured at 22°C can be made as low as 650-700 μmho/cm. Electric fields of 95V/cm are obtained at 200 mA which corresponds to a current density of 65 mA/cm^2. The period of circulation is 37.5 sec. The temperature varies between 20 and 23°C in the buffer belt under these conditions depending upon the distance from the cooling surfaces. It has been found (11) that the rise in temperature from the 0° cooling water to 23° in the center of the buffer belt takes place primarily in the walls separating the cooling water from the buffer. This can amount to a linear rise in temperature of 18-20°C. The temperature distribution in the buffer is parabolic with its vertex halfway between the surfaces of core and mantle and ranges over 3-5°C. The variation of these parameters with current is shown in Table 1 (11).

A buffer system which has proved convenient in the pH range from 3.5 to 9.3 is tris-acetic acid adjusted to conductivity

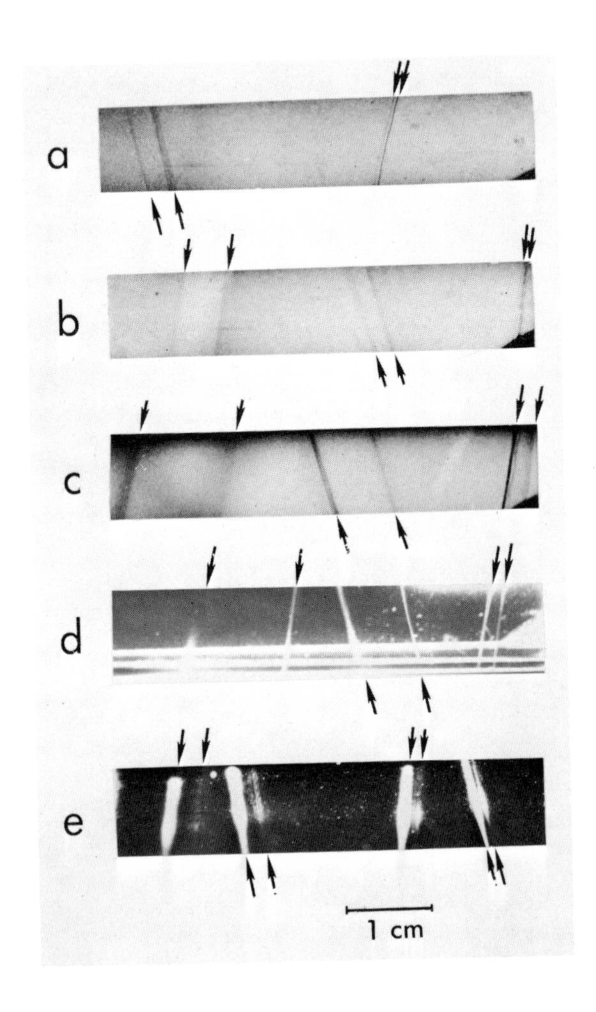

Fig. 21. Photographs of streaks in the electrophoresis apparatus, a and b, photographed by ultraviolet light, c by visible light absorption and d and c by light scattering. Current was 150 ma in example a and 200 ma in all the others.

a. TMV strains U1 left and U2 right pH 7.1.
b. Human serum proteins: albumin, left, γ globulin, right, pH 8.75.
c. A mixture of bovine hemoglobin, left, and horse heart, cytochrome-c, right, pH 9.3.
d. <u>Rhodotorula</u> sp. left and <u>Eschericia coli</u> right, pH 7.
e. <u>Erythrocytes</u> left and mainly granulocytes right, pH 7.1. The streak pattern descending from the injector is out of the field of view. (From Kolin & Luner 10)).

Table 1

Operating Parameters at Various Currents

i ma	E V/cm	σ μmho/cm	η cp	T °C	t sec	10^{-4} μ cm^2/V sec
220	97.7	692	0.91	24°	31	2.9
200	94.9	649	0.98	21°	37.5	2.4
180	87.2	633	1.00	20°	40.7	2.6
160	84.2	585	1.08	17°	50	2.12
140	77.5	557	1.14	15°	63	2.2
120	71.4	517	1.23	12°	76	2.1

T and η are average values of temperature and viscosity in the buffer belt calculated from the average electrical conductivity, σ, as determined from current and voltage measurements. The period of circulation of buffer around the core is t. The mobility of Brilliant Blue determined from voltage, time, and streak separation distance, is μ.

650-700 μmho/cm (11) Figure 22). Since the buffer solutions are adjusted to a constant conductivity for all values of pH and since tris H^+ and acetate$^-$ are the only ions contributing appreciably to conductivity above pH 4, the concentrations of tris H^+ and acetate$^-$ are almost equal to each other and constant over the whole pH range. The only concentrations varying are those of the un-ionized species and, of course, H^+ and OH^-. Changes in mobility will be dependent only upon pH, therefore, and not on ionic strength, which is always approximately 0.01. As the extremes of the pH range are reached there will also be changes in the dielectric constant of the buffer solution which may affect the dissociation constants of the sample material. This system can be seen from Figure 22 to have a minimum of buffering capacity in the range between pH 6 and 6.75; however, it can still be used to establish the pH of the sparse suspensions of cells or subcellular particles used to form streaks visualized by light scattering for determination of a pH-mobility curve.

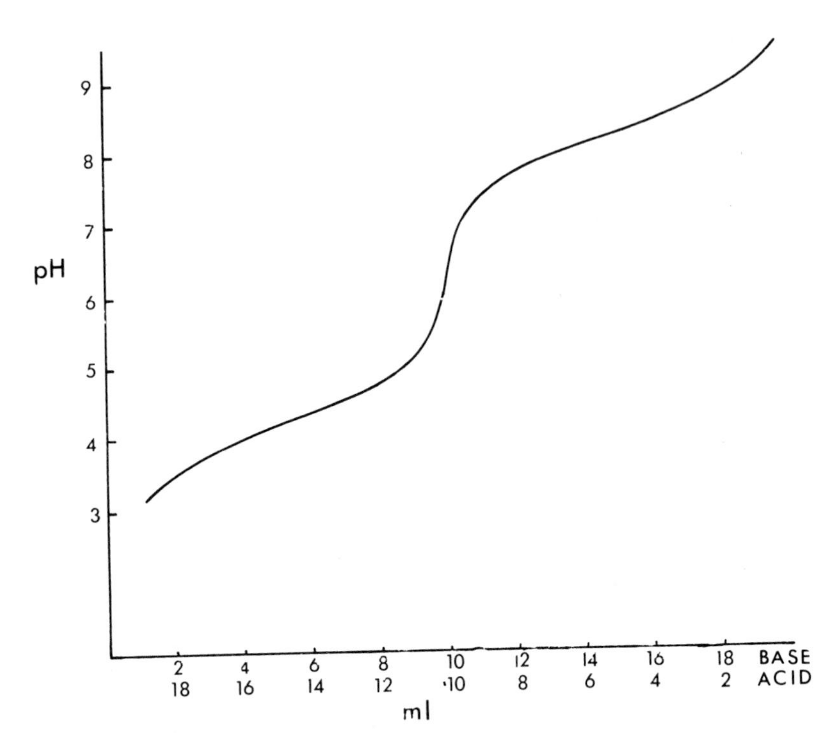

Fig. 22. pH vs. composition of Tris-acetic acid buffer. (From S. Luner (11)).

VIII. EXAMPLES OF APPLICATIONS

Initial illustrations of the application of endless belt electrophoresis to protein chemistry have been the fractionation of artificial mixtures of proteins obtained commercially. Figure 21b is an ultraviolet photograph of the separation of a mixture of human serum albumin and gamma globulin. In order to obtain a distinct streak the gamma globulin was put in as a saturated solution. The wide separation after one revolution indicates that only one or two revolutions would be necessary to resolve a number of components of intermediate mobility present in normal serum. An extremely wide spread in mobility is shown between positively charged horse heart cytochrome c and bovine hemoglobin charged negatively at pH 9.3 in Figure 21c.

The distance traversed by the buffer belt around the core is proportional to time, so that the electrophoretic mobilities can be calculated from graphs of separation between streaks plotted against distance circulated. Such a graph is shown in Figure 23 taken from the data in Figure 21b and c.

The ultraviolet photograph of tobacco mosaic virus strains U1 and U2 (Figure 21a) has been used to calculate the mobility

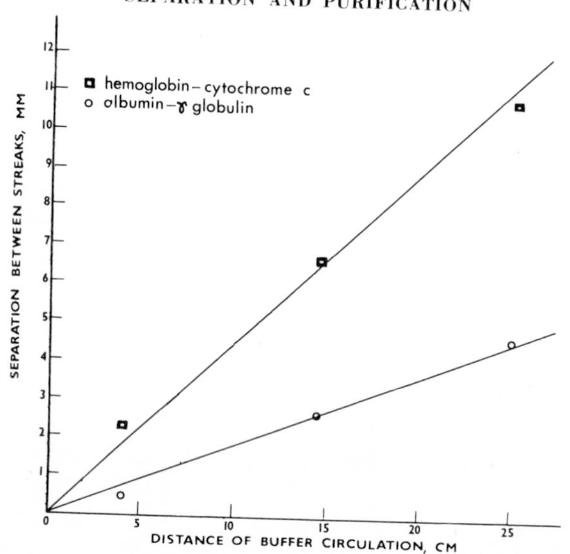

Fig. 23. Separation between streaks as a function of distance circulated by the sample around the core. Taken from the streak patterns of Fig. 21. (From Kolin & Luner (10)).

difference between these two strains (10). Considerably larger mobilities are obtained using this method than are measured in the Tiselius moving boundary method. This is so because the endless belt employs buffers of lower ionic strength 0.01 compared to 0.1 and lower viscosity 0.91 centipoise at 24°C compared to 1.67 at 2°C.

Figure 21e shows the fractionation of human blood cells obtained from a washed buffy-coat preparation. Phosphate-EDTA buffer containing (14) 0.2% glucose was used with 9.5% sucrose added to make the buffer isotonic with the cells. Centrifuging the cells onto round cover-glasses permitted the preparation of fixed and stained slides from which the histogram of Figure 24 was prepared.

The viability of mammalian cells following electrophoresis was checked by using sterile buffer and culturing L and HeLa cells emerging from the collector in Eagles medium containing calf serum and antibiotics. Large colonies were formed showing that the cells retained the ability to undergo an unlimited number of divisions.

Figure 21d illustrates a separation of a yeast _Rhodotorula_ from a strain of _E. coli_. The concentration of microorganism in the collector output was determined using a Petroff-Hausser counting chamber to give the histogram of Figure 25.

A survey of 22 bacterial species and strains currently being completed in the authors' laboratory (16) has yielded mobility values at several pH's which can be used to characterize a particular strain as well as to yield information on the bacterial wall. An additional useful identifying characteristic is the

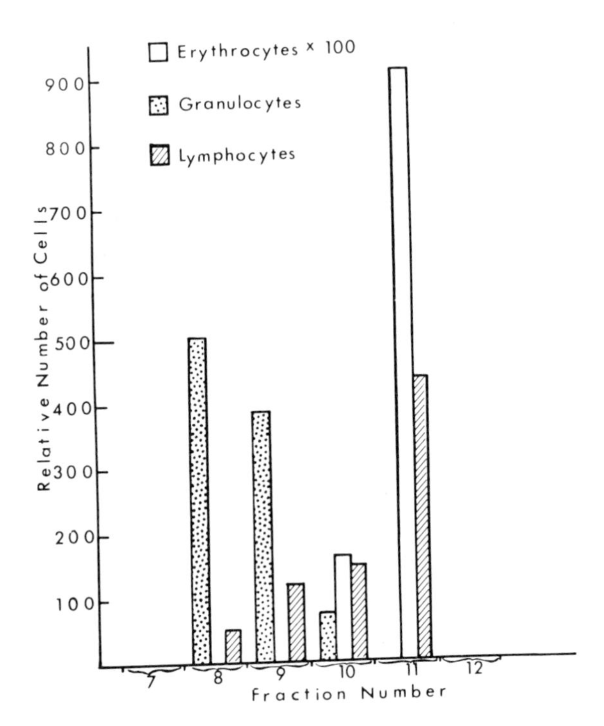

Fig. 24. Histogram of collection patterns of blood cells. (From Kolin & Luner (10)).

number of distinct lines or number and width of continuous bands in the mobility spectrum of a bacterial population of a single strain grown under well-defined conditions.

Figure 26 (16) shows differences in streak pattern of different individual strains of bacteria grown on nutrient agar and washed twice in buffer. A and B show two strains of the same species, B. subtilis and B. subtilis VN, at pH 4. In contrast to the single streak of B. subtilis shown in A, the VN strain can be seen to split into two clearly defined separate streaks of different electrophoretic mobility. The multiple streak pattern of P. fluorescence at pH 6.9 is shown in C, while D illustrates the broad diffuse bands characteristic of A. viscolatis at pH 10. Further study is underway to determine whether these components of different mobilities represent different phases of the bacterial life cycle or are genetically distinct variant strains.

Endless belt electrophoresis has been applied to the study of metaphase chromosomes isolated from the HeLa and PtK1 cell lines (11). After 15 hours' exposure to Vinblastine to accumulate them in metaphase the cells were washed and swollen with a hypotonic solution. Mechanical agitation or shearing was used to break open the cells and release the metaphase chromosomes. In the

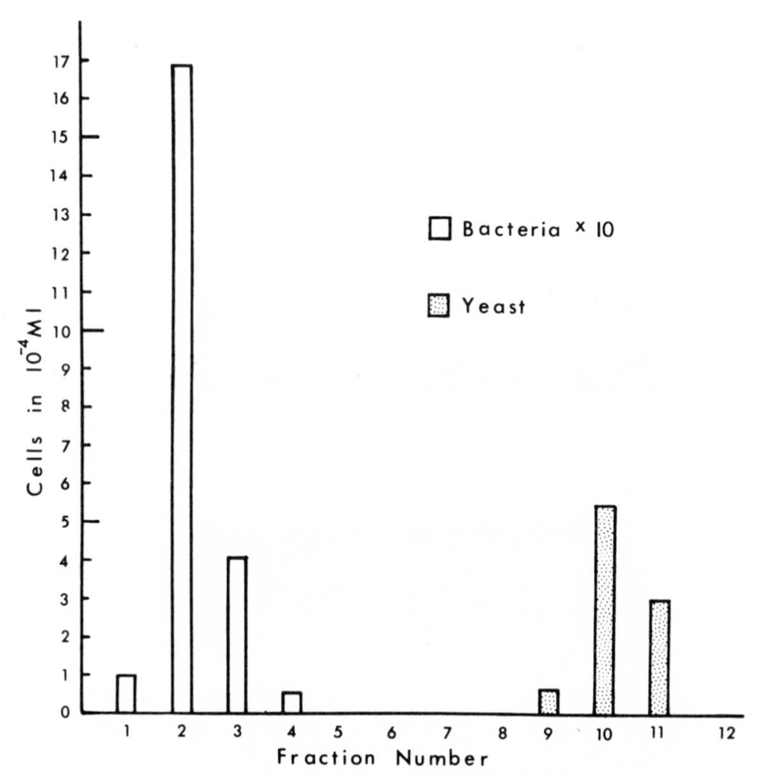

Fig. 25. Collection pattern from Rhodotorula-Escherichia coli separation. (From Kolin & Luner (10)).

case of PtKl cells this was done at pH 3 using the cationic detergent cetyltrimethyl-ammonium bromide. HeLa chromosomes were released from the cells at pH 8.6 in the presence of saponin, a non-ionic detergent.

The pH mobility curve for HeLa chromosomes (Figure 27) shows an isoelectric point near pH 4.6. The high positive mobility of the PtKl chromosomes and its insensitivity to change of pH (Figure 28) suggest that the cationic detergent has coated the chromosomes, contributing a considerable amount of positive charge.

The success of these initial applications indicate that endless belt electrophoresis could prove a valuable research tool as apparatus becomes more generally available.

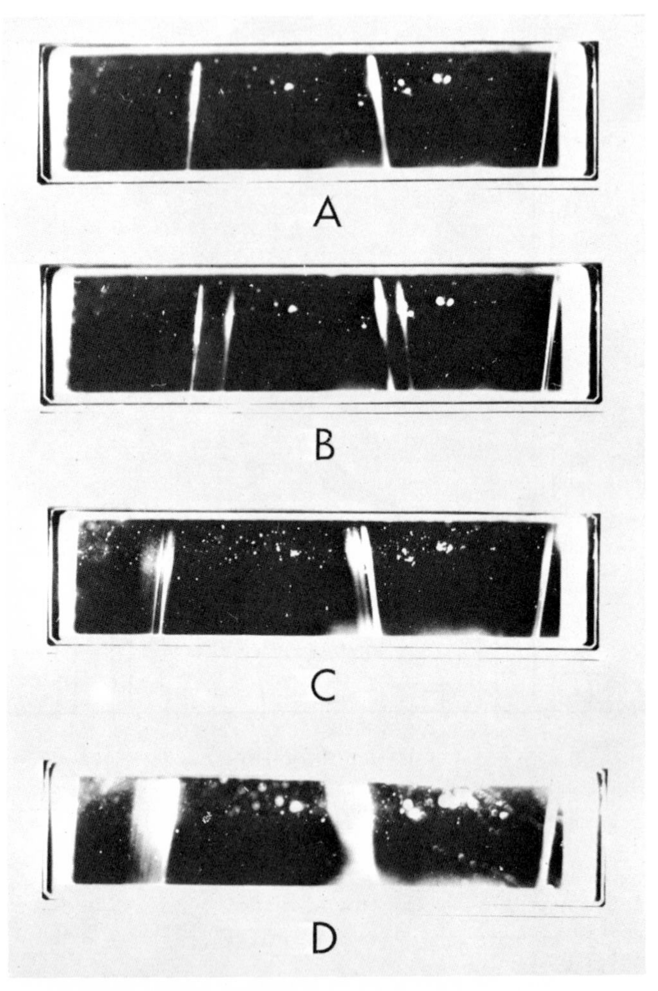

Fig. 26. Mobility spectra of single strains of bacteria: (A) B. subtilis, pH 4; (B) B. subtilis VN, pH 4; (C) P. fluorescence pH 6.9; (D) A. viscolatis pH 10. (From Kolin, et al (16)).

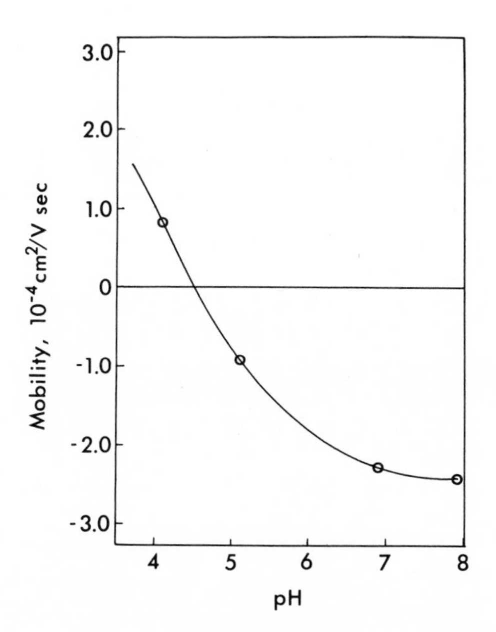

Fig. 27. Mobility vs. pH of HeLa chromosomes. (From S. Luner (11)).

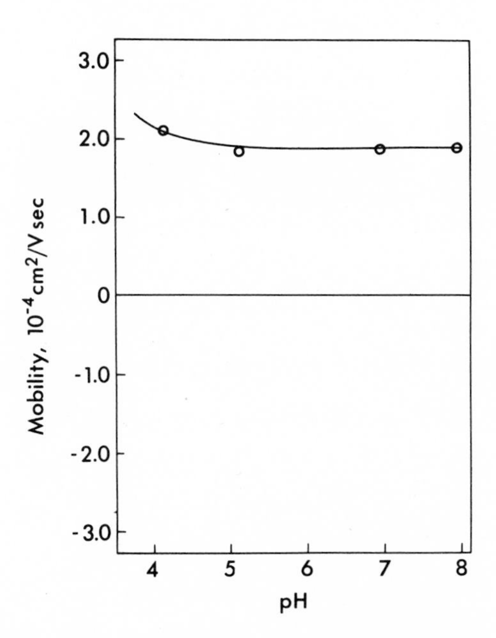

Fig. 28. Mobility vs. pH of PtK1 chromosomes isolated at pH 3 using cetyltrimethylammonium bromide. (From S. Luner (11)).

REFERENCES

1. Grassman, W. and K. Hannig, Naturwiss. 37, 496 (1950); Svensson, H. and I. Brattsten, Arkiv Kemi, 1, 401 (1949).

2. Barrollier, J., E. Watzke, and H. Gibian, Z. Naturforsch., 136, 754 (1958).

3. Hannig, K., Z. Physiol. Chem., 338, 211 (1964).

4. Kolin, A., Proc. Natl. Acad. Sci. U.S., 51, 1110 (1964).

5. Kolin, A., Proc. Natl. Acad. Sci. U.S., 46, 509 (1960).

6. (a) Kolin, A., J. Chromatog., 26, 164 (1967).

   (b) Kolin, A., J. Chromatog., 26, 180 (1967).

7. Lamb, H., Hydroanimics (Dover, N.Y., 1945), p. 582.

8. Smoluchowski, N., in Handbuch der Elektrizitaet und der Magnetismus (L. Graetz, ed., A. Barth, Leipzig), v. 2, p. 366.

9. Kolin, A., Proc. Natl. Acad. Sci. U.S., 56, 1051 (1966).

10. Kolin, A. and S. J. Luner, Anal. Biochem. 30, 111 (1969).

11. Luner, S. J., "Electrophoretic Studies of Isolated Mammalian Metaphase Chromosomes," Thesis University of California, Los Angeles (1969).

12. Kolin, A., To be published.

13. Werum, L. N., H. T. Gordon, and W. Thornburg, J. Chromatog. 3, 125 (1960).

14. Luner, S. J., Anal. Biochem. 23, 357 (1968).

15. Hannig, K., Z. Physiol. Chem., 349, 161 (1968).

16. Kolin, A., K. S. Kwak, and A. Kolangian, In preparation.

HYDROCARBON SEPARATIONS WITH SILVER(I) SYSTEMS

H.W. Quinn
Dow Chemical of Canada, Limited
Sarnia, Ontario, Canada

I. INTRODUCTION

For several decades researchers have been investigating the interaction of olefins with the transition metals and have been evaluating transition metal compounds as agents for removal of olefins from hydrocarbon systems. In recent years many new transition metal-olefin complexes have been discovered; although study of these has lead to a better understanding of the nature of metal-olefin interactions and some of them have shown activity as catalysts for reactions of olefins, most cannot be considered to have potential for application to hydrocarbon separation processes because of the irreversibility of the complex formation step.

Of the transition metal compounds which do interact reversibly with olefins, those which have been investigated most thoroughly for hydrocarbon separations are compounds of monovalent copper or silver. There is a voluminous patent literature reading upon the use of these compounds but no attempt will be made herein to review it in detail. Although some of the earlier patents (e.g. 1,2) read, in general, on the use of the monovalent salts of copper, mercury and silver including the halides, nitrates, sulfates, phosphates, formates, acetates, propionates, carbonates and lactates, they do not record as examples the use of any salt other than cuprous chloride.

Dating from the pioneering investigation of Gilliland and coworkers (3), the use of cuprous chloride as a separation agent has been thoroughly studied (4,5). Employed as a solid, it has been slurried in an aqueous system (6), ground in a ball mill (7), supported on an inert substrate (8) and activated by the cyclic formation-decomposition of an intermediate complex (9). It has been used also as a solution in alcoholic HCl (10) or organic amines such as orthophenetidine (11). The use of solutions of cuprous nitrate in ethanolamine (12), cuprous sulfate in propionitrile (13) and cuprous salicylate in aqueous ammonia (14) has been described as well. However, despite the diverse studies which have been made, as far as this author is aware, the only commercially successful application of the cuprous ion system for olefin separation has been the use of aqueous ammoniacal cuprous acetate for removal of 1,3-butadiene from C_4 hydrocarbon fractions (15).

Most of the early studies of the complexing of olefins with silver salts have involved aqueous silver nitrate solutions. Subsequently, aqueous solutions of silver tetrafluoroborate caught the attention of investigators because the much higher water solubility of that salt yields solutions with greater capacity for olefin absorption. A commercial unit employing aqueous $AgBF_4$ for ethylene extraction has been operated by Farbwerke Hoechst (4,12).

While this review will provide a summary of the research on the use of solutions of silver salt for olefin separations including their application in gas chromatography, it will be concerned primarily with the study of the interaction of olefins with solid silver salts and other solid systems incorporating a significant concentration of silver ion.

II. NATURE OF THE METAL-OLEFIN COMPLEX

The generally accepted model for silver ion-olefin complexing is that described in 1951 by Dewar using molecular orbital terminology (16). As illustrated by Figure 1, Dewar postulated a double bonding between the olefin molecule and the silver ion, the σ-component of the double bond being formed by overlap of the occupied bonding σ-molecule orbital of the olefin with the vacant 5s orbital of the ion and the π-component by overlap of an occupied 4d orbital of the ion with the vacant antibonding π-molecular orbital of the olefin. With the Cu(I) species, it is the vacant 4s and occupied 3d orbitals of the metal that are involved. The Dewar postulate accounts for the tendency of metals near the ends of the transition series to form stable olefin complexes since it is these that have the larger number of filled d orbitals available for overlap with the antibonding π-orbital of the olefin.

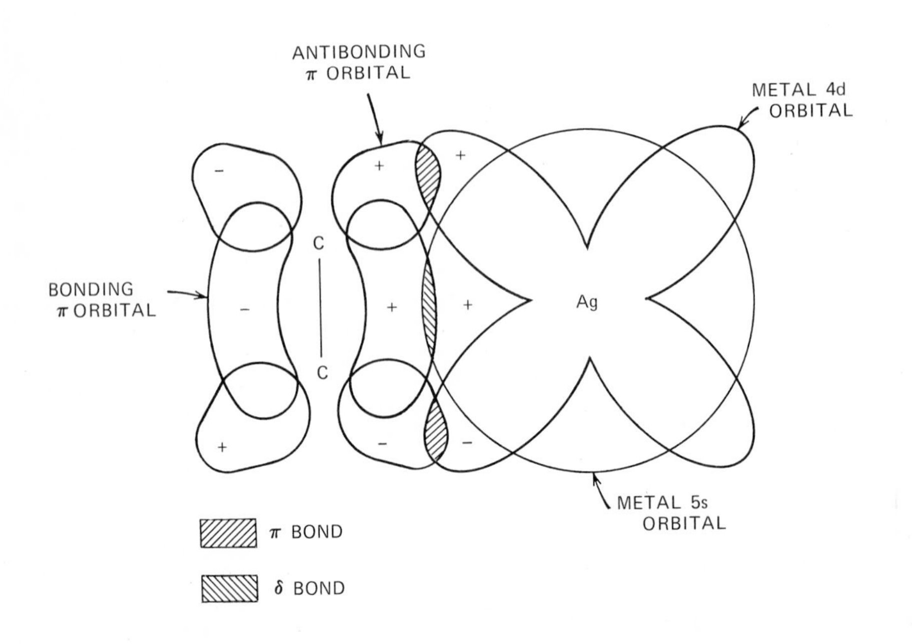

Figure 1. Dewar model for silver ion-olefin complex.

Upon variation of the olefin or of the metal one would anticipate a change in the relative magnitudes of the σ- and π-components of the coordinate bond. Despite earlier reports to the contrary (17), some recent studies (18,19) have indicated that an increase in basicity of an olefin results in an increase in stability of the silver ion complex which may be a reflection of an increase in the strength of the coordinate bond. Infrared studies (20) also show a stronger interaction of the more basic olefins with silver. Although it has been suggested (16) that there may be a weakening of the π-component of the coordinate bond concurrent with a strengthening of the σ-component by an electron-donating alkyl substituent, it is not known how significant the effect would be. A qualitative estimate (21) indicates that with silver the σ-component should predominate. The dependence of complex stability upon the electron-donating or -withdrawing power of substituent groups on the olefin tends to confirm that suggestion (22,23).

III. AQUEOUS SOLUTIONS OF SILVER SALTS

A. Silver Nitrate

The earliest studies of the interaction between unsaturated hydrocarbons and silver ion involved the use of aqueous solutions. In 1922, Hill (24) observed a salting-in effect of $AgClO_4$ upon benzene in an aqueous system. Then, in 1938 Winstein and Lucas (17) published the results of a comprehensive study of the equilibrium distribution of a variety of olefins between carbon tetrachloride and aqueous silver nitrate. The equilibrium constants for the complexes became known as the argentation constants for the olefins. These indicated a decrease in complex stability with alkyl substitution of ethylene, an effect which is probably attributable largely to increased entropy change upon complexing of the substituted ethylenes. For the butenes, the equilibrium constants for formation of the monosilver complex in aqueous solution are: 1-butene 119.4, isobutene 71.5, *cis*-2-butene 62.3 and *trans*-2-butene 24.6 (25). The complex with a *cis*-olefin is always more stable than that with its *trans* analogue. It has been suggested (26) that the differences between the solubilities of individual butene isomers in $AgNO_3$ solutions can be considered to be sufficient for separation of the butenes from each other on a technical scale. However, this author is not aware that a commercial unit employing aqueous $AgNO_3$ has even been constructed.

With monoolefins, the aqueous complex (17,27) consists largely of one olefin molecule and one silver ion, the enthalpy change for complex formation amounting to about 6kcal./mole, while with diolefins the complex can consist of one diene molecule and two silver ions (17). At high silver ion concentrations, there is some evidence for disilver complexes of monoolefins as well (17,27). Although the monoolefin complexes do not tend to precipitate readily from aqueous solutions, those of the dienes are more stable and much less soluble. Accordingly, Kraus and Stern (28a) have isolated both $(C_4H_6)AgNO_3$ and $(C_4H_6)(AgNO_3)_2$ from aqueous silver nitrate solutions of 1,3-butadiene and Traynham (29) has isolated complexes of the same stoichiometry with norbornadiene.

Cope and co-workers, as well as other researchers, have made extensive use of aqueous silver nitrate solutions for separation of cyclic mono-, di- and oligo-olefins from synthetic mixtures (see Table IV, reference 23). Separation was effected, dependent upon complex solubility, either by concentration of the desired olefin in the aqueous phase or by precipitation of the complex. Although Kraus and Stern have patented the use of aqueous $AgNO_3$ solution for recovery of 1,3-butadiene as a solid complex from admixture with the butenes (28b), the commercial acceptability of this process would be dubious since Long (30) has observed that the solid complex is about as explosive as black powder. He suggests that other solid $AgNO_3$-olefin complexes may show similar behavior.

In general, separation of monoolefins is effected by concentration of the olefin in the aqueous phase. The absorption capacity of the $AgNO_3$ solution depends upon the pressure, the solution concentration and the mole fraction of olefin in the hydrocarbon mixture (31). For example, the variation of propylene solubility in saturated (71 weight percent) aqueous $AgNO_3$ at 25° with mole fraction of olefin in the liquid hydrocarbon is illustrated by the data of Table I.

TABLE I

Solubility of Liquid Propylene in Saturated Aqueous Silver Nitrate at 25°C (31)

Mole Fraction (C_3H_6 in C_3H_8)	Propylene Solubility gm./100 gm. soln.
1.000	18.7
0.78	14.1
0.405	10.4
0.33	4.7
0.22	4.4

The solubility of liquid propylene at 25°C in 39 percent silver nitrate solution is only 9.6g./100g. soln. (31). Although the most practical approach would be the use of a saturated solution, Francis (32) has described a system employing less water than that required to completely dissolve the salt which has, as one would expect, greater capacity for olefin absorption than does the saturated solution. It may be, however, more difficult to handle because of the presence of a solid phase. It has been reported (33) that the addition of an amine salt, such as monobutylamine nitrate, to aqueous silver nitrate also markedly increases its capacity for olefin absorption.

B. Silver Tetrafluoroborate

Although there have been some studies made of the solubility of olefins in aqueous solutions of silver perchlorate (34) and silver trifluoroacetate (35), the only other silver salt that has been extensively investigated is silver tetrafluoroborate. In general, it has been found that olefins are more soluble in aqueous $AgBF_4$ than in aqueous $AgNO_3$; this is especially the case when highly concentrated solutions are employed (Table II). As shown by Figure 2 for ethylene, olefin solubility increases with increasing concentration of $AgBF_4$ but decreases with increasing concentration of $AgNO_3$ (34,36). It is probable (23) that this solubility dependence is a reflection of the change in the degree of association of the ions with increasing salt concentration. The solubility data suggest (12, 34) that the highest ratio of olefin: Ag^+ approaches 1:1 with aqueous $AgNO_3$ and 2:1 with aqueous $AgBF_4$. These ratios are similar to the stoichiometry of the corresponding anhydrous complexes described in Section V; the stoichiometry differences can also be rationalized in terms of the differences in the cation-anion interaction in the solid salt. In this connection, Baker (36) has satisfactorily correlated the solubility data with the values of the activity coefficient for each salt. He has reported also that the addition of other fluoroborates, including fluoroboric acid, increases the solubility of C_2H_4 in aqueous $AgBF_4$ to an extent proportional to the tendency (related to the charge: radius ratio) of the added

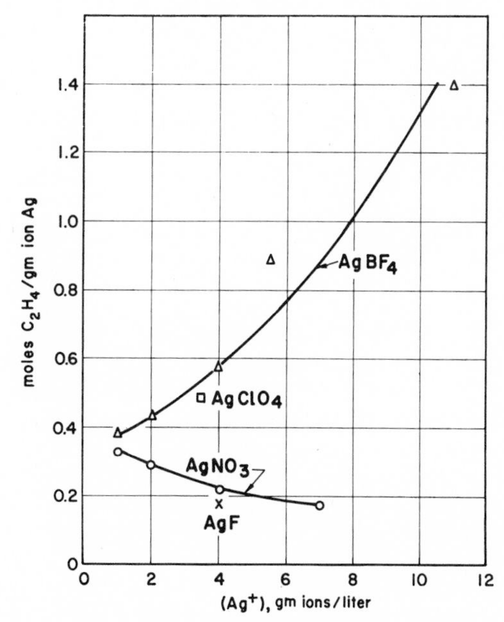

Figure 2. Ethylene absorption by aqueous solutions of various silver salts (36).

TABLE II

Solubility of Olefins in Aqueous Silver Tetrafluoroborate and Nitrate (34)

Olefin	Solubility mole/mole Ag^+ $AgBF_4$ [a]	$AgNO_3$ [b]	Olefin Vapor Pressure atmos. at 20°C
Ethylene	0.68	0.225	1
Propylene	0.78	0.23	1
1-Butene	0.976	0.374	1
cis-2-Butene	1.04	0.39	1
trans-2-Butene	0.61	0.09	1
Isobutene	0.94	0.21	1
1-Pentene	1.066	0.29	0.7
3-Methyl-1-butene	0.94	0.35	1
2-Methyl-2-butene	0.9	0.245	0.51
1-Hexene	1.03	0.036	0.20
4-Methyl-1-pentene	0.83	0.03	0.29
2-Methyl-1-pentene	0.57	0.025	0.21
1-Octene	0.97	0.009	0.017
Cyclohexene	1.03	0.26	0.088
Styrene	1.15	0.18	0.006

a - 650 g. Ag/1.

b - 600 g. Ag/1.

cation to reduce the degree of hydration of the silver ion and thus make it more readily available for olefin complexing. In addition, Krekeler and colleagues (12) have shown that the ethylene solubility is directly proportional to the product of the concentrations of silver and fluoroborate ions, $[Ag^+] \times [BF_4^-]$, Figure 3.

Much attention has been devoted to evaluation of the technical and commercial feasibility of using aqueous $AgBF_4$ for olefin separations. These investigations have resulted in the issuance of a number of patents to duPont (37), I.C.I. (38) and Farbewerke Hoechst (39). Although a number of patents read upon the use of silver hexafluorosilicate solutions as well for olefin separation, this compound has not been extensively investigated and those studies which have been made indicate that its solutions do not absorb as much olefin as those of silver tetrafluoroborate (12).

During an investigation of the extraction of styrene from admixture with xylenes by aqueous $AgBF_4$, Featherstone and Sorrie (34) observed that the solubility of the xylenes in the aqueous phase was enhanced by the presence of the complexed styrene (Figure 4). Further study of the effect of olefins on the solubility of benzene and m-xylene in $AgBF_4$ solutions showed an increase in the order 1-octene > 1-pentene > styrene indicating that greater olefin chain length results in greater aromatic solubility. This is not surprising when one considers the effect

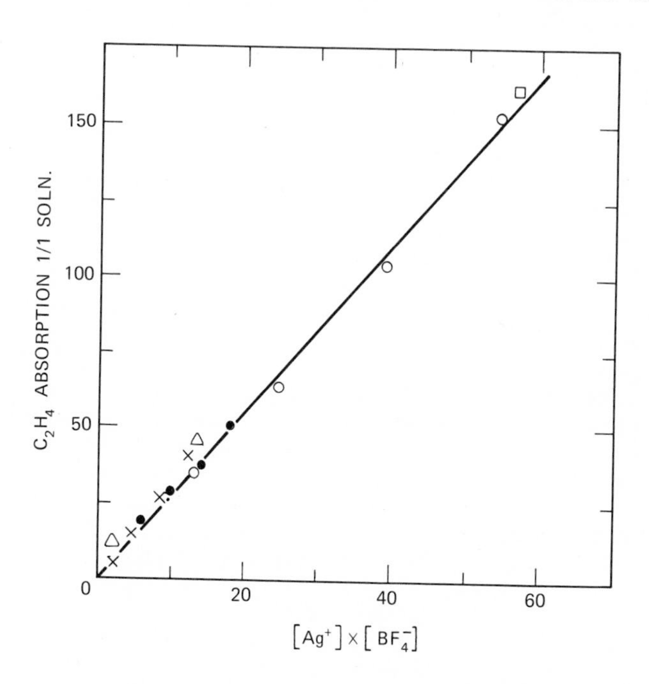

Figure 3. Dependence of ethylene absorption in aqueous $AgBF_4$ upon the product $[Ag^+] \times [BF_4^-]$. Concentrations expressed in moles/1 (12).

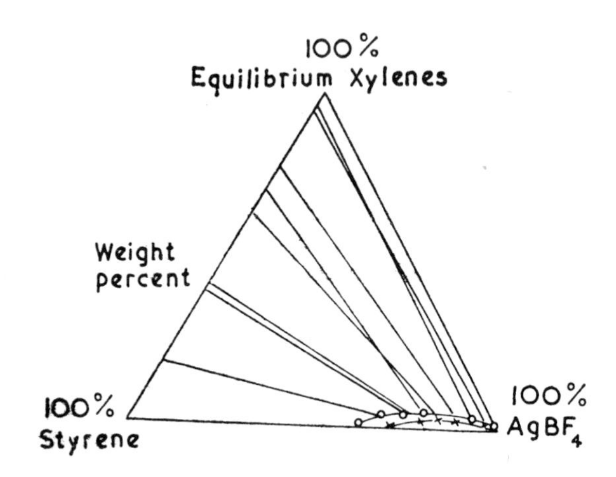

Figure 4. Ternary system: styrene-xylenes-$AgBF_4$ (34).
O $AgBF_4$ contained 810 g Ag/l at 23°C.
X $AgBF_4$ contained 590 g Ag/l at 5°C.

of olefin chain length upon the organic character of the complex $AgBF_4$.2 olefin. This is further demonstrated by decreasing hygroscopicity of solid complexes with increasing molecular weight of 1-olefins (20).

An exhaustive study of olefin, especially ethylene, separation with aqueous $AgBF_4$, containing HBF_4, has been conducted by researchers at Farbewerke Hoechst (12). The data presented appear to have been collected by operation of both experimental and production units. In addition, the aqueous $AgBF_4$ solution has been compared with an ethanolamine cuprous nitrate solution for ethylene or propylene extraction. Although the silver salt solution is very much more costly than the ethanolamine cuprous nitrate system (cost ratio of perhaps 60:1), it shows an ethylene absorption capacity that is 6 to 10 times greater, thus offering the possibility of effecting a given olefin separation with a much smaller extraction unit.

The design of the extraction unit employing aqueous $AgBF_4$ would be determined by the nature of the gas stream to be treated. With those streams containing no reducing gas such as hydrogen or carbon monoxide which could cause precipitation of metallic silver, it is necessary to have only an absorption and a desorption tower. Ethylene absorption can be easily effected at about 1 atmosphere pressure and 20°C while desorption proceeds at about 0.3 atmosphere pressure and 80°C. In the presence of reducing gases, however, the extraction conditions and thus the design of the extraction unit are quite different. In order to prevent the reduction of silver to the metallic state, hydrogen peroxide is added to the extent of about 1 g/kg of ethylene absorbed. Because of H_2O_2 instability at higher temperatures, it is necessary to operate the desorption stage at a temperature below about 30°C when the peroxide is employed. In this unit, a stripping tower employing ethylene as a stripping gas to remove residual gases from the $AgBF_4$ solution is inserted between the absorption and desorption stages. This ethylene is recovered by subsequent extraction with fresh $AgBF_4$ solution introduced near the head of the stripping tower. Operation of the desorption stage at reduced pressure results in some loss of water from the aqueous solution; consequently, additional water must be added before the solution returns to the absorption tower. A schematic diagram for this latter extraction unit is shown in Figure 5.

Although the presence of other gases such as carbon dioxide or oxygen does not detrimentally affect the silver salt solution, one must be very concerned about the presence of acetylene. Although silver acetylide is quite soluble in the concentrated $AgBF_4$ solution, upon dilution with water it is precipitated as a highly explosive solid. It is consequently desirable to remove it from the gas stream prior to the olefin ammoniacal contaminants because of their tendency to complex strongly with the silver ion.

With the $AgBF_4$ solutions one can employ stainless steel as the material of construction for operation at temperatures up to about 30°C; when, however, the operating temperature is higher, plastic or resinous materials must be used (12). The extraction

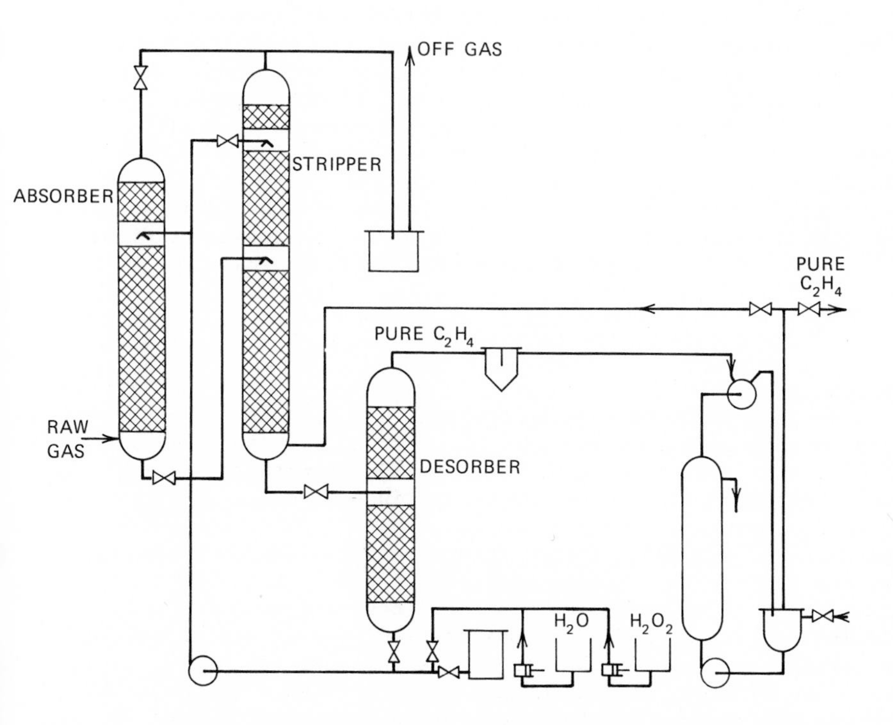

Figure 5. Schematic diagram of a processing unit for extraction of ethylene with aqueous $AgBF_4$ from a hydrocarbon mixture containing reducing gases (12).

tower can be packed with Raschig rings made of polyvinyl chloride or polyethylene. The HBF_4 present in the solution would attack silica surfaces in the system resulting in some precipitation of SiO_2 and some generation of SiF_4 which would come out with the stripping gas. This solution is less hazardous to handle than is concentrated H_2SO_4 (12). When the solution is contacted with the skin, however, a dark brown stain is produced, the intensity of which can be reduced by washing with a potassium thiocyanate solution.

IV. GAS CHROMATOGRAPHIC SEPARATION

The interaction of olefins with silver ions has been applied for analytical purposes by preparation of gas chromatographic (g.c.) columns having as the liquid phase a solution of silver nitrate in some non-volatile polar substrate such as ethylene glycol (40,41) or benzyl cyanide (42,43). In general, paraffins are not strongly retained on these columns. Unlike their retention on the more conventional columns, the retention of the alkenes on the $AgNO_3$ column bears no relationship to olefin volatility; the retention order is determined instead by the relative stability of the silver nitrate complexes. Table III records the retention data relative to isoprene for a number of the lower alkanes and alkenes on a silver nitrate-benzyl cyanide column (43). The maximum temperature at which the silver nitrate columns can be satisfactorily operated is about 50°C.

The silver nitrate-ethylene glycol columns have been employed, as well, for determination of the equilibrium constants for formation of a large number of silver nitrate-olefin complexes (44,45). Using the terminology of Muhs and Weiss (45), the partition coefficient, H, for partitioning of the olefin between the liquid phase and the carrier gas can be expressed by equation [1] wherein K_L is the partition coefficient for the

$$H = K_L + K_L K_1 [AgNO_3] \qquad [1]$$

solvent glycol in the absence of $AgNO_3$ and K_1 is the equilibrium constant for the $AgNO_3$-olefin complex. If K_L were not dependent upon the concentration of $AgNO_3$, i.e. there were no salting-out effect, a plot of H vs $AgNO_3$ would be linear with slope $K_L K_1$ and intercept K_L. In practice, a salting-out effect is observed at $AgNO_3$ concentrations greater than about 1 molar and, with monoolefins, a smooth curve with decreasing slope is obtained for the above plot. Since, however, at low concentration of $AgNO_3$ the salting-out effect is small compared with the complexing effect, satisfactory values of K_L and K_1 can be obtained by taking the slope of the tangent to the curve as $[AgNO_3] \longrightarrow 0$. The value of H can be determined from a knowledge of the inlet and outlet pressure and the flow rate of the carrier gas, the volume of the liquid phase at the column temperature employed and the retention time for the olefin relative to air or some other species that is not strongly retained.

TABLE III

Retention Data Relative to Isoprene for Various Hydrocarbons on a Benzyl Cyanide - $AgNO_3$ Column[a] at 22°C (43)

Hydrocarbon	Relative Retention Time
n-Butane	0.036
Ethylene	0.041
Propylene	0.095
n-Pentane	0.103
Propadiene	0.103
trans-2-Butene	0.147
Isobutene	0.188
n-Hexane	0.254
1-Butene	0.287
cis-2-Butene	0.330
Cyclopentane	0.342
1,3-Butadiene	0.353
trans-2-Pentene	0.355
2-Methyl-2-butene	0.381
3-Methyl-1-butene	0.457
2-Methyl-2-butene	0.602
1-Pentene	0.609
n-Heptane	0.633
Cyclohexane	0.766
cis-2-Pentene	0.832
Isoprene	1.00
2-Methyl-1-pentene	1.155
1,4-Pentadiene	1.648
Cyclopentene	2.611

a - 11.4' x 1/4" column with 3.806 g $AgNO_3$ in 9.660 g benzyl cyanide on 33.31 g 30-50 mesh Chromosorb.

The equilibrium constants obtained in this way for a series of olefins, while of different absolute value, are, relative to one another, very similar to those obtained from earlier distribution studies with aqueous $AgNO_3$ (17,25). This technique for determination of complex equilibrium constants has been used also (46) for determination of secondary deuterium isotope effects upon the stability of silver ion-olefin complexes.

V. ANHYDROUS SILVER SALTS

Francis (47) was the first to report the formation of complexes by interaction of monoolefins with solid $AgNO_3$. He found that contact of the salt with liquid propylene of 1-butene yielded liquid complexes containing about 1.3 moles of olefin per mole of salt which were stable only under pressures approaching the olefin vapor pressure and not at all above 36° and 25°C for C_3H_6 and C_4H_8, respectively. More recently, Long (30,48) has prepared the solid complexes $(AgNO_3)_2$·olefin by contacting the monoolefin vapor with anhydrous highly porous $AgNO_3$ obtained by carefully controlled dissociation of the complex with 1,3-butadiene. This 2:1 complex stoichiometry is identical to that reported (49) for the ethylene complex precipitated from aqueous $AgNO_3$. Long (48) has described the use of the porous $AgNO_3$ for separation of ethylene from admixture with ethane and has noted that finely ground commercial $AgNO_3$ is not at all effective as a solid for olefin separations.

Upon contact of excess cyclohexene with either $AgNO_3$ or $AgClO_4$, the complexes $AgNO_3 \cdot 2C_6H_{10}$ or $AgClO_4 \cdot 2C_6H_{10}$ are formed (50). Similarly, with α- and β-pinene, $AgClO_4$ forms the 1:2 complexes. It has been used (50) for separation of a 1:1 mixture of the pinenes into a non-complexed fraction containing 78 percent α-piene and a complexed fraction containing 92 percent β-pinene. However, because of the tendency of the perchlorate complexes to explode upon heating, $AgClO_4$ has not been extensively employed for olefin separations.

Detailed studies of the olefin complexes formed with solid anhydrous $AgBF_4$ and of the use of the solid salt for extraction of olefins from mixtures of hydrocarbons have been conducted by Quinn and co-workers (20,51-55). With the exception of ethylene, the most stable complex formed by $AgBF_4$ with all monoolefins has the 1:2 stoichiometry. Those obtained with the olefins through the C_4 series, the normally gaseous olefins, are shown in Table IV along with their equilibrium dissociation pressures at 25°C. Ethylene is exceptional in that it forms the 1:1 and 2:3 complexes, both of which are much more stable than the 1:2 complex. The much lower stability of the 1:2 ethylene complex than of the other 1:2 complexes is presumably the result of a weaker coordinate bond between the less basic ethylene molecule and the silver ion. The existence of the 1:1 and 2:3 ethylene complexes probably results from the small size and high symmetry of the ethylene molecule by comparison with the other olefins. Most of these olefins will form also very much less stable 1:3 complexes (Table IV) which could be encountered when

TABLE IV

Equilibrium Dissociation Pressures at 25°C for Complexes of Normally Gaseous Olefins with $AgBF_4$ (51)

	Dissociation Pressure of Complex (torr)				
Olefin	Complex	1.1	2:3	1:2	1:3
Ethylene		12.8	41.3	208	2033
Propylene				51.1	2539
1-Butene				9.4	888
Isobutene				8.8	
cis-2-Butene				4.1	706
trans-2-Butene				20.8	

effecting extractions at higher pressures. The absence of the 1:3 complexes with such olefins as isobutene and *trans*-2-butene is considered to be the result of steric factors (51).

With the exception of the complex with 1-butene, all of the 1:2 complexes listed in Table IV are solids which upon heating dissociate without melting. The 1-butene complex melts at about 44°C. A study of the phase diagram for the system $AgBF_4$-1-pentent, Figure 6, has shown not only the 1:2 and 1:3

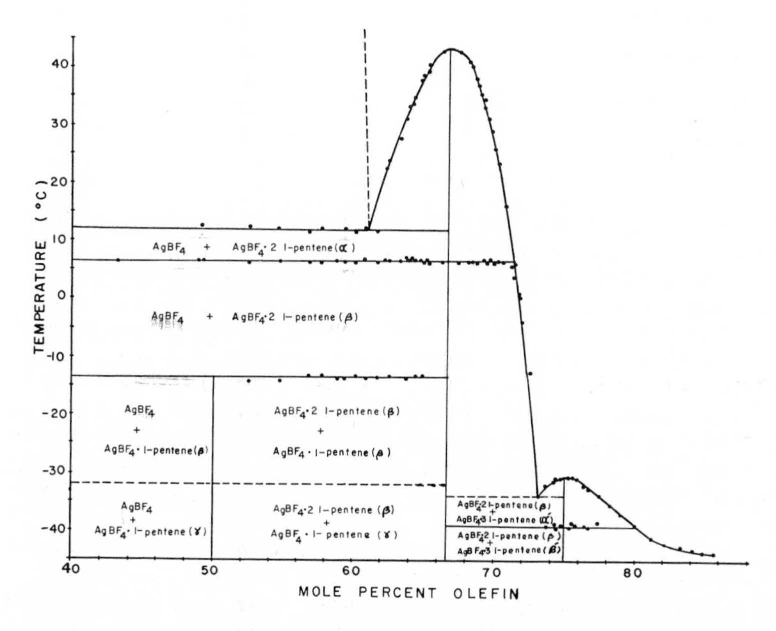

Figure 6. Phase diagram for the system $AgBF_4$-1-pentene (53).

complexes as compounds melting congruently at 43.3°C and -30.3°C, respectively, but also a 1:1 complex stable only in the solid state at temperatures below about -10°C. The 1:2 complexes of many other olefins are also congruently melting compounds, the melting point being markedly dependent upon the position and degree of branching of the hydrocarbon chain (Table V) (56).

TABLE V

Melting Point of $AgBF_4$.2 Olefin Complexes (56)

Olefin	M.P. (°C)
1-Butene	44
2-Methyl-1-butene	60
3-Methyl-1-butene	61
3,3-Diemethyl-1-butene	106
3-Methyl-1-pentene	17
4-Methyl-1-pentene	98
4,4-Diemethyl-1-pentene	143
1-Hexene	32
2-Methyl-1-hexene	-89
3-Methyl-1-hexene	-32
4-Methyl-1-hexene	91
5-Methyl-1-hexene	61
3,5,5-Trimethyl-1-hexene	52
1-Heptene	38

The stability of the complex as defined by its equilibrium dissociation pressure curve is a measure of the lowest level to which a given olefin could be extracted from its solution in a non-complexing solvent such as a paraffin. Assuming zero solubility of the complex in the paraffin phase, (this is not a good assumption with the higher olefins), one can use the equilibrium dissociation pressure and the Raoult's Law approximation to calculate the concentration of olefin in the mixture in equilibrium with the complex at a given temperature. From the stability data for the propylene complexes shown in Figure 7, it is obvious that the order of extraction efficiency for the silver salts reported would be $AgBF_4 > AgClO_4 > AgNO_3$. At 0°C, the dissociation pressure of the $AgBF_4$ complex is about 1/1000th of that of the $AgClO_4$ complex and about 1/20th of that of the $AgNO_3$ complex. The $AgNO_3$ complex of propylene is somewhat less stable than the CuCl complex (3). The greater stability of the $AgBF_4$ complexes may account for the high solubility of olefins in aqueous $AgBF_4$

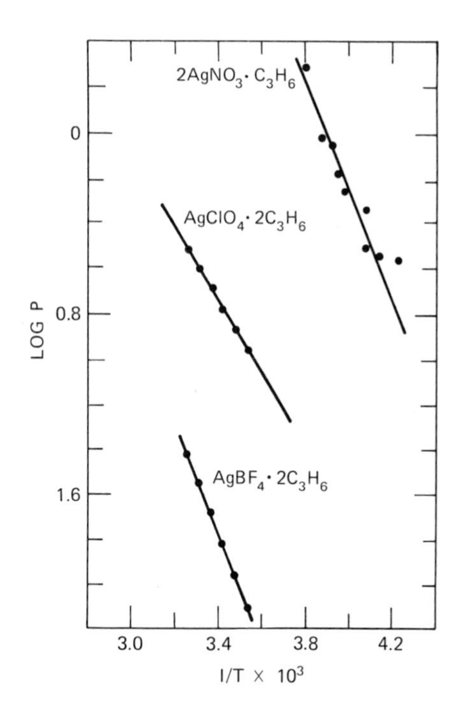

Figure 7. Stability lines for propylene complexes of $AgNO_3$ (30), $AgClO_4$ and $AgBF_4$ (51). Stoichiometry of $AgNO_3$ complex not firmly established. (P in atmospheres, T°K).

solutions as discussed earlier. Both of these facts may be related to a weaker interaction between the silver ion and the anion in $AgBF_4$ than in the other silver salts.

Evaluation (57) of a number of silver salts as extractants of the normally gaseous olefins from solution in n-pentane has produced the data of Table VI. It is quite obvious that no extraction is effected by silver fluoride or silver trifluoroacetate. Although $AgClO_4$ extracts some olefin, it is much poorer as an extractant than $AgBF_4$ which is in turn much poorer than $AgSbF_6$. Although $AgSbF_6$ will extract olefins to a very low level, it is not a practical extractant because the very low dissociation pressure of its complexes makes the olefin recovery very difficult.

The $AgBF_4$ complexes of the olefins of Table VI are solids at 25°C, even in the presence of the paraffin. The concentration of olefin in the paraffin phase after extraction, as determined by G.C. analysis, is quite close to the minimum value calculated from the equilibrium pressure data for ethylene, propylene, 1-butene and cis-2-butene but it is significantly higher than that minimum for isobutene and trans-2-butene. The much poorer extraction of these latter olefins indicates a slower approach to equilibrium in the complex formation with these. With all of the olefins, the extraction time was five minutes.

TABLE VI

Extraction of Olefins from n-Pentane at 25°C

Olefin	Silver Salt	Olefin Concentration (Wt. %)		
		Initial	Final	
			Observed	Calculated[a]
Ethylene	$AgBF_4$	0.173	0.013	0.011
Propylene	$AgBF_4$	0.73	0.156	0.104
	$AgSbF_6$	1.15	0.019	
	$AgClO_4$	1.37	1.12	
	AgF	1.37	1.37	
	$AgCOOCF_3$	1.37	1.37	
1-Butene	$AgBF_4$	6.13	0.439	0.432
	$AgSbF_6$	5.22	0.041	
Isobutene	$AgBF_4$	9.70	1.43	0.302
	$AgSbF_6$	9.70	0.564	
cis-2-Butene	$AgBF_4$	8.47	0.225	0.204
trans-2-Butene	$AgBF_4$	7.52	4.10	0.938
	$AgSbF_6$	7.52	0.101	

a - Using dissociation pressure and Roalt's Law

A detailed investigation (57) of the extraction of 1-heptene from n-heptane has shown, as one would anticipate from a knowledge of the 1:2 complex stoichiometry, that there is no dependence of extraction efficiency upon the $AgBF_4$:olefin ratio provided that it is greater than 0.5:1. For many extractions, however, a ratio of 1.4:1 was used in order to reduce the period of time necessary for optimum extraction. With this olefin, as with all 1-olefins from 1-pentene through 1-dodecene, the complex phase is liquid at 25°C in the presence of the paraffin. At

13°C, the 1:2 complex freezes out from the liquid complex phase. The ready attainment of equilibrium conditions over the temperature range from 13° to 75°C is indicated by the linear plot of the logarithm of the olefin concentration in the paraffin phase as determined by infrared spectroscopy and the reciprocal of the absolute temperature (Figure 8).

Extractions with $AgBF_4$ of the linear 1-olefins from C_5 to C_{18} from solutions in n-heptane are reported in Table VII. As observed above, the olefins through C_{12} form a liquid complex phase at 25°C in n-heptane and with increasing chain length the extractions become less efficient due to increasing miscibility of the complex and paraffin phases. At 25°C, no extraction of 1-decene or 1-dodecene is obtained because of this miscibility; however, at 0°C, crystallization of the complex from the paraffin solution yields a very good extraction of 1-dodecene. The complexes of the olefins beyond C_{12} show, even at 25°C, increasing crystallization from the n-heptane solution with increasing chain length. Consequently, 1-octadecene is extracted much more efficiently at 25°C than is 1-tetradecene.

A comparison of the extraction of a number of olefins by various silver salts is shown by Table VIII. The olefins with more than one substituent group at the double bond are much less

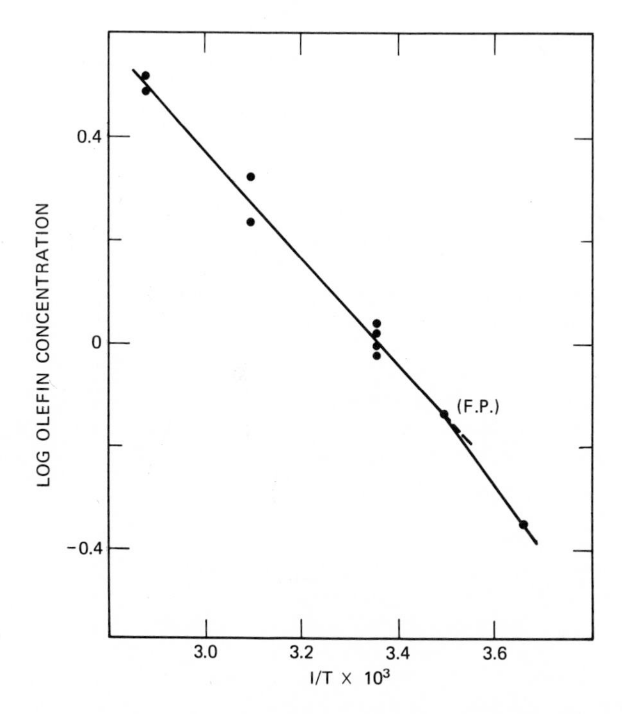

Figure 8. Temperature (°K) dependence of equilibrium concentration (wt. %) of olefin for extraction of 1-heptene from n-heptane with anhydrous $AgBF_4$.

TABLE VII

Extraction of Linear 1-Olefins from n-Heptane with $AgBF_4$[a]

Olefin	Complex Phase at 25°C	Initial Conc. (wt. %)	Final Conc. (wt. %)	
			25°C	0°C
1-Pentene	liquid	9.07	0.68	
1-Heptene	"	9.11	0.93	0.43
1-Octene	"	9.16	2.83	1.44
1-Decene	"	9.09	9.05	4.36
1-Dodecene	"	9.34	9.30	0.72
1-Tetradecene	solid	8.98	5.56	
1-Hexadecene	"	8.88	1.91	
1-Octadecene	"	9.10	0.75	

a - $AgBF_4$:olefin mole ratio = 1.4:1

TABLE VIII

Extraction of Olefins from n-Heptane with Silver Salts[a] at 25°C

Olefin	Final olefin concentration (wt. %)			
	$AgBF_4$	$AgPF_6$	$AgSbF_6$	$AgClO_4$
1-Heptene	0.93	0.10	nil	7.5
1-Decene	9.05		0.15	
1-Tetradecene	5.56	3.58	1.20	
1-Octadecene	0.75		0.44	
2-Pentene	2.52		nil	
2-Methyl-2-butene	5.84		5.34	

a - Ag^+:olefin mole ratio = 1.4:1

efficiently extracted than are the mono-alkyl substituted 1-olefins. This table demonstrates very decisively that, for the silver salts employed, the order of extraction efficiency is $AgSbF_6 > AgPF_6 > AgBF_4 > AgClO_4$. For the series of salts with perfluorinated anions, the extraction efficiency, therefore the complex stability, increases with increasing size of the anion. This stability differences may reflect the difference in the change of free energy of the salt lattice when the olefin molecule is incorporated into the various lattices. The difference between the stability of the $AgBF_4$ and $AgClO_4$ complexes may result also from a difference in the change of lattice free energy due not to difference in anion size but, more likely, to different cation-anion interactions in the salts. Because oxygen is a better electron donor than fluorine, ClO_4^- will interact more strongly with Ag^+ than does BF_4^-. Furthermore, the observation (58) that, in solid silver nitrate-olefin complexes, there is a very definite bonding between the Ag^+ and NO_3^- ions may account for the relatively low stability of the $AgNO_3$ complexes.

Extraction of an aryl-substituted olefin from its paraffinic analogue also can be effected with solid $AgBF_4$ (57). When a 1:1 mixture of o-chlorostyrene and o-chloroethylbenzene, diluted with n-hexane, is treated for 10 minutes with $AgBF_4$ ($AgBF_4$;olefin = 1:1), about 85 percent of the o-chlorostyrene is separated as a white crystalline complex. The olefin can be displaced from the complex by treatment of a suspension of the solid in n-hexane with propylene vapor.

Because the utility of the solid silver salts as olefin extractants depends not only upon the complex stability but also upon the rate of complex formation, a study has been made (52) of the kinetics of formation of the $AgBF_4$-propylene complex upon interaction of the gaseous olefin with the solid salt, equation [2]. Although these conditions are not identical to those used

$$\underset{(s)}{AgBF_4} + \underset{(g)}{2C_3H_6} \rightleftharpoons \underset{(s)}{AgBF_4 \cdot 2C_3H_6} \qquad [2]$$

for the study of propylene extraction, i.e. C_3H_6 in n-pentane solution, the fact that the complex is solid in both systems suggests that the mechanism of transfer of propylene into the salt particle might be the same in both systems.

When $AgBF_4$ is exposed to propylene for the first time, complex formation proceeds much more slowly than it does when the salt has been taken previously through a complex formation-dissociation cycle. In order to obtain reproducible kinetic data it is necessary to activate the $AgBF_4$ by subjecting it to several such cycles with a resultant 40-fold increase in its surface area and a marked reduction in its particle size. There is, however, no change in its lattice structure. This process is somewhat analogous to that described by Long (48) for generation of active, highly porous $AgNO_3$.

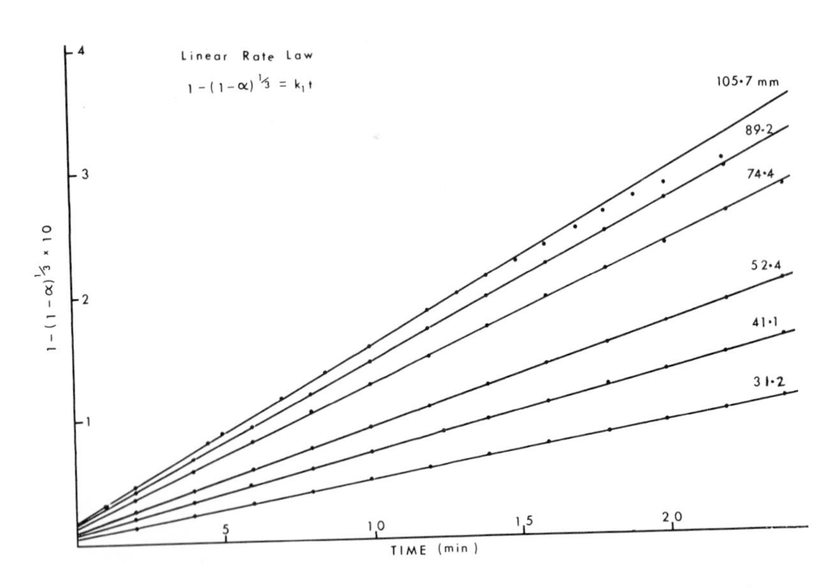

Figure 9. Linear rate law plot for a reaction of propylene with activated $AgBF_4$ at various pressures (52). (Weight of $AgBF_4$:552 mg.; temperature: 25.0°C).

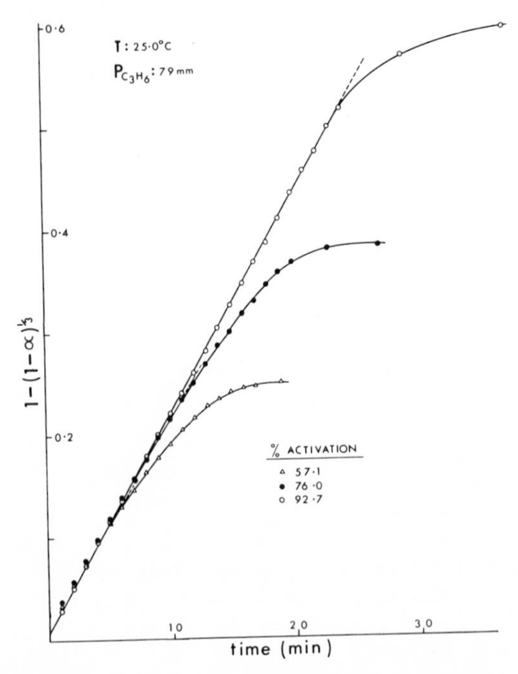

Figure 10: Effect of varying cyclohexene activation upon range of applicability of linear rate law (52).

The primary rate data, obtained by gravimetric determination of the rate of propylene absorption by $AgBF_4$ are best correlated (Figure 9) by the linear rate law

$$1-(1-\alpha)^{1/3} = k_1 t \qquad [3]$$

wherein α is the fraction of $AgBF_4$ complexed, t is the time and k_1, the slope of the linear plot, is proportional to the radius, and thus the area, of the uncomplexed particle. The linear rate law indicates a kinetically controlled reaction proceeding by movement of the reaction interface from the surface toward the center of the particles. The extent to which the process of activation of the salt affects the degree of kinetic control of the propylene complex formation is shown by Figure 10 for a sample of $AgBF_4$ activated by cyclic formation-decomposition of the cyclohexene complex, $AgBF_4.2C_6H_{10}$. While the rate, k_1, from the linear rate law plot increases only slightly upon increasing the extent of activation from 57.1 to 92.7 percent, there is a very marked effect on the value of α at which deviation from the linear rate law occurs. This increases from 0.34 at 57.1 percent activation to 0.88 at 92.7 percent activation.

At a given temperature, the slope, k_1, of the linear rate plot is directly proportional to the logarithm of the propylene pressure P and the rate expression for constant temperature operation is in essence

$$k_1 = k_2 \log P/Pe \qquad [4]$$

wherein k_2 is the temperature dependent rate constant and Pe is

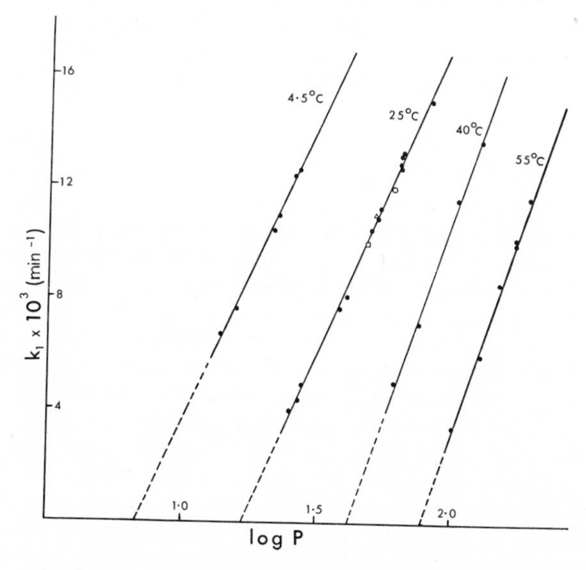

Figure 11. Variation of the linear rate law slope k_1 with log P at several temperatures. (P in torr) (52).

the equilibrium dissociation pressure of the complex. As shown by Figure 11, k_2 is not very dependent upon the temperature. An Arrhenius plot of the k_2-temperature data indicates an activation energy for complex formation of less than 2 kcal/mole which tends to support the suggestion (59) that, for a truly reversible reaction such as this at a solid-solid interface, the activation energy is zero.

The dependence of the reaction rate on log P/Pe rather than on the simple pressure differential P-Pe suggests that an adsorption process which determines the concentration of olefin at the surface of the solid particles is important. This process might be expected to be operative, as well, when the solid salt is employed to extract the olefin from a liquid paraffin solution. A consideration of the various adsorption isotherms which might apply has indicated that the Freundlich isotherm,

$$\log \theta = AT \log P + BT \qquad [5]$$

wherein θ is the surface coverage, is the most likely. Since the Freundlich isotherm is based on the assumption that the heat of adsorption decreases logarithmically with increasing θ, one is lead to conclude from equation [4] that the reaction rate is not as much dependent upon the surface coverage of the salt by olefin molecules as upon the heat of adsorption. If the mechanism by which the olefin is transferred from the surface of the salt particle to the reaction interface involves successive complex decomposition-formation reactions, the heat of adsorption might be important for initiation of the decomposition step.

It has been observed (52) that, if the activated sample of $AgBF_4$ is exposed to sintering conditions for a period of time, the kinetic data deviate from the linear rate law at a rather low value of α and at some point begin to adhere very closely, Figure 12, to the parabolic rate law, equation [6], which infers

$$[1-(1-\alpha)^{1/3}]^2 = k_d t \qquad [6]$$

a diffusion controlled process. Interestingly, this same rate law describes also the kinetics of the first reaction of a sample of $AgBF_4$ with propylene. The pressure dependence of k_d, Figure 13, is very similar to that which has been found (60) for the variation of the diffusion coefficient with amount of adsorption for transport of a gas through a porous plug on which it is adsorbed. The diffusion control shown by $AgBF_4$ samples suggests then the existence of a pore structure in the solid. The process of activation of the salt apparently produces a porous structure with pore diameters sufficiently large that, even after adsorption of olefin on the surface, there is little hindrance to gaseous diffusion and the kinetically-controlled reaction proceeds for a significant part of the total reaction. The dependence of the degree of kinetic control upon the extent of activation shown by Figure 10 tends to confirm this suggestion.

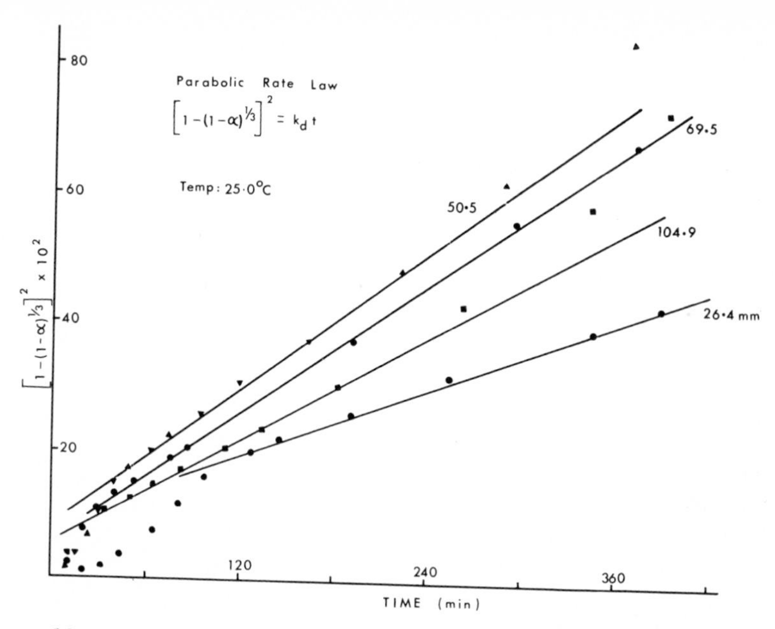

Figure 12. Parabolic rate law plot for reaction of propylene with a sintered sample of $AgBF_4$ (weight of $AgBF_4$:513 mg.) (52).

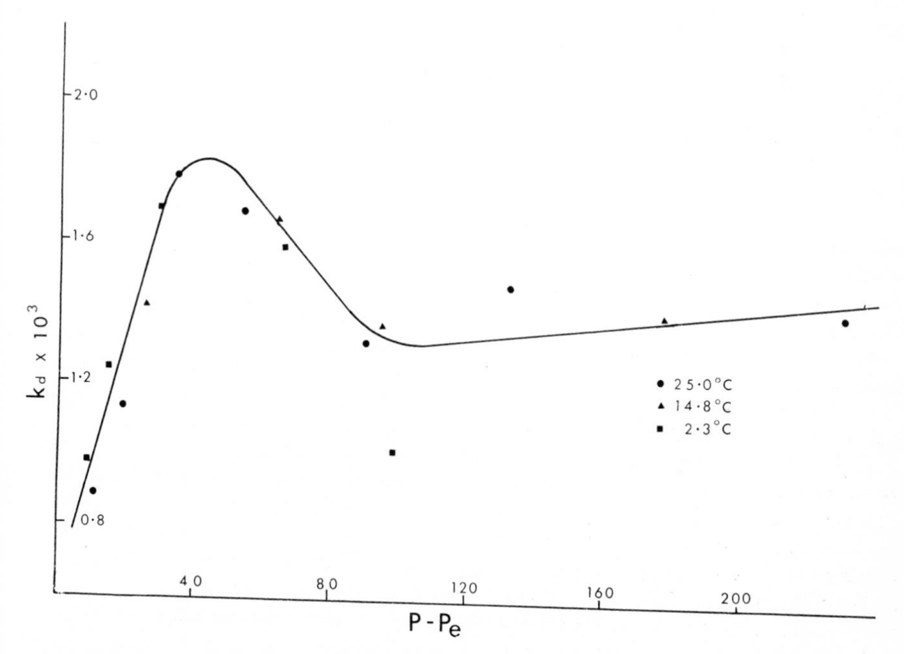

Figure 13. Pressure dependence of slope k_d of parabolic rate law plots (52).

When the pore diameter is decreased, either by sintering or by expansion upon complex formation of the salt particles, a point is reached where, upon formation of an adsorbed film, gaseous diffusion into the salt bed is hindered. Subsequently, the rate of complex formation is determined by the rate of surface diffusion of olefin molecules.

The kinetic studies have shown the advantage to be gained by introduction of a large pore volume into the solid salts (48,52). In addition, it has been shown that silver salts with the larger perfluorinated anions are better extractants than those with smaller anions. These observations lead one to think that silver ion systems which have large permanent pore structures, such as zeolites, or macromolecular anions, such as ion exchange resins, could be good olefin extractants. Furthermore, they are often thermally more stable and more easily dried than are salts such as $AgBF_4$.

VI. SILVER ION EXCHANGE RESINS

Silver ion exchange resins can be easily prepared by conventional ion exchange techniques employing a cation exchange resin in the H^+ or Na^+ form and an aqueous solution of the silver salt. The resins are normally dried before use as extractants of unsaturated hydrocarbons.

An alternative approach to preparation of the silver resin has been to use an anion exchange resin containing amine groups which complex with the metal ion and retain it at the resin surface. A resin of this nature, derived from Amberlite IR4B, containing $AgNO_3$ has been used (61) at ambient temperature for separation of piperylene from n-hexane.

With these silver ion resins, the olefin separation or enrichment is usually effected by passing the hydrocarbon mixture, either as a vapor or a liquid, through a bed of resin at temperatures at or below 30°C (62). The olefin can be recovered from the resin by evacuation or inert-gas purging at somewhat higher temperatures. Care must be excercised to assure that the temperature does not reach that, at which resin decomposition will occur.

Very often it is found that the capacity of the silver resin for olefin adsorption is quite low. Niles (62) has observed a sharp breakthrough of 1-pentene when one bed volume of a 1 percent solution of olefin in n-pentane has passed through a column of silver Amberlite IRC-50, a carboxylate resin, at -26°C. The breakthrough occurred after use only of about 3.5 percent of the theoretical capacity (based on Ag content) of the resin. With silver Dowex 50W-X8, a sulfonate resin, the breakthrough is not sharp but occurs at ambient temperature after use of only about 1.2 percent of the theoretical capacity.

Small (63) has described the use of highly porous cation exchange resins in the silver ion form for separation of polar molecules, including olefins, from less polar molecules such as paraffins. The porous resins offer higher sorption capacity, selectivity and efficiency even though they do not have an ion exchange capacity significantly different from that of the conventional resins.

An evaluation of silver ion exchange resins for olefin-paraffin separations also has been made by using the resin as a gas chromatographic column packing (57). Surprisingly, a dry Dowex 50-X4 resin containing 43 weight percent of Ag^+ gives no resolution of olefin-paraffin mixtures. When, however, the resin is coated with about 20-25 weight percent of ethylene glycol, good resolution is obtained with the olefin appearing as a rather broad peak on the chromatogram (Figure 14a). As shown by the data of Table IX,

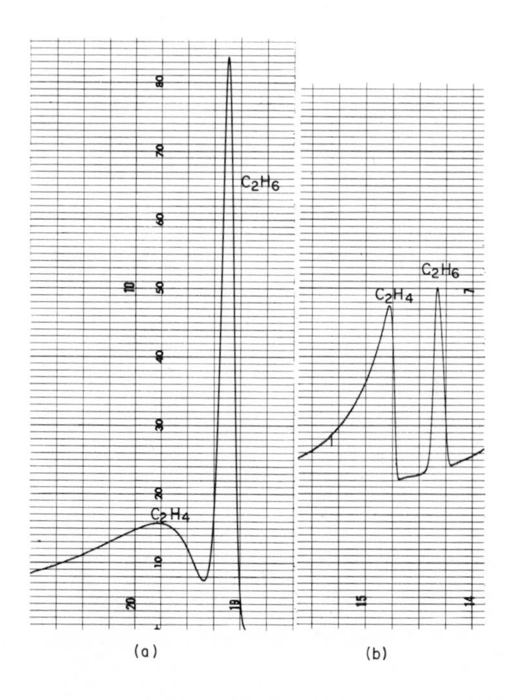

Figure 14. (a) Resolution of ethane-ethylene mixture on glycol-coated silver Dowex 50-X4 resin at 55°C.

(b) Resolution of ethane-ethylene mixture on dry silver Amberlite IRC-50 resin at 24°C.

this resolution increases initially with increasing temperature and then begins to decrease. The increase is thought to be due to decreasing viscosity of the glycol film which permits more rapid diffusion of the olefin to the silver ion sites. At some temperature, however, between 55 and 86°C for ethylene, this effect is offset by decreasing stability of the silver-olefin complex. A similarly coated sodium ion form of the resin gives no olefin-paraffin resolution. The effect of the glycol may be to reduce the donor-acceptor interaction between the silver ions and sulfonate groups and thus permit stronger interaction of the silver ions with the olefin molecules.

TABLE IX

Resolution of C_2 and C_3 Mixtures on Silver Ion Exchange Resins

Mixture	Resin	Temperature (°C)	R.T.O./R.T.P.[a]
C_2H_4/C_2H_6	A[b]	0	1.33
		25	1.58
		55	1.76
		86	1.54
	B[c]	0	1.87
		24	1.44
		50	1.24
C_3H_6/C_3H_8	A	0	1.34
		25	1.54
		55	1.48
	B	0	1.90
		24	1.86
		50	1.63

a - retention time of olefin/retention time of paraffin.

b - Ag^+ Dowex 50W-X4 coated with ethylene glycol.

c - Dry Ag^+ Amberlite IRC-50; 0.4 cm^3 of olefin-paraffin mixture.

With the silver ion form of the carboxylate resin Amberlite IRC-50 containing 50 weight percent of Ag^+ quite the opposite behavior is observed. When coated with glycol, this resin gives no olefin-paraffin resolution. The bare dry resin effects, on the other hand, a very good resolution with a fairly sharp olefin peak (Figure 14b). The resolution decreases with increasing temperature (Table IX), a reflection of decreasing stability of the silver-olefin complex. With this resin the resolution decreases also with increasing size of the olefin-paraffin sample suggesting that only a relatively small proportion of the silver ions are readily accessible to the olefin molecules. The resolution demonstrated by the dry Amberlite resin relative to that of the dry Dowex resin suggests that there is less interaction of the silver ions with the carboxylate groups than with the sulfonate groups.

Silver resins have also been found (64,65) to be effective for removal of acetylenes from olefin streams. The extent of acetylene adsorption increases with increasing temperature. That fact, in conjunction with the observation (64) that the acetylene cannot be removed by purging with inert gas at temperatures below the decomposition point of the resin, suggests that silver acetylide formation may occur. The acetylene-loaded resin shows, however, no evidence of shock sensitivity even at higher temperatures (64).

The silver carboxylate resin has been shown (64) to perform very efficiently for removal of trace quantities of acetylene from ethylene. When an ethylene stream containing 6 p.p.m. C_2H_2 is passed at 300 ml/min through 1 ml of the resin at 64°C, the acetylene is removed to a level below 0.2 p.p.m. from more than 7×10^5 ml of the ethylene. This corresponds to an acetylene adsorption of about 4.4 ml but represents still an acetylene:silver ion ratio of only about 1:25. Although it effects good acetylene extractions, the silver resin probably will not be commercially acceptable because its thermal instability prevents regeneration by oxidative removal of the adsorbed species at higher temperatures. The silver zeolites do not have that deficiency.

VII. SILVER ZEOLITES

Zeolites, or molecular sieves, are often employed for separations involving unsaturated hydrocarbons (66,67). These metal aluminosilicates have a crystal structure such that there is a large internal surface for adsorption. Access to this surface is by way of openings or pores in the lattice.

The zeolites consist basically of a three-dimensional network of SiO_4 and AlO_4 tetrahedra. Because of the trivalency of the aluminum, each alumina tetrahedron introduces one negative charge to the lattice which is neutralized by the positive charges of associated cations. The tetrahedra are joined through the oxygen atoms to form the so-called sodalite unit containing 24 (Si, Al) atoms interconnected by 36 oxygen atoms. The sodalite units are then packed in a simple cubic array to form Type A[a] zeolites with four bridge oxygen atoms between neighboring units or in a tetrahedral (or diamond) array to form Type X[a] or Type Y[a] zeolites with six bridge oxygen atoms. The main adsorption cavity of the A zeolite is a space surrounded by eight sodalite units and entry into the cavity is by means of a channel bounded by four sodalite units, the channel having a diameter of about 4Å. The adsorption cavity of the X zeolite is a space surrounded by ten sodalite units and the channels are bounded by six sodalite units resulting in a channel diameter of about 9Å.

Whether the cubic or diamond array is obtained depends upon the SiO_4:AlO_4 ratio in the system. The Type 4A zeolite has a chemical composition approximating $Na_2O.Al_2O_3.2SiO_2.xH_2O$ resulting in about 12 monovalent cations per sodalite unit (68) while the Type 13X zeolite has a composition approximately $Na_2O.Al_2O_3.2.8SiO_2.yH_2O$ with about 10 monovalent cations per sodalite unit (69).

a. Terminology employed by Linde Division, Union Carbide Corporation.

Crystallographic investigations have established that one-third of the sodium ions per unit cell of zeolite 4A (68) and about one-half of the sodium ions per unit cell of zeolite 13X (69) are located near the walls of the main cavity. While these ions will be those most easily exposed to adsorbate molecules, some or all of the other ions may be exposed as well, dependent upon the adsorbate molecular dimensions. The hydrated zeolites are activated for adsorption by driving off most of the water of crystallization at temperatures of 300-400°C.

With conventional ion-exchange techniques, the sodium ion of the crystalline zeolite can be replaced by many other metal ions (70) such as Ag^+. Although many separations with the zeolites, especially Type 4A, are based on molecular size, others result from differences in such physical properties as polarity or polarizability. Because olefins are, in general, more polarizable than paraffins, they are adsorbed more strongly on the zeolite surface. The cations exposed at the adsorption surface act as sites of strong localized charge which induce dipoles in the polarizable molecules and then electrostatically interact with them.

When the zeolite carries a cation such as Ag^+ which is known to coordinate with an olefin molecule, very strong adsorption would be expected. The adsorption of ethylene on Type X zeolites (71, 72) has shown a marked irreversibility on AgX but complete reversibility on NaX. In addition, calorimetric determination of the heat of ethylene adsorption has shown a value of about 18 kcal/mole for AgX and only about 8-9 kcal/mole for NaX (73). Yates (72) has observed two types of adsorbed ethylene molecules on AgX, one weakly adsorbed that can be removed by evacuation at 25°C (about 45 percent of the total) and another strongly adsorbed that can be removed only by evacuation at elevated temperature. The quantity of strongly adsorbed ethylene corresponds approximately to adsorption of one ethylene molecule at each silver ion in the large cavity. Ethylene adsorption on AgA has shown a similar relationship (57). It is probable that the ethylene molecules cannot approach sufficiently closely to the other silver ions in the zeolite lattice to permit a strong interaction.

The relative ability of AgX and NaX zeolites to extract 1-pentene from a 2.9 percent solution in n-heptane employing a zeolite bed 45 cm. in length and 1.5 cm in diameter is shown in Figure 15. Although both zeolites adsorb significant quantities of the olefin, breakthrough of 1-pentene occurred first and the initial concentration of the olefin in the effluent was achieved more rapidly on the NaX. Upon desorption of the olefin by flushing of the loaded zeolite with paraffin, the greater irreversibility of adsorption on AgX (containing 35.4 wt. percent Ag^+) is again observed, Figure 16. Flushing with 150 gm. of paraffin results in recovery of 66 percent of the olefin adsorbed on NaX but of only 19 percent of that adsorbed on AgX.

The temperature dependence of the adsorption of ethylene on a number of zeolites is shown in Figure 17. The data were obtained by passing a stream containing 5.7 volume percent of

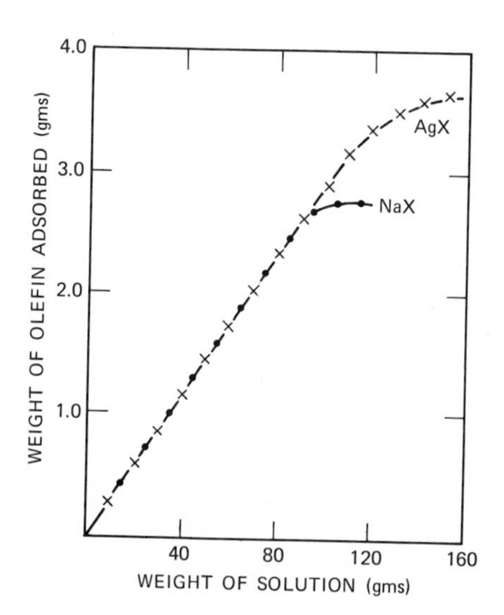

Figure 15. Extraction of 1-pentene from n-heptane with NaX and AgX zeolites at 25°C.

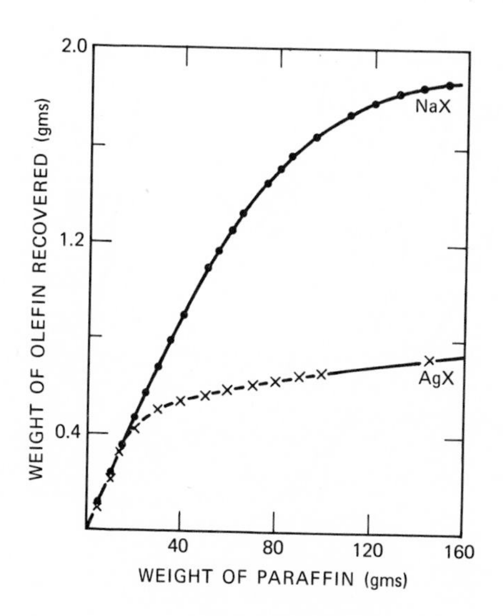

Figure 16. Recovery of adsorbed 1-pentene from NaX and AgX zeolites by flushing with n-heptane at 25°C.

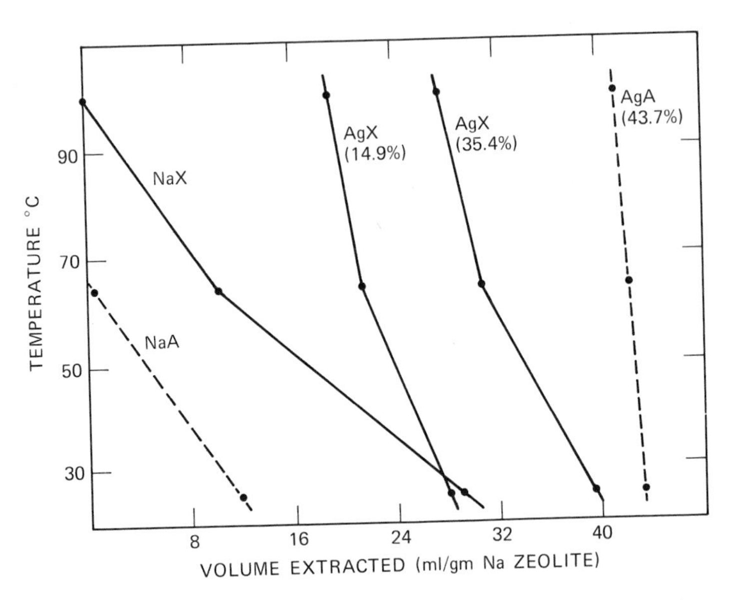

Figure 17. Temperature dependence of the extraction of ethylene from helium with various zeolites.

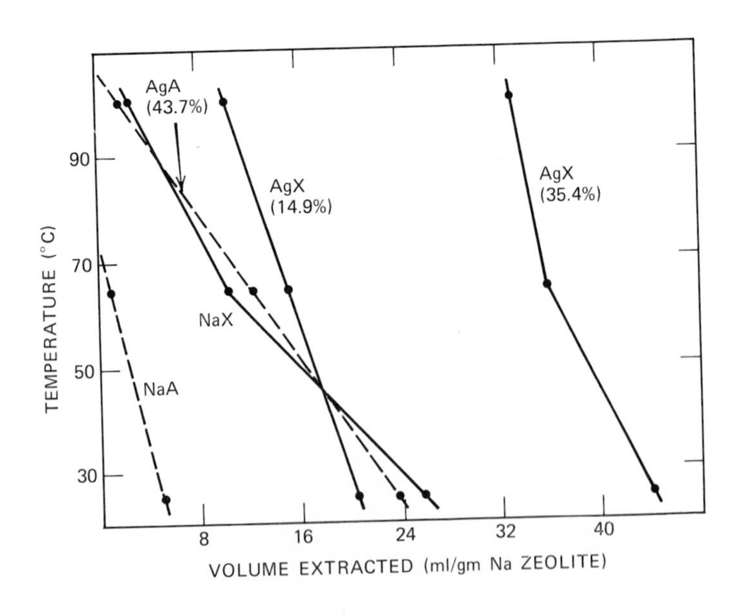

Figure 18. Temperature dependence of the extraction of acetylene from helium with various zeolites.

C_2H_4 in helium at 25 ml/min through a 3 ml bed of zeolite and recording the volume passed before the olefin concentration in the effluent exceeded 1 ppm.

The capacity for ethylene adsorption appears to increase with increasing silver content of the zeolite. The small temperature dependence of the adsorption on Ag^+ zeolites relative to their Na^+ counterparts is again evidence for much stronger adsorption on the silver system. Furthermore, the ethylene may be even somewhat more strongly adsorbed on AgA than on AgX.

A similar investigation of the extraction of acetylene from a 0.23 volume percent solution in a helium stream passed at 75 ml/min through a 3 ml bed of zeolite has demonstrated a marked difference in behavior between AgA and AgX (Figure 18). Although acetylene is very strongly held on AgX, its adsorption on AgA is completely reversible. The capacity of AgA for acetylene adsorption is almost identical at all temperatures from 25° to 100°C to that of NaX. With AgX, acetylene adsorption appears to be more dependent upon silver content than does ethylene adsorption.

Acetylene is one of the most common trace level contaminants in commercial ethylene. When a stream of ethylene containing 6-9 p.p.m. of acetylene is passed at a rate of 300 ml/min through a 1 ml sample of zeolite held at 65°C in a 6 mm tube with a bed depth of about 3.5 cm, the extraction data of Table X are obtained (74). Acetylene is removed to a concentration below 0.2 p.p.m. from more than 300 liters by the partially-exchanged AgX but from only 68 liters by the partially-exchanged AgX. There is, however, no effective removal of acetylene by the AgA or the sodium zeolites. The acetylene-loaded AgX shows no evidence of either thermal or shock sensitivity, despite the possibility that acetylide formation may occur, and it can be regenerated for subsequent use by flushing with air at 300°C.

VIII. SUMMARY

Hydrocarbon separations, in particular the separation of olefins from paraffins, by formation of complexes with silver ions have involved the use of silver salts, either as solids or as aqueous or organic solutions, silver ion exchange resins and silver zeolites. The only commercial separation to date has employed an aqueous solution of silver tetrafluoroborate. Although silver nitrate solutions could be employed, their capacity for olefin absorption is lower because of lower aqueous solubility of the salt. Solid $AgBF_4$ is also much better than solid $AgNO_3$ for olefin separations, primarily because of greater stability of its complexes.

When the silver ion is placed on a chromatographic substrate as an ethylene glycol solution of $AgNO_3$ or by exchange with an ion exchange resin, good olefin-paraffin separations can be accomplished for either analytical or preparative purposes. A silver carboxylate resin shows a tendency to strongly adsorb acetylene and can be used for removal of trace quantities of acetylenes from olefin streams.

TABLE X

Extraction of Acetylene from Ethylene at 65°C with Silver Zeolites

Zeolites	Ag^+ Content (wt. %)	Feed Conc. (p.p.m. C_2H_2)	Volume Treated (liters)	Final Conc.[a] (p.p.m. C_2H_2)	C_2H_2 Removed[b] (ml/g)
AgX	35.4	6.0	320[c]	<0.2	3.2
AgX	14.9	6.5	68	0.2	0.68
NaX	0	8.4	2	4.7	0.01
AgA	43.7	8.4	2	0.7	0.03
NaA	0	8.4	2	7.8	0.01

a. Exit concentration after treating indicated volume of ethylene.

b. Ml C_2H_2 adsorbed/g zeolite, Na form.

c. No evidence of C_2H_2 breakthrough when experiment ended.

Silver zeolites will adsorb olefins sufficiently strongly to permit their removal as contaminants from paraffin streams. Some of them also show a very strong affinity for acetylenes and thus can be employed for purification of olefin streams. The zeolites are much more thermally stable than are silver salts and the silver zeolite loaded with olefin or acetylene can be repeatedly regenerated by oxidative removal of the hydrocarbons at elevated temperature.

Because of the large capital investment associated with use of the silver ion systems on a commercial scale, the greater thermal stability of zeolite than of salts or ion exchange resins makes its use more attractive. Silver losses resulting from reduction to the metallic state by hydrogen or some other reducing species in the system are lower with the zeolites because oxidative regeneration also results in re-oxidation of Ag(0) to Ag(I).

REFERENCES

1. N.K. Chaney, U.S. Patent 2,395,954 (1946).
2. F.J. Soday, U.S. Patent 2,395,958 (1946).
3. E.R. Gilliland, J.E. Seebold, J.R. FitzHugh, and P.S. Morgan, J. Am. Chem. Soc. 61, 1960 (1939).
4. K.R. Gray and P. Thurairajan, British Chem. Eng. 13, 824 (1968).
5. E.R. Atkinson, D. Rubinstein, and E.R. Winiarczyk, Ind. Eng. Chem. 50, 1553 (1958).
6. R.C. Scofield, U.S. Patent 3,080,437 (1963).
7. F.J. Soday, U.S. Patent 2,423,414 (1947).
8. E.W. Stern, U.S. Patent 3,243,471 (1966).
9. W.A. Knarr and E.A. Hunter, U.S. Patent 3,413,085 (1968).
10. A.W. Francis, U.S. Patent 2,735,878 (1956).
11. G.C. Ray, U.S. Patent 2,589,960 (1952).
12. H.W. Krekeler, J.M. Hirschbeck, and U. Schwenk, Erdöl Kohle 16, 551 (1963).
13. G.C. Blytas, E.R. Bell, and A.K. Dunlop, U.S. Patent 3,449,240 (1969).
14. M.I. Bogdanov and E.P. Krushinskaya, U.S.S.R. Patent 137,114 (1962).
15. C.E. Morrell, W.J. Paltz, J.W. Packie, W.C. Asbury, and C.L. Brown, Trans. Am. Inst. Chem. Eng. 42, 473 (1946).
16. M.J.S. Dewar, Bull. Soc. Chim. France 18, C79 (1951).
17. S. Winstein and H.J. Lucas, J. Am. Chem. Soc. 60, 836 (1938).
18. T. Fueno, O. Kajimoto, and J. Furukawa, Bull. Chem. Soc. Japan 41, 782 (1968).
19. T. Fueno, T. Okuyama, T. Deguchi,and J. Furukawa, J. Am. Chem. Soc. 87, 170 (1965).
20. H.W. Quinn, J.S. McIntyre, and D.J. Peterson, Can. J. Chem. 43, 2896 (1965).
21. J.G. Traynham and J.R. Olechowski, J. Am. Chem. Soc. 81, 571 (1959) (footnote 22).
22. R. Cramer, J. Am. Chem. Soc. 89, 4621 (1967).

23. H.W. Quinn and J.H. Tsai, Advan. Inorg. Chem. Radiochem. 12, 217 (1970).
24. A.E. Hill, J. Am. Chem. Soc. 44, 1163 (1922).
25. F.R. Hepner, K.N. Trueblood, and H.J. Lucas, J. Am. Chem. Soc. 74, 1333 (1952).
26. W. Gabler, A. Orlicek, and H. Poell, Erdol Kohle 17, 10 (1964).
27. K.N. Trueblood and H.J. Lucas, J. Am. Chem. Soc. 74, 1338 (1952).
28. J.W. Kraus and E.W. Stern, (a) J. Am. Chem. Soc. 84, 2893 (1962); (b) U.S. Patent 3,070,642 (1962).
29. J.G. Traynham, J. Org. Chem. 26, 4694 (1961).
30. R.B. Long, private communication (1969).
31. A.W. Francis, U.S. Patent 2,463,482 (1949).
32. A.W. Francis, U.S. Patent 2,673,225 (1954).
33. C.P. Strand, U.S. 2,515,140 (1950).
34. W. Featherstone and A.J.S. Sorrie, J. Chem. Soc. 5235 (1964).
35. P. Brandt, Acta Chim. Scand. 13, 1639 (1959).
36. B.B. Baker, Inorg. Chem. 3, 200 (1964).
37. duPont de Nemours, U.S. Patents 3,007,981 (1961); 3,218,366 (1965).
38. I.C.I., U.S. Patent 3,304,341 (1967).
39. Farbewerke Hoechst, U.S. Patent 2,913,505 (1959).
40. B.W. Bradford, D. Harvey, and D.E. Chalkley, J. Inst. Petrol. 41, 80 (1955).
41. E. Gil-Av, J. Herling, and J. Shabtai, J. Chromatog. 1, 508 (1958).
42. F. van de Craats, Anal. Chim. Acta. 14, 136 (1956).
43. F. Armitage, J. Chromtog. 2, 655 (1959).
44. E. Gil-Av and J. Herling, J. Phys. Chem. 66, 1208 (1962).
45. M.A. Muhs and F.T. Weiss, J. Am. Chem. Soc. 84, 4697 (1962).
46. R.J. Cvetanovic, F.J. Duncan, W.E. Falconer, and R.S. Irwin, J. Am. Chem. Soc. 87, 1827 (1965).
47. A.W. Francis, J. Am. Chem. Soc. 73, 3709 (1951).
48. R.B. Long, U.S. Patent 3,395,192 (1968).
49. K. Tarama, M. Sano, and K. Tatsuoka, Bull. Chem. Soc. Japan 36, 1366 (1963).
50. A.E. Comyns and H.J. Lucas, J. Am. Chem. Soc. 79, 4339 (1957).
51. H.W. Quinn and D.N. Glew, Can. J. Chem. 40, 1103 (1962).
52. C.A. Jull, R.J. Kominar, N.K. Mainland, and H.W. Quinn, Can. J. Chem. 44, 2663 (1966).
53. H.W. Quinn and R.L. VanGilder, Can. J. Chem. 46, 2707 (1968).
54. H.W. Quinn and R.L. VanGilder, Can. J. Chem. 47, 4691 (1969).
55. H.W. Quinn, U.S. Patent 3,189,658 (1965); Canadian Patent 691,251 (1964).
56. H.W. Quinn and R.L. VanGilder, unpublished data.
57. H.W. Quinn, unpublished data.
58. N.C. Baenziger, H.L. Haight, R. Alexander, and J.R. Doyle, Inorg. Chem. 5, 1399 (1966).
59. W.E. Garner, Chemistry of the Solid State, Buttworths London, 1955, pp. 228.

60. G.J. Field, H. Watts, and K.R. Weller, Rev. Pure Appl. Chem. 13, 1 (1963).
61. C.L. Swarthmore, U.S. Patent 2,865,970 (1958).
62. E.T. Niles, U.S. Patent 3,219,717 (1965).
63. H. Small, U.S. Patent 3,409,691 (1968).
64. H.W. Quinn, U.S. Patent 3,273,314 (1966).
65. W. Featherstone, British Patent 1,060,424 (1967).
66. S. Danatos, Chem. Eng. 71 (25), 155 (1964).
67. D.R. Silbernagel, Chem. Eng. Progr. 63 (4), 99 (1967).
68. T.B. Reid and D.W. Breck, J. Am. Chem. Soc. 78, 5972 (1956).
69. L. Broussard and D.P. Shoemaker, J. Am. Chem. Soc. 82, 1041 (1960).
70. H.S. Sherry, J. Phys. Chem. 70, 1158 (1966).
71. H.W. Habgood, Can. J. Chem. 42, 2340 (1964).
72. D.J.C. Yates, J. Phys. Chem. 70, 3693 (1966).
73. J.L. Carter, D.J.C. Yates, P.J. Lucchesi, J.J. Elliott, and V. Kevorkian, J. Phys. Chem. 70, 1126 (1966).
74. D.N. Glew and H.W. Quinn, U.S. Patent 3,331,190 (1967).

PARAMETRIC PUMPING

Norman H. Sweed
Department of Chemical Engineering
Princeton University
Princeton, New Jersey 08540

I. INTRODUCTION

Parametric pumping is a dynamic principle of separation based on periodic, synchronous, coupled transport steps. Both theory and experiment have demonstrated its ability to separate fluid mixtures, gas or liquid, at the expense of energy in almost any form: thermal, mechanical, electrical, or chemical.

This chapter summarizes the theory of the principle of parametric pumping, and then presents several examples of its implementation. Attention is focused on mathematical models of parapump processes and their use in predicting and optimizing separations. Experimental methods are discussed for determining model parameters.

However, before discussing its details, let us put parametric pumping into perspective with other separation methods. We begin by developing a new way of looking at separation processes based on a two-step sequence of transport actions.

II. TWO-STEP APPROACH TO SEPARATIONS

Traditionally, a student in the field of separations studies either a number of specific processes such as distillation, chromatography, and reverse osmosis; or a few broad groupings of processes such as staged and phase change operations. When this student is confronted later with a new process from the literature, he tries to fit the unfamiliar ideas into the framework of what he learned previously. Unfortunately, he often finds that the old framework just won't support the new concepts. Its underlying foundation is insufficient because it is not based on a general enough theory of separations.

Parametric pumping is quite different from conventional separations, both in theory and practice, and so might cause this student some difficulty. To help him better understand parametric pumping (and other separations as well), we have developed a simple, unifying way of viewing all separations.

This approach is based on the following idea: all separation processes contain only two essential steps. In the first, chemical species separate themselves on the microscopic level by migrating from a region of one chemical potential (1) to a region of lower chemical potential. In the second step, either simultaneously or subsequently, these regions are separated macroscopically.

In the microscopic step species migrate so as to reduce the gradients of chemical potential. Within a single phase, in the absence of external fields, this migration is what we usually call ordinary diffusion. The concentration distribution formed by this process is less separated than before the diffusion occurred.

1. We will assume that the chemical potential can be defined in this non-equilibrium situation.

At equilibrium the gradients of both concentration and chemical potential disappear.

However, if a second phase or an external field is allowed to interact with the original phase, then the species migrating under the chemical potential driving force may distribute themselves into regions more separated than before. Indeed, at equilibrium, these systems may exhibit concentration gradients even though chemical potential gradients are absent.

The spontaneous migration of chemical species at the microscopic level is at the heart of all separation processes.

It should be obvious that we engineers and scientists never actually separate mixtures at the microscopic level. Rather we merely adjust the environmental conditions around the system so that the desired separations can occur naturally, in accordance with the Second Law of Thermodynamics.

Completion of the separation process requires that the regions of different concentration be divided up. In a multiphase system, for example, the macro step might involve segregating the various phases. In a single phase, the system could be divided by partitions, valves, and the like. The separated products may be directed toward other similar separators thus staging the process.

No matter how complicated a separation method appears on the surface, no matter how complex its mathematical description, it is designed primarily for performing these two transport steps.

One important consequence of the two-step approach is that processes behave similarly if their microsteps and macrosteps are similar. This is irrespective of the phase(s) present, the form of energy used, or the details of the two steps. The differences that occur in similar separations are due entirely to differences in the relative rates of the steps.

A. Applications of the two-step approach

Instead of using the two-step approach to analyze existing processes, let us create new processes with it. For example, let us formulate the following process and then examine its properties (see Figure 1a).

> For the macroscopic step, divide the material in the system into two portions and cause one portion to move steadily past the other. Simultaneously, for the microscopic step, allow the various chemical species to transfer from one portion to the other according to their chemical potential gradients.

Is this process familiar? It should be. It is the two-step description of all of the following: steady state distillation, extraction, gas absorption, chromatography, as well as several others. Since the properties of these processes are well known, we need not discuss them here.

It is important to point out the value of the two-step approach. Nowhere in the process description is mention made of the details of how the step is to be carried out. This is left to the imagination of the person performing the separation, subject only to the restrictions imposed by Nature.

Thus by thinking in terms of microscopic and macroscopic steps rather than chemistry, physics, fluid mechanics, and thermodynamics, we were able to create a great many different yet similar processes from a very simple description. The two step approach gives us an unencumbered view of the entire separation field, unprejudiced by processes we are already familiar with.

Let us formulate another separation process. In this case, allow both steps to be periodic and synchronous (Figure 1b).

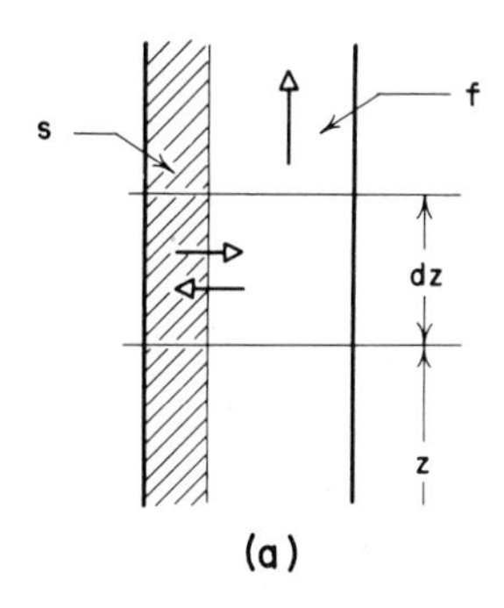

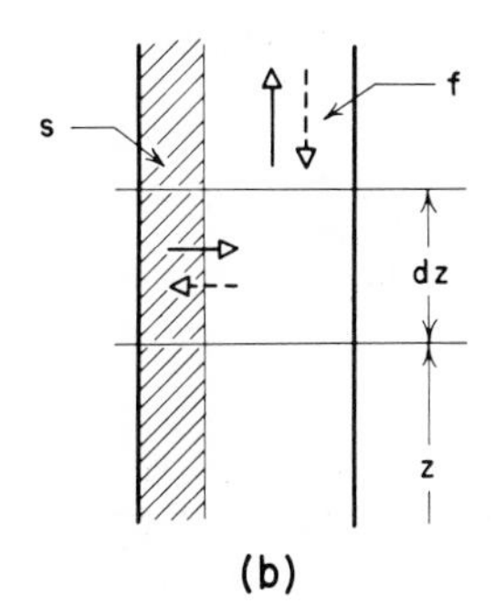

Figure 1. Two-step view of two separation processes: (a) distillation, extraction, etc.; (b) parametric pumping. Part (b) is also the Tinkertoy Model of parametric pumping.

> For the macroscopic step, divide the system into two portions and cause one portion to move past the other so that the relative velocity between the portions is periodic and alternating in direction. Synchronously, cause the gradient of chemical potential between the portions to be periodic and alternating in sign (direction).

This process is probably less familiar than the one discussed above. It is the two-step description of the Principle of Parametric Pumping (PPP), a method of separation originally conceived of by R. H. Wilhelm.

III. PARAMETRIC PUMPING PRINCIPLE

The PPP is purposely stated in the general two-step form rather than by referring it to a specific implementation, to make it clear that the Principle is not limited to the few parapump processes which have been investigated to date. Any states of matter and forms of energy can be used in parapumping. Any equipment configuration which allows the two steps to occur as defined in the PPP is acceptable.

Figure 1 depicts schematically the micro and macro steps for the two processes described above. While it is well known that separations do result from the process described in Figure 1a, it is not so clear that separation occurs by applying the PPP, Figure 1b.

A. Tinkertoy model

To show that the PPP is a generally valid means of separation, Wilhelm, Rice, Rolke, and Sweed (1968) developed a very simple mathematical model of the principle*. They called it the "Tinkertoy" model, since it represents the building block for all parapump systems. The model, Figure 1b, may be viewed as a differential element in a parapump process, which contains all the features required by the PPP. The figure shows the two portions into which the system is divided (one stationary [s], and the other flowing [f]), the alternating velocity of f (relative to s), and the alternating inter-portion mass flux which results from the changes in the direction of chemical potential gradient. For the present we omit details of how the gradient is forced to alternate.

A material balance around the flowing portion gives

$$V(\omega t)\,\frac{\partial \phi_{fi}}{\partial z} = -\,\frac{\partial \phi_{fi}}{\partial t} + \frac{\partial \phi_{si}}{\partial t} \qquad [1]$$

where

$V(\omega t)$ = Alternating velocity of portion f relative to s
ϕ_{fi} = Concentration of the ith species in f
ϕ_{si} = Concentration of the ith species in s
ω = Frequency of periodic velocity
t = time
z = position measured parallel to the flow direction

* Horn and Lin (1969) present a simple geometric interpretation of the PPP.

Both ϕ_{si} and ϕ_{fi} are periodic with frequency ω. Although no formal proof will be given, it is not unreasonable to assume that a system with no natural frequency of its own will respond synchronously when subjected to periodic forcing.

Because the state variables ϕ_{si} and ϕ_{fi} are periodic, their derivatives are also periodic at the same frequency. We may therefore replace the right-hand side of Equation 1 by a new periodic function, $q\ (\omega t)$. For one component thus

$$V\ (\omega t)\ \frac{\partial \phi_f}{\partial z} = q\ (\omega t)$$

Solving for the concentration gradient we obtain:

$$\frac{\partial \phi_f}{\partial z} = \frac{q(\omega t)}{V(\omega t)} \qquad [2]$$

This gradient represents the instantaneous separation over the differential element.

We now ask whether the local gradient, when averaged over a cycle, is non-zero, indicating a useful separation, or is zero, indicating that no net separation exists. In the former case we might expect the possibility that the second cycle would build on the first akin to a staged operation in time instead of position.

Although they are synchronous, q and v need not be in phase; they may be displaced by an angle ε, thus we replace $q\ (\omega t)$ by $q\ (\omega t + \varepsilon)$. We examine now the integral time average concentration gradient over the differential element. Thus

$$\overline{\frac{\partial \phi_f}{\partial z}} = \frac{\omega}{2\pi} \int_0^{2\pi/\omega} \frac{q(\omega t + \varepsilon)}{V(\omega t)}\, dt \qquad [3]$$

where the overbar indicates time averaged.

If V is zero for at most a finite number of points, and if q is piecewise continuous, then the above integral exists. Is its value zero or non-zero?

By choosing a simple example

$$V(\omega t) = A_1 \cos \omega t$$

$$q(\omega t + \varepsilon) = A_2 \cos\ (\omega t + \varepsilon)$$

where the A's are arbitrary constants it can be shown that

$$\overline{\frac{\partial \Phi_f}{\partial z}} = \frac{A_2}{A_1} \cos \varepsilon \qquad [4]$$

which is clearly not usually zero. Moreover, the time-averaged value (over a cycle) of the ratio of any two synchronous periodic functions is usually not zero. Thus in terms of parametric pumping the time-averaged local separation is usually non-zero.

Therefore, this model proves that the PPP can produce separations. It also provides much additional information: if the amplitude of the relative velocity, A_1, is decreased, or if A_2 is increased, separation increases. Also, Equation 4 shows the important effect of phase angle: a shift of 180° produces a separation unchanged in magnitude but opposite in sign. A phase angle of 90° or 270° produces no separation at all. Many features of the Tinkertoy Model are apparent in the process implementions of the PPP.

Now that we know the PPP produces separations on the differential level, we must consider how to implement it. For example, what wave form should we choose for V? How should q be made periodic? What equipment configuration is desirable? How should the differential elements be connected together?

B. Implementing the principle

To begin to answer these questions, let us consider a specific application. Suppose we want to separate into its components a liquid solution of toluene and n-heptane. The PPP requires that the system first be divided into two portions in such a way that the chemical potential in each portion can be made different (and so give rise to micro separation). We could use a second phase for one of the portions, such as the vapor of toluene and heptane--and thereby parapump distillation. However, it is difficult, though not impossible, to alternate experimentally the relative velocity of a vapor-liquid system because of gravitational effects. It is far easier to use a solid second phase, such as the adsorbent silica gel. With the adsorbent fixed in a column, the liquid can be made to flow back and forth over the bed with no difficulty. We select this fixed bed configuration.

Though the PPP is not restricted to systems with two or more phases, it is difficult in most situations to move one portion of a single fluid back and forth past another without substantial mixing and loss of separation. Therefore, further attention will be focussed only on two-phase systems.

The next requirement of the PPP is that the chemical potential gradient between the phases be made to alternate. But how? The chemical potential of toluene in each phase depends on temperature, on the concentrations of all the chemical species which might be present, and to a lesser extent on all other thermodynamic variables. By periodically varying one or more of these parameters the potential can be altered in one phase more than in the other, producing an imbalance and hence the periodic microstep. For the present we choose to vary temperature alone.

We have satisfied the requirements of the Principle of Parametric Pumping. But to make the process workable we must consider additional problems in implementation. For example, how should the temperature of each differential element in the fixed bed be varied? Should they all be treated the same way, or should some lag the others? The process could be operated, for example, by directly heating and cooling the bed with a jacket. Or the liquid entering one end of the bed might be heated, the thermal energy being transported into the bed by the moving fluid. Similarly, cooling could be achieved by introducing cold fluid at the other end. In this case temperature oscillations within the bed would depend on the thermal properties of the bed, the rate and duration of flow, and the rate of interphase heat transfer.

Both of these "modes" of implementing thermal parametric pumping --introducing heat directly through column walls (Figure 2a), or recuperatively at column ends (Figure 2b) -- will produce separations. However, their operational characteristics are quite different.

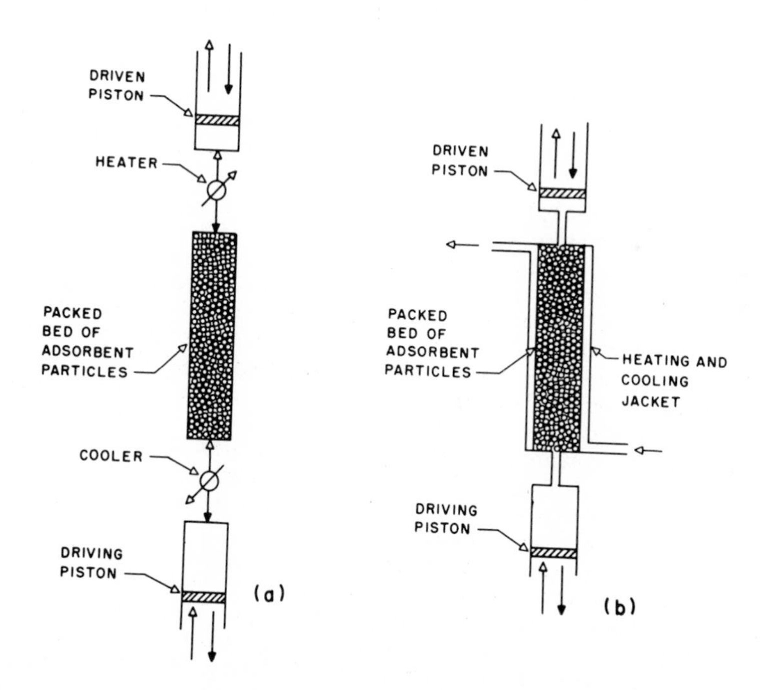

Figure 2. Two implementations of thermal parametric pumping: (a) direct mode; (b) recuperative mode.

To better understand these two modes of operation, let us examine both their experimental behavior and mathematical descriptions. For the latter we obtain general, analytical solutions for idealized parapump systems, and specific numerical solutions to real, nonideal cases. We couple the mathematical models with laboratory experimentation to obtain parameters for the model equations. Finally, we use the models to predict experimental separations and optimize processes.

IV. EXPERIMENTAL THERMAL PARAMETRIC PUMPING SEPARATIONS

A. Direct mode

Experimental separations are presented in this section for three different implementations of direct thermal parapumping. These are the batch, semi-batch, and continuous methods of operation. In the batch system, no product is removed during the process; in the semi-batch, one product is removed in a transient process; and in the continuous system, two (or more) products are removed continually.

Several different adsorbent-solution systems are described: NaCl-H_2O- ion retardation resin, urea-H_2O-activated carbon, toluene-n-heptane-silica gel. This is done to show that parametric pumping is not limited as to what can be separated. Any adsorbent-solution system could have been used with any implementation.

1. Batch separations

Consider first a batch direct-coupled thermal parapump separation of a liquid-phase toluene-heptane mixture on a silica gel adsorbent, a system previously reported by Wilhelm and Sweed (1968) and Wilhelm et al (1968).*

The experimental apparatus consists of a jacketed glass column (100 x 1 cm i.d.) packed with particles (30- to 60-mesh) of chromatographic grade silica gel (Matheson Coleman Bell), a constant rate, positive displacement, dual-syringe infusion-withdrawal pump (Harvard Apparatus Co.), sources of hot and cold water for the jacket (not shown in Figure 2a), and a programmed cycle timer. The timer is adjusted to reverse the direction of the fluid stream periodically, and also to cycle the jacket temperature by connection to hot or cold sources. Both alternations have the same frequency and are in phase. Pump and timer adjustments establish the alternating velocities to be uniform at all times and displacements to be equal in each half cycle.

To compare experiments, each run was started under identical conditions: the interstitial fluid was a mixture of 20 per cent (volume) toluene in n-heptane and was in equilibrium with the silica gel at ambient temperature. The bottom syringe was filled initially with 30 ml. of the same 20 per cent solution; the top syringe initially was empty. Temperature limits between hot and cold sources are shown in Figure 3.

At the beginning of a run, the syringe pump is started and hot water is circulated through the jacket. When 30 ml. have been displaced upward, the pump reverses direction and the jacket is switched to the cold water source. This process continues until 30 ml. are pumped back downward through the column, completing one cycle. Since no product is removed, we call this method "total

*(Portions of the following are reprinted from Wilhelm, et al (1968) by permission of the American Chemical Society.)

reflux". This analogy with distillation is explored more fully below.

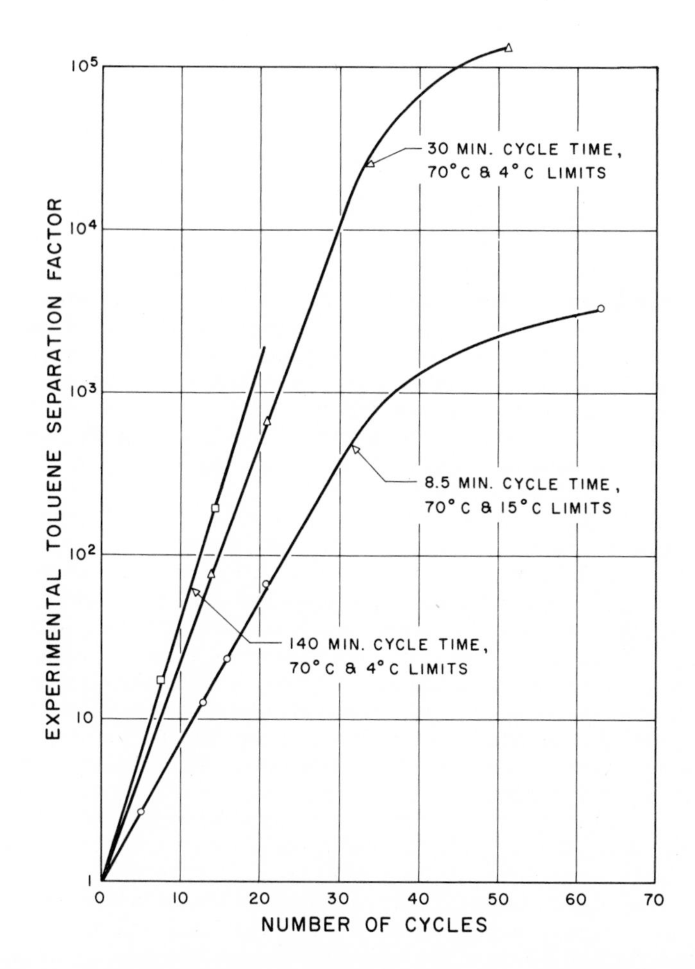

Figure 3. Experimental parametric pumping separation of toluene-heptane liquid mixture with silica gel adsorbent showing effect of cycle time, τ. [Reprinted with kind permission of the American Chemical Society from IEC Fundamentals 7, 340 (1968)].

Separation factor (SF) is defined as the ratio of mixed average toluene volume concentrations in the syringes, compared top to bottom.*

*All products containing more than about 1 per cent toluene are analyzed by means of refractive index with a sodium D line source; those containing less, by ultraviolet absorption (Cary 14 recording spectrophotometer, 2685-Å wave length).

Figure 3 shows the results of three experimental separation series, each performed with a different cycle time. Experimental points represent fresh starts from initial conditions to the number of cycles noted on the abscissa. The separation factor is observed first to increase approximately exponentially with number of cycles and finally to lean toward a limiting separation. The potential for still further separation by increasing the number of cycles is evident in each sequence.

The largest separation attempted yielded an average volumetric toluene fraction of 0.59 in the upper piston cylinder and 4.7×10^{-6} in the lower (separation factor > 10^5) after 52 cycles. Higher separations were not tried because of the difficulty in analyzing the low concentration product.

The separation increase each cycle is greater for longer cycles. This effect may be interpreted as follows: if the cycle time, T, is very short, little transfer takes place, and there is no separation. In such a situation the bed appears to be inert. As T increases, more time is allowed for interphase mass transfer during a half cycle, hence more material is transferred, up to the limit imposed by equilibrium. This effect was predicted by the Tinkertoy Model through factor A_2.

Additional experimental results, shown in Figure 4, depict the effect of phase angle, ε, between flow and heating cycles. Each experiment lasted 23 cycles, each being 8.5 minutes. Qualitatively these phase-angle separation responses over the experimental column reflect the corresponding characteristics at the local, differential level, as discussed earlier for the Tinkertoy model--i.e., the separation gradient equals cos ε. As predicted, the direction of separation changes sign experimentally, higher concentration being at the top of the column at one phase angle and at the bottom at a 180° phase shift. Further, as predicted, there is no separation (SF = 1.0) at two values of ε, 180° apart. However, the separation maxima are displaced slightly from the predicted 0° and 180° locations. This small shift of the wave is due to thermal and mass diffusive lags inherent in the experimental system.

In another batch separation, Sweed and Gregory (1970) investigated direct thermal separations of aqueous NaCl solutions using an ion retardation resin (Bio-Rad AG11A8) as the solid phase. With equipment and procedures similar to those above, but also introducing mixing in each reservoir to keep their concentrations uniform*, they obtained the separations shown in Figures 5 and 6.

Figure 5 shows the effect of dimensionless fluid displacement volume, α, on separation. Clearly if α is greater than one (i.e., more than one column volume of fluid displaced in a half cycle), fluid moves from one reservoir, through the column, and into the other reservoir with mixing and less separation as the result. When α is less than one, mixing no longer occurs directly between reservoirs. Rather there is an indirect mixing which depends

*The reservoirs are mixed to satisfy the boundary conditions in the mathematical model.

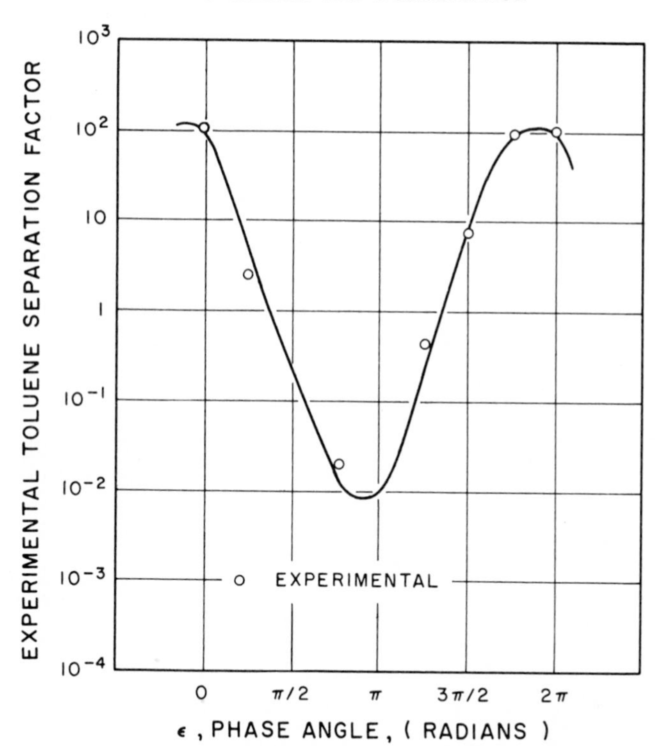

Figure 4. Experimental effect of phase angle ε between flow and heating cycles on toluene separation factor. [Reprinted with kind permission of the American Chemical Society from IEC Fundamentals 7, 340 (1968)].

on the existence of non-sharp axial concentration profiles, such as occur with finite interphase mass transfer rates or axial dispersion. Because of these profiles, the concentration of column effluent changes continuously during a half cycle. In most cases, effluent concentration at the high concentration end decreases as the less concentrated regions of the axial profile leave the column. The reverse occurs at the other end. Thus, at larger α, more of the profile leaves the column, decreasing separation. Remarkably, the effect of α was also predicted by the Tinkertoy Model where the corresponding quantity is A_1.

Figure 6 is similar to Figure 3, showing the important effect of cycle time, τ.

Similar batch separations have been reported by Wakao, et al (1968) for benzene and hexane on silica gel.

It is obvious from the experimental data presented here that spectacular separations are possible with direct, thermal, batch parametric pumping. It will be shown below that in certain situations there is no theoretical limit to the separation; the low concentration end approaches pure solvent.

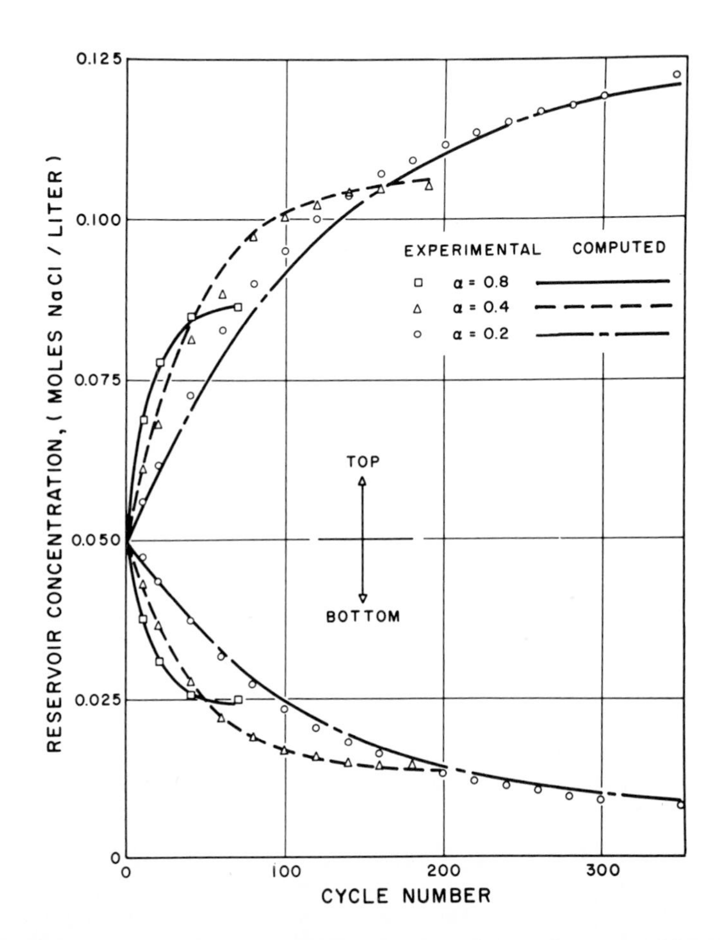

Figure 5. Experimental and calculated batch separations of a $NaCl-H_2O$ solution on Bio-Rad AG-11A8 resin, showing the effect of displacement volume α. [Reprinted with kind permission from the A.I. Ch. E. Journal 17, 171 (1971)].

2. Semi-batch separations

The direct thermal parapump systems above were all operated in the total reflux, batch, or closed manner. Columns and reservoirs were initially loaded with solute and solvent and then closed to mass input or removal. Product was removed only after the run was terminated.

However, there is no reason to restrict parapumping to batch units. Goldstein (1969) and Bringhurst (1970) demonstrated the applicability of semi-batch parametric pumping to separating aqueous urea solutions.

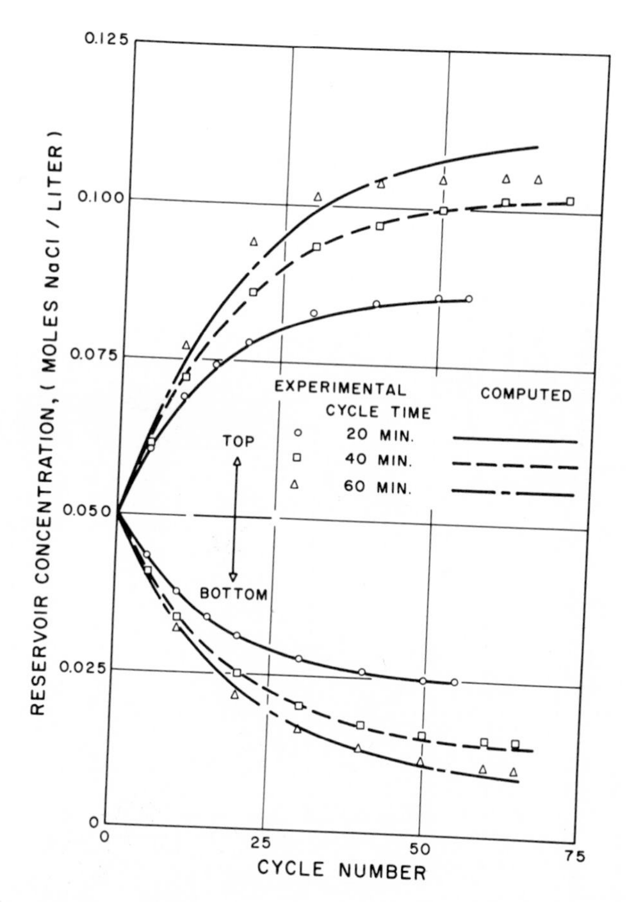

Figure 6. Experimental and calculated separations of a $NaCl-H_2O$ solution on Bio-Rad AG-11A8 resin, showing the effect of cycle time, τ. [Reprinted with kind permission from the A.I.Ch.E. Journal 17, 171 (1971)].

The semi-batch system introduces solute into the column continually through an open feed arrangement which replaces one reservoir of the batch system, as shown in Figure 7. All fluid which enters the top of the column comes from a large reservoir of constant composition; fluid leaving this end is diverted into a fraction collector and is not recycled back into the column. Other than this and the 180° phase shift (downflow occuring on the hot half cycle), the semi-batch and batch systems are identical in construction and operation. The adsorbent consists of 20 - 48 mesh particles of Type LC activated carbon (Union Carbide Corp.); the volume flowed each half cycle, 25 ml., is equivalent to $\alpha = 0.83$.

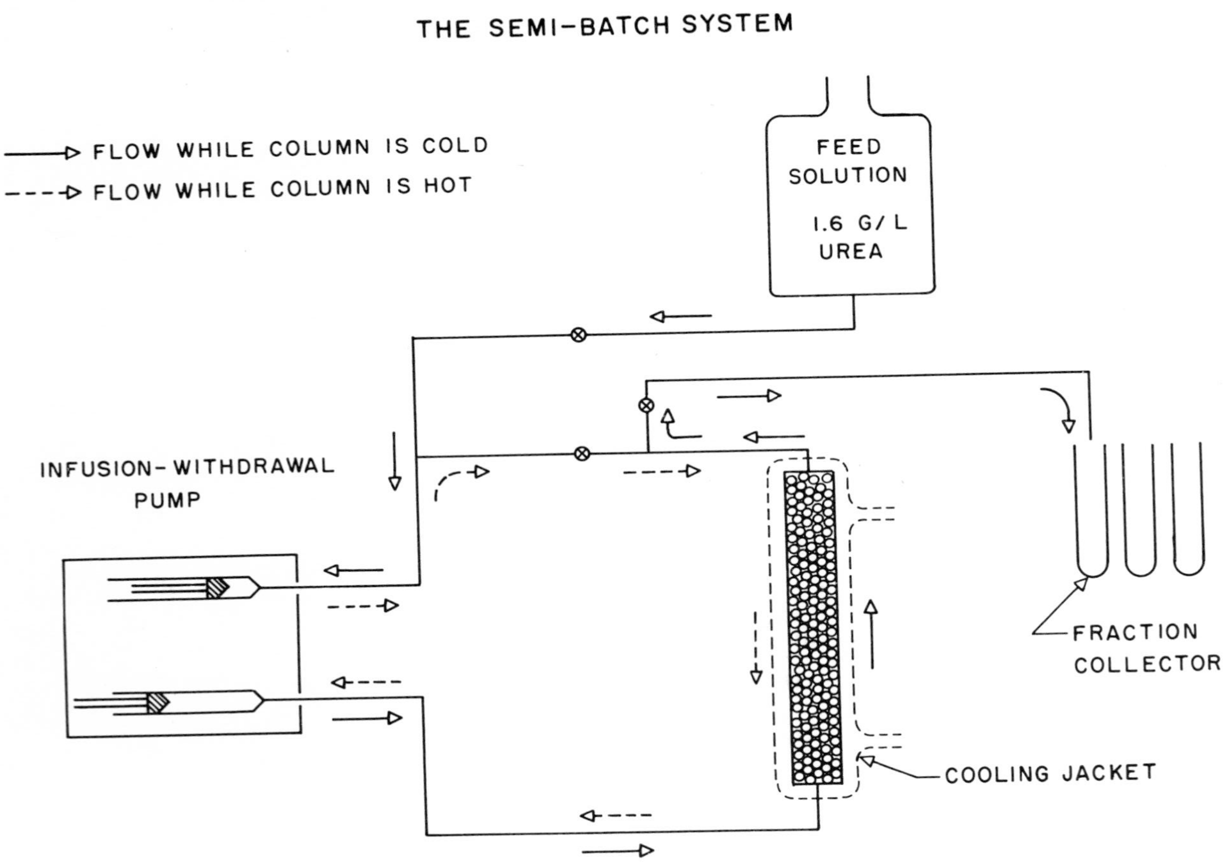

Figure 7. Schematic of a semi-batch, direct thermal mode parametric pump.

The results of a semi-batch experiment are shown in Figure 8.

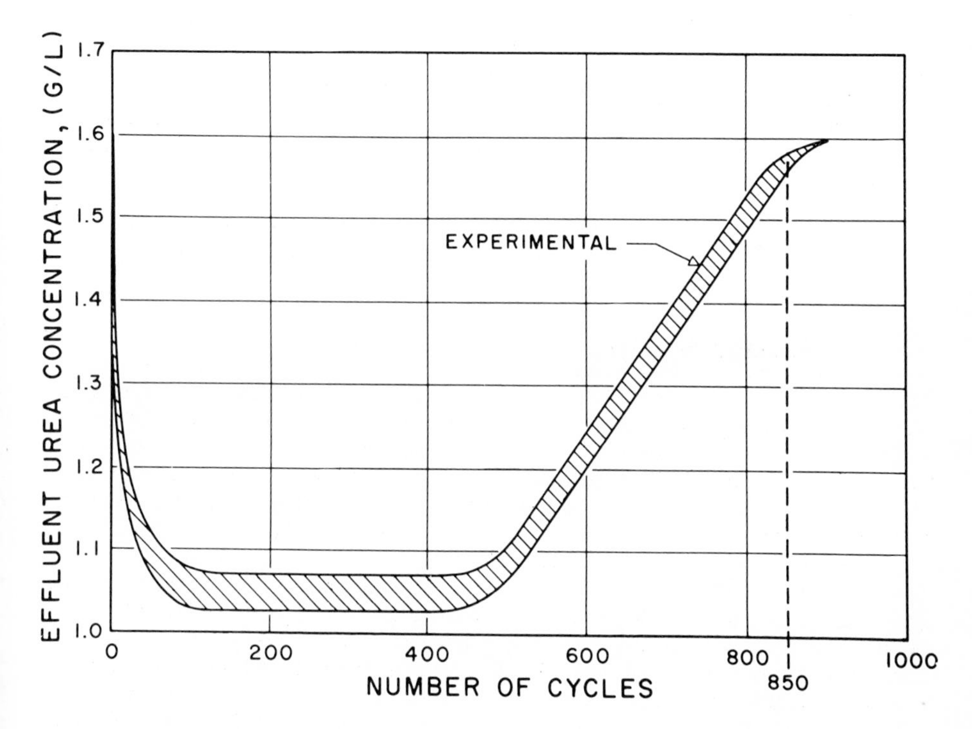

Figure 8. Experimental urea concentration from the semi-batch system of Figure 7.

The initial fluid concentration in the column and reservoirs was 1.60 g urea/l.; the operating temperatures were 4° and 37° C. The column effluent was sampled every cycle; the same could not be done for the closed reservoir because removing a sample for analysis would disturb column operation. Hence samples were taken only twice: after 499 and 850 cycles.*

As shown in Figure 8, the initial 1.60 g/l solution in the column dropped after a very few cycles to give an effluent of 1.05 g/l. For over 400 cycles (8.6 minutes each) --almost 2-1/2 days-- the column still accepted 25 ml of 1.60 g/l solution and expelled 25 ml at 1.05 g/l per cycle, the difference being taken up by the column and closed reservoir.

After about 400 cycles the effluent concentration rose slowly, indicating that the limit of the column's ability to concentrate urea was being approached. Finally -- after 850 cycles -- the column could no longer remove solute from the feed and the effluent

*Urea concentrations below about 20 g/l were measured by the colorimetric diacetyl monoxime method of Coulombe and Favreau (1963). Higher concentrations were determined by refractive index at 20 C with the D-line of sodium.

left the column with composition unchanged from the feed.

The concentrated reservoir after 499 cycles contained a solution of 110 g/l. After 850 cycles it had risen to 120 g/l. This is 75 times as concentrated as the feed, and is 20 per cent of the concentration of a saturated solution at 20° C.

Clearly, the semi-batch system is very useful for concentrating a solute. It has an advantage over the batch in that more solute is available because of the large volume of feed. Higher concentration product is the result. However, it does not remove as much solute from the feed and in this respect it is inefficient.

The choice between batch and semi-batch depends on the objective of the separation.

3. Continuous separations

Separations in batch and semi-batch parametric pumps are inherently transients. After operating for some time they reach a limit, either in the extent of separation or in the volume of product. Once the limit is reached additional cycles consume energy but to no productive end. For separating small volumes of solution both of these processes are quite useful; but for large volume streams, a continuous process is essential.

A section above dealt with a batch parapump: feed is not introduced; product is not removed. In analogy with distillation, this process can be thought of as operating at "total reflux" because all of the fluid leaving a column end is sent back into that end on the next half cycle. If feed were introduced and product withdrawn, then the system would be open and would operate at "partial reflux". The semi-batch system is also semi-open. It has total reflux at one end, zero reflux at the other.

The number of open systems which can be devised is very large. Not only do we have available the degrees of freedom usually found in most column separations (reflux ratios, temperatures, feed location, feed rate), but also the freedom of when in the cycle to feed, when to withdraw product. It is not at all necessary that feed be introduced continuously.

Let us consider here two possible open systems to illustrate this behavior (Gregory and Sweed, 1970). Although any feed location is acceptable, we will look at systems fed only at one end, this being the state of the literature on this subject to date (1970). A column fed in the middle may, of course, be treated as two columns, one fed at the top, the other at the bottom.

Each open system is made up of several distinct cycle portions, each having its own temperature and flow pattern. For example, one portion might consist of feeding the column at one end and withdrawing product at the other. Another portion might have liquid flowing from one reservoir through the column and into the other reservoir with no feed anywhere. The kind and sequence of these cycle portions completely specifies the particular open system.

The cycle portions need not be of equal duration, except as indicated below.

The two open systems considered here are shown in Figures 9 and 10. The former is termed "non-symmetric" since it has a net flow of solution from one end to the other. The latter (Figure 10) is "symmetric", having no net flow of solution within the column per cycle.

To simplify mathematical analysis later, several restrictions are imposed.

1. There is no reservoir dead volume, i.e., all the fluid in a reservoir is flowed each cycle.

2. There is no change in reservoir volume from cycle to cycle.

3. All flow rates are constant and of the same magnitude.

4. Initially the column is uniform with fluid concentration the same as the feed, and the solid in equilibrium with it at the hot temperature.

a. Non-symmetric flow system

The non-symmetric flow system has five distinct portions in each cycle (Figure 9):

When the column is cold and flow is down, there are three portions:

1. Feed into the top of the column, flow through the column, and remove product from the bottom, for time t_I.

2. Feed into the top of the column, flow through the column and into the bottom reservoir, for time t_{II}, and

3. Flow from the top reservoir through the column into the bottom reservoir, for time t_{III}.

When the flow is upward and the column is hot, there are two portions:

4. Flow from the bottom reservoir through the column and remove top product, for time t_{IV}, and

5. Flow from the bottom reservoir through the column and into the top reservoir for time t_V.

With the constraint of constant reservoir volumes this system produces a net solution flow down the column each cycle.

b. Symmetric flow system.

The symmetric flow system has four distinct portions in each

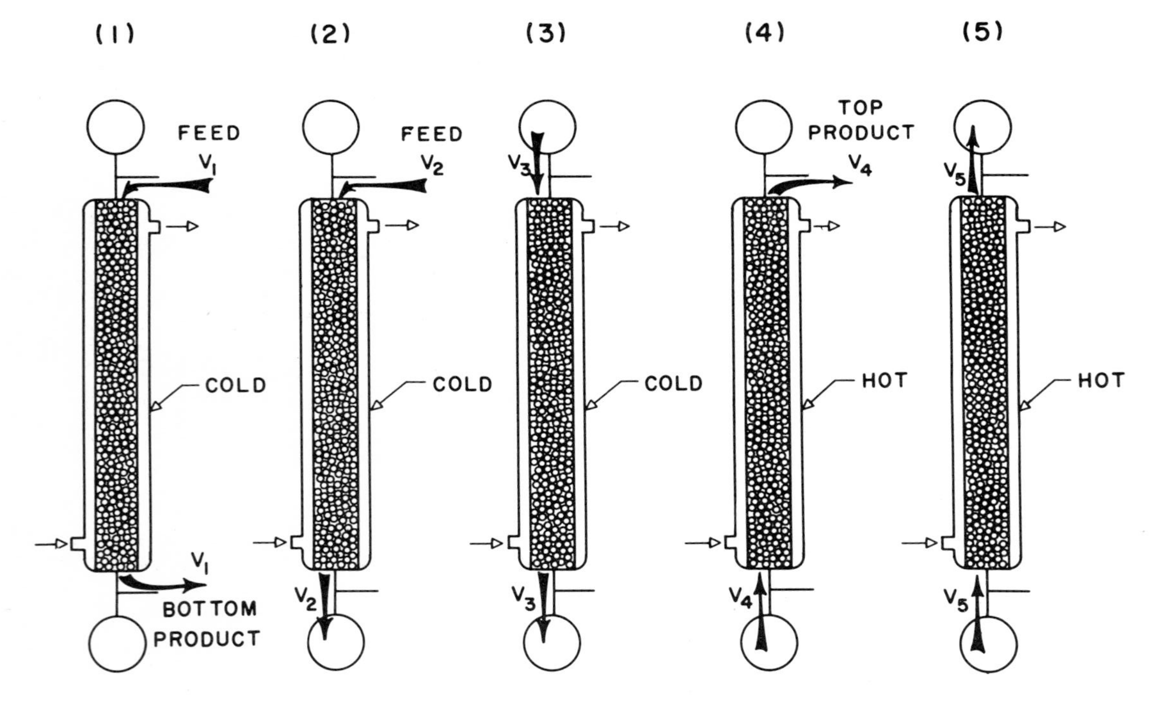

Figure 9. Schematic description of the non-symmetric open system. Each column represents a separate cycle portion. [Reprinted with kind permission from the A.E.Ch.E.Journal 17, 171 (1971)].

cycle (Figure 10). The symmetry results in one less degree of freedom, forcing the top reflux ratio to equal the reciprocal of the bottom reflux ratio. On the cold downflow, there are two parts:

1. Flow from the top reservoir through the column to the bottom product, for time t_I, and

2. Feed at the top and flow through the column to the bottom reservoir, for time t_{II}.

While the column is hot and flow is up, there are two parts:

3. Feed at the top into the top reservoir with no flow in the column, for time t_{III}, and

4. Flow from the bottom reservoir through the column to the top product, for time t_{IV}.

Gregory (1970) studied these two continuous direct mode processes using the same NaCl-H_2O-ion retardation resin system as in the batch. Figure 11 shows schematically the apparatus used for both open systems. Two pumps are required, one to introduce feed and the other to provide reciprocating flow.

The apparatus was designed to satisfy the assumptions of the model equations developed below. Circulation pumps were installed in the reservoirs to keep their contents well mixed. Short periods with no fluid flow (1 minute) were introduced at each temperature change to better approximate a square temperature wave within the bed in the presence of thermal lags. Tubing and connections were kept short to minimize unwanted and unmixed dead space between the reservoir and column.

Auxillary reservoirs provide temporary expansion-contraction relief during each cycle; their net volume does not change from one cycle to the next.

Concentration of feed in all continuous runs was 0.05 N NaCl; concentration of both top and bottom products was monitored each cycle using electrical conductivity. For all runs the cycle time was 24 minutes. Figures 12, 13, and 14 show the top and bottom product concentrations vs. cycle number for three runs of the non-symmetric open system, each with a different reflux ratio. Figure 15 shows the results of a symmetric system run. Note the initial transient and the final steady periodic separation. Product of steady composition is removed continually from both ends of the column once limiting conditions are reached.

For the non-symmetric runs, the bottom product concentration decreases as the reflux ratio at that end increases. The top product concentration increases when the top reflux ratio increases. Thus, the overall separation improves as either or both of the reflux ratios are increased. This is exactly the same response found in distillation. The analogy extends to the effect of reflux

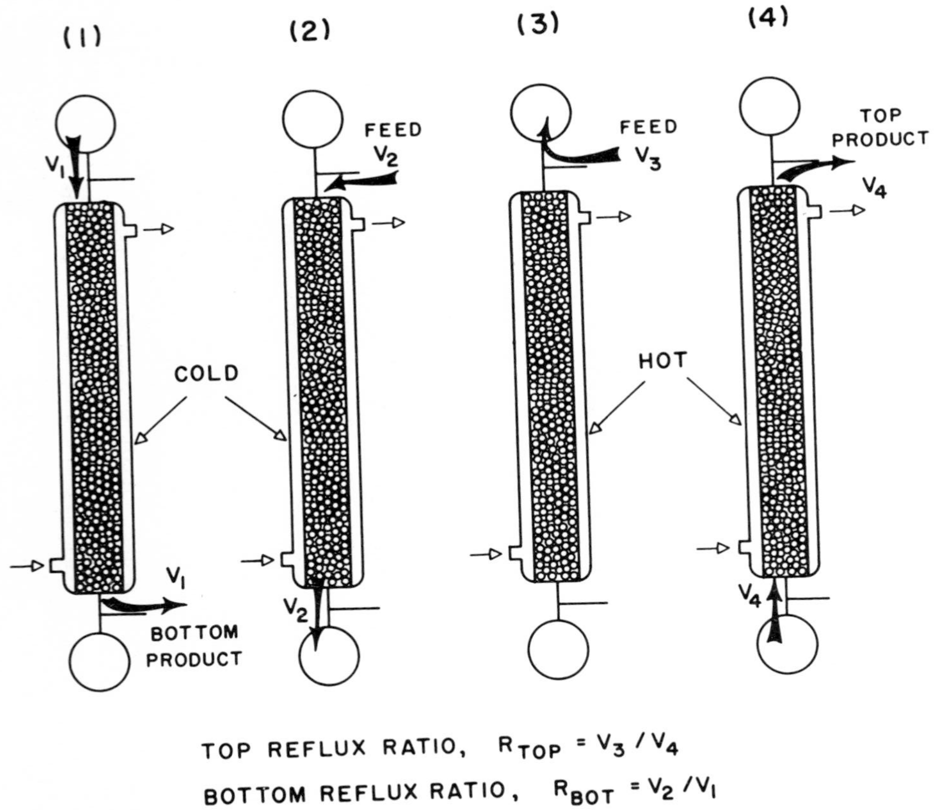

(1) (2) (3) (4)

FEED V_2

FEED V_3

TOP PRODUCT V_4

V_1

BOTTOM PRODUCT V_1

V_2

V_4

COLD

HOT

TOP REFLUX RATIO, $R_{TOP} = V_3 / V_4$

BOTTOM REFLUX RATIO, $R_{BOT} = V_2 / V_1$

Figure 10. Schematic description of the symmetric open system. Each column represents a separate cycle portion. [Reprinted with kind permission from the Chemical Engineering Journal, 1, 207 July 1970].

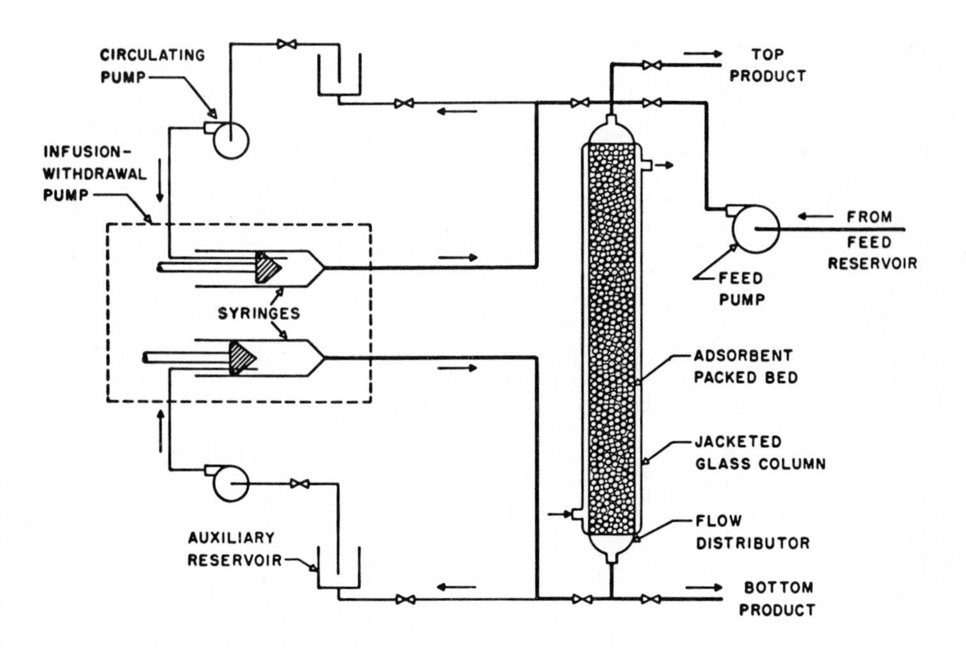

Figure 11. Apparatus for implementing the open system. [Reprinted with kind permission from the A.I.Ch.E. Journal 17, 171 (1971)].

ratio on product volume. As the ratio increases, the volume of more separated product decreases.

B. Recuperative mode

Recuperative thermal parametric pumping was first introduced by Wilhelm, Rice, and Bendelius (1966), and was later treated extensively by Wilhelm *et al* (1968) and Rolke and Wilhelm (1969) for liquid-adsorbent systems. Experimental separations of gases were studied by Jenczewski and Myers (1968) and McAndrew (1967).

In this mode, fixed but different temperatures in the reservoirs (Figure 2b) cause the moving fluid to carry not only the

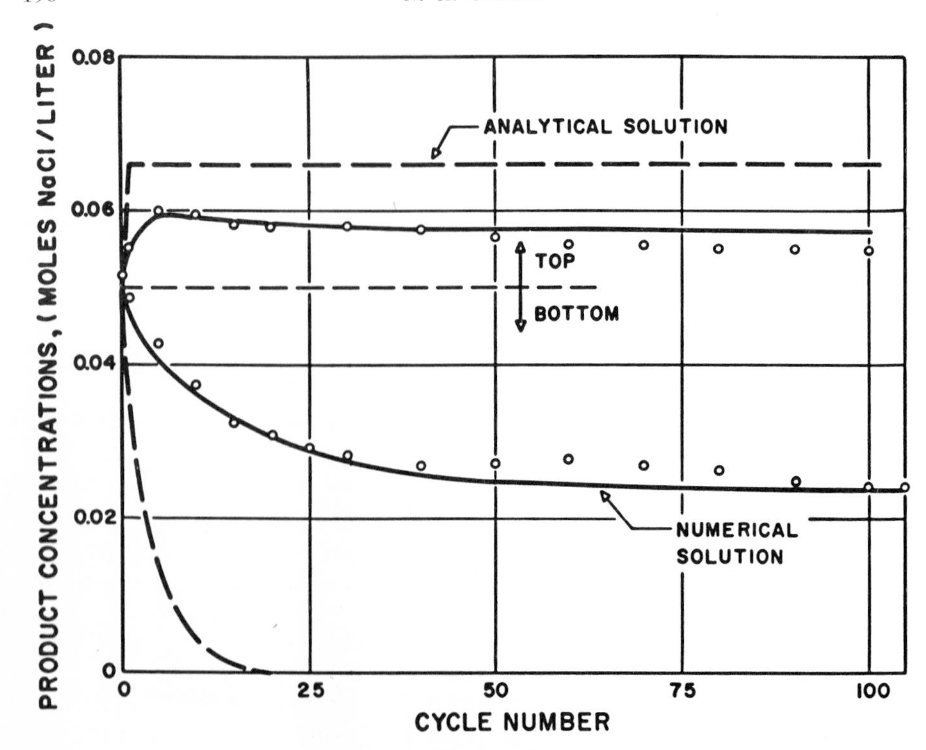

Figure 12. Experimental and computed separations of an aqueous NaCl solution for the non-symmetric open system on Bio-Rad AG-11A8. $R_{TOP} = 0.19$. $R_{BOT} = 4.4$. Computed curves are based on the STOP-GO algorithm for finite λ and on Tables I and II for infinite λ. [Reprinted with kind permission from the Chemical Engineering Journal, July 1970].

mixture components but also thermal energy into and out of the column as the flow direction alternates. The temperatures within the bed depend on the heat capacity of both phases, fluid flow rate, end temperatures, and the rate of interphase heat transfer. The recuperative process may be described qualitatively as follows (after Wilhelm, et al (1968)), with the top of the bed hot, the bottom cold: fluid is displaced downward from a warmer region, raising the temperature of the adjacent adsorbent. As a result of the temperature change, the adsorbent transfers solute to the fluid (the microstep). The enriched (and cooled) fluid next is displaced axially upward to come in contact with warmer adsorbent (the macrostep). The cycle is completed as the fluid cools the adsorbent now adjacent and loses solute to it. The difference between the fluxes of solute in the two directions of motion is the net flux due to parametric pumping.

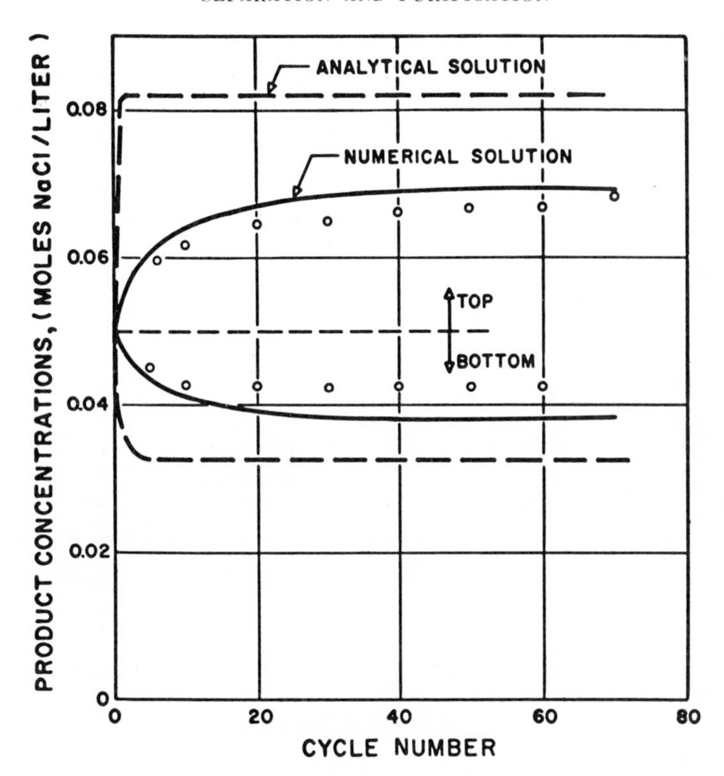

Figure 13. Experimental and computed separations of an aqueous NaCl solution for the non-symmetric open system on Bio-Rad AG-11A8. $R_{TOP} = 1.0$, $R_{BOT} = 1.09$. Computed curves are based on the STOP-GO algorithm for finite λ and on Tables I and II for infinite λ. [Reprinted with kind permission from the Chemical Engineering Journal, July 1970].

As with the direct mode, the product may be withdrawn completely, returned as reflux, sent to another column, or a combination of these. Rolke and Wilhelm reported separations for a NaCl-H_2O-ion exchange resin system (equicapacity mixture of Rohm & Haas IR-45 and IRC-50) for a zero reflux arrangement. Using an insulated column to reduce heat leak to the surroundings, with cycle time, τ = 80 minutes and displacement α slightly greater than 1.0, they found that the steady periodic column effluents, averaged over a cycle, these gave separation factors of 1.11, a value substantially lower than the direct mode. This is not surprising.

The ion exchange resins used here transfer mass to the fluid at a much slower rate than the silica gel, activated carbon, or ion retardation resin used with the direct mode. This makes the bed almost "inert" compared to the other adsorbents.

Another reason for the low separation is the total feed (no

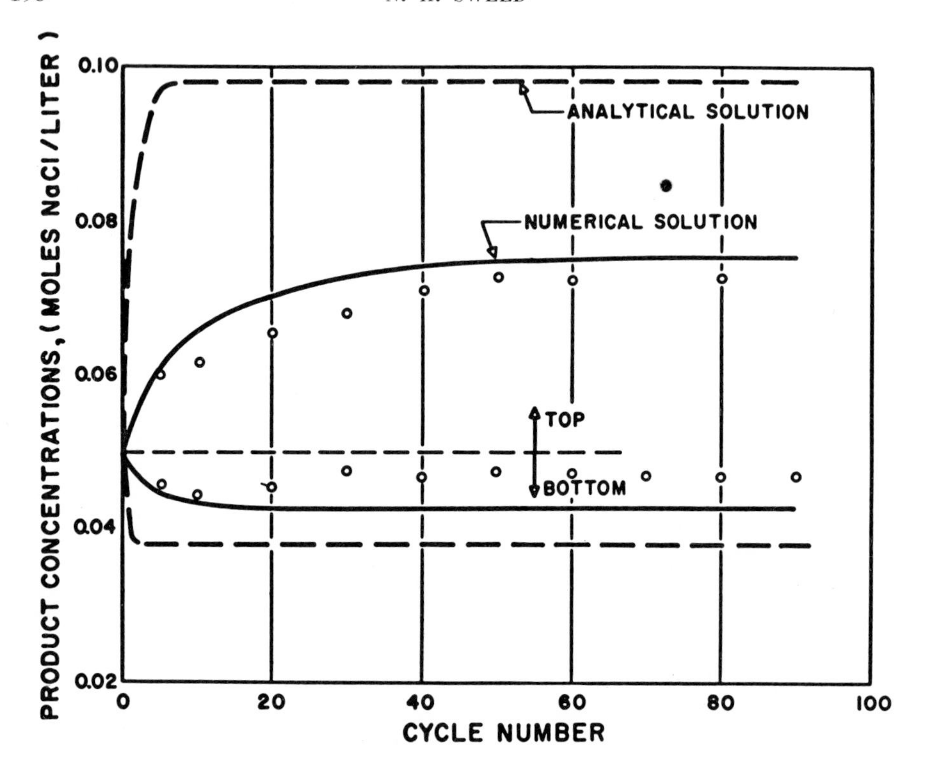

Figure 14. Experimental and computed separations of an aqueous NaCl solution for the non-symmetric open system on Bio-Rad AG-11A8. $R_{TOP} = 2.0$, $R_{BOT} = 0.75$. Computed curves are based on the STOP-GO algorithm for finite λ and on Tables I and II for infinite λ. [Reprinted with kind permission from the Chemical Engineering Journal, July 1970].

reflux arrangement. Any tendency for larger separation is immediately diminished when feed is introduced at the same point that product is removed. Potentially high concentration product is diluted by the feed; low concentration product is mixed with more concentrated feed.

A third cause of the low separation is that, although the bed operates between two widely separated temperatures, no point in the column experiences the full ΔT over a cycle. For a bed operating between 97° C and 11° C, Rolke and Wilhelm show that each location in the column feels a ΔT of only a fraction of this range. In the direct mode, the column would experience the entire range over its whole length. In addition, the phase angles between heating and flow cycles are not uniform throughout the bed. As shown in the Tinkertoy model, an angle other than 0° or 180° reduces separation. Thus part of a recuperative column operates at other than the optimum phase angle.

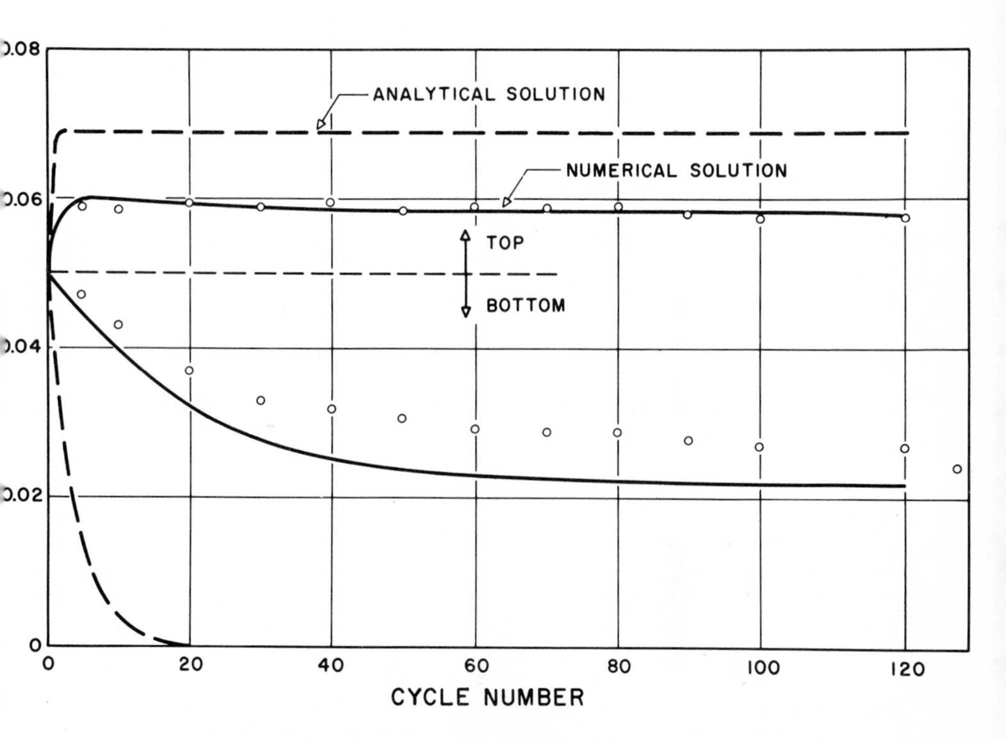

Figure 15. Experimental and computed separations of an aqueous NaCl solution for the symmetric open system on Bio-Rad AG-11A8. $R_{TOP} = 0.296$, $R_{BOT} = 3.38$. Computed curves are based on the STOP-GO algorithm for finite λ; and on Table III and Eq. 46 for infinite λ. [Reprinted with kind permission from the Chemical Engineering Journal, July 1970].

Though direct parapumping gives much better separation here, the difference between modes would not have been so great had the same adsorbents been used in each case. The recuperative mode has the advantage of better thermal contact between phases while the direct mode requires radial heat conduction in the bed, an inefficient process in large diameter beds. Also, the recuperative mode is potentially better for recovery of process heat.

Parametric pumping is clearly a powerful method for separating mixtures. We have seen it applied to several different classes of compounds with four experimental implementations. In one case, the process has reduced the concentration of one hydrocarbon in another by a factor of 100,000: 1. In another situation it has concentrated an organic solid from dilute aqueous solution to 20 percent of saturation without exceeding human body temperature. This latter point indicates the possibilities of using parapumping where temperature-sensitive compounds must be separated.

Many useful conclusions about parapumping can be drawn from these few experiments, but we still cannot predict *a priori* how each parameter will affect separation. What effect, quantitatively, would obtain if we were to vary displacement or cycle time, or, more importantly, change the chemical composition of solution and adsorbent? We cannot extrapolate with confidence from our experiments to vastly different conditions.

However, if we model accurately the parapump process, we can explore without experiments the entire domain of operation. The next section presents the general mathematical description of these thermal parametric pumping systems.

V. MATHEMATICAL MODELS OF PARAMETRIC PUMPING

For the thermal parapumping processes depicted in Figure 2, there are two sets of model equations. One set describes the distribution of each chemical species in the bed. These equations are exactly those of nonequilibrium chromatography. The other set describes the distribution of thermal energy in the bed. In the direct mode this latter set consists of a single equation stating how temperature is programmed to vary with time. Recuperative parapumping, on the other hand, requires a set of partial differential equations, similar to those of chromatography, to define the temperature profile.

The equations presented below are in dimensionless form, i.e., all variables have been normalized relative to characteristic length, time, mass, composition, and temperature terms.

form.

The model uses the following assumptions:

1. The fluid and solid phases have constant and uniform densities.
2. Bed cross section and fluid fraction are constant in time and uniform in position.
3. The packed bed may be modelled using the dispersion model with constant axial diffusivity.
4. The fluid velocity profile is flat (plug flow).
5. Material enters and exits the bed at the ends only.
6. Bed properties are radially uniform.

A fluid phase material balance is required for n-1 of the n components, the remaining balance being the total continuity equation. With the assumptions above, the continuity equation reduces to

$$v(z,t) = v(t)$$

i.e., velocity is independent of position in the bed. (If gases were used rather than liquids, this might have to be modified.)

The fluid phase balance on the ith component is

$$\alpha\, f(t)\, \frac{\partial \phi_{f_i}}{\partial z} + \frac{\partial \phi_{f_i}}{\partial t} + \frac{\partial \phi_{s_i}}{\partial t} - \eta\, \frac{\partial^2 \phi_{f_i}}{\partial z^2} = 0 \qquad [5]$$

The first term represents the transport of species i by convection, where $\alpha f(t)$ is the dimensionless velocity. The second and third terms are the fluid and solid concentration transients, respectively, and the last term describes the axial transport of species i by dispersion.

An interphase mass transfer rate expression relates concentrations between the mobile and stationary phases. In the two step view, this equation details the microstep. An approximate form of the rate expression, convenient for its simplicity yet exact for many real situations is

$$\frac{\partial \phi_{s_i}}{\partial t} = \lambda\, (\phi_{f_i} - \phi_{f_i}^*) \qquad [6]$$

where $\phi_{f_i}^*$ is the concentration of i in the fluid phase which would obtain if the adsorbent were at equilibrium with its current composition and temperature. Thus

$$\phi_{f_i}^* = \phi_{f_i}^*\, (\phi_{s_j},\ \Theta_s) \qquad j = 1, 2, \ldots n-1 \qquad [7]$$

is an equilibrium function.

Note that this rate expression lumps all resistance to mass transfer at the particle surface; the intraparticle compositions are assumed uniform. If this were not the case, as with some ion exchange resins, a distributed parameter model could be substituted to account for intraparticle profiles. Thus

$$\frac{\partial \phi_{s_i}}{\partial t} = \lambda \left(\phi_{f_i} - \phi_{f_i}^* \,\Big|_{r = R} \right) \qquad [8]$$

where $\phi_{f_i}^*|_{r = R}$, is the value of $\phi_{f_i}^*$ at the particle surface, determined in part from the intraparticle diffusion equation (i.e., Fick's Law).* Rolke and Wilhelm (1969) discuss the distributed system in detail. We will not treat it here since, except for ion exchange resins, most parapump adsorption rates can be modelled adequately by Equation 6 or some other lumped parameter expression.

If the interphase mass transfer rate becomes very large, or

*If electrical effects are important, the Nernst-Plank equation could be used instead of Fick's Law.

equivalently, if the time allowed for mass transfer becomes very large, the column approaches equilibrium operation, and $\lambda = \infty$. This special or ideal case will be explored in detail later.

The energy balance equations for the direct mode are quite simple. If radial heat transfer is rapid, and the column is uniformly jacketed, then

$$\Theta\ (z,t) = \Theta\ (t)$$

i.e., temperature is a prescribed periodic function of time alone.

For the recuperative mode, $\Theta(z,t)$ is given by the solution of

$$\alpha f(t)\ \frac{\partial\Theta_f}{\partial z} + \frac{\partial\Theta_f}{\partial t} + \beta\ \frac{\partial\Theta_s}{\partial t} - \xi\frac{\partial^2\Theta_f}{\partial z^2} = 0 \tag{9}$$

and

$$\frac{\partial\Theta_s}{\partial t} = \gamma\ (\Theta_f - \Theta_s) \tag{10}$$

where Equation 9 is the thermal counterpart of Equation 5; Equation 10 is similar to Equation 6.

The effect of the heat of adsorption has been neglected in the energy equation as have all heat losses to the surroundings. Thus oscillations in the energy equation drive the composition equations through the term ϕ_f^* but not vice versa. This one-way coupling simplifies enormously the process of solving the equations, since the energy set is completely independent of composition and can be solved separately. Were the coupling two-way, all equations would have to be solved simultaneously.

In nonthermal forms of parametric pumping, the thermodynamic variable is likely to be two-way coupled (pressure, chemical energy, etc.), making model solution in those cases increasingly difficult.

Model specification is completed with a consistent set of initial and boundary conditions. For parapumping, any physically realistic initial conditions are acceptable. The boundary conditions are not so simple. At the outlet end of the bed, the boundary condition is

$$\frac{\partial\phi_{f_i}}{\partial z} = 0 \tag{11A}$$

and, for the recuperative mode,

$$\frac{\partial \Theta_f}{\partial z} = 0 \qquad [11B]$$

This is the zero gradient condition of Wehner and Wilhelm (1956)*. Recall that the outlet end of the column switches position every half cycle. Thus the above condition applies only half of the time at either end.

At the inlet end, again from Wehner and Wilhelm,

$$\phi_{f_i}\Big|_{inlet} = \phi^{\circ}_{f_i} + \frac{1}{Pe}\frac{\partial \phi_{fi}}{\partial z}\Big|_{inlet} \qquad [12]$$

and for the recuperative mode,

$$\Theta_f\Big|_{inlet} = \Theta^{\circ}_f + \frac{1}{Pe}\frac{\partial \Theta_f}{\partial z}\Big|_{inlet} \qquad [13]$$

where Pe is the Peclet number, and the superscript zero refers to either the reservoir or the feed, depending on the cycle portion. These inlet conditions ** also flip from one end of the bed to the other each half cycle.

The reservoir concentrations usually depend on previous cycles in some specified way. This specification is external to the solution of the model equations each half cycle. However, it is required for connecting one cycle to the next.

The model equations and their initial and boundary conditions have now been given for the general case. To make them specific to any real situation we need the following experimental data:

1. An equilibrium function, ϕ^*_f, with its temperature dependence.
2. Values for λ, γ, η, ξ, and β (including their temperature, composition, and flow rate dependence).

In addition, as designer or user of a parapump system we must specify several more items to satisfy the remaining degrees of freedom. These design variables are:

1. The temperatures at the column ends, Θ (0,t) and Θ (1,t) in the recuperative mode, or the function $\Theta(t)$ in the direct mode.
2. The disposition of fluid leaving the bed at each end, either removed as product, returned as reflux, sent to another column, or some combination of these.

*It is usual practice to use these boundary conditions in transient problems even though they are strictly correct only for steady state.
**See previous footnote.

3. The velocity function, $\alpha f(t)$, with its wave form, amplitude, and phase displacement from Θ (t).
4. Cycle time, τ.

With the model now completely specified, let us examine several methods of solution. Then we will look at typical system responses.

VI. ANALYTICAL SOLUTIONS OF MODEL EQUATIONS

A. Direct mode

An analytical solution of the parametric pumping equations, although highly desirable, is not possible in general. If the equilibrium function is nonlinear, there exists no such solution. Even if it is linear, solutions in general are difficult to obtain for all possible initial and boundary conditions. Any "solution" of the parametric pumping PDE's can be applied only one half-cycle at a time because the initial and boundary conditions for any half-cycle depend on the previous half-cycle.

With certain assumptions, however, we can obtain a complete analytical solution and with it a better understanding of para-pumping.

If we assume instantaneous interphase equilibration ($\lambda = \infty$), no axial diffusion ($\eta = 0$), square wave variation in temperature (via direct mode), and a linear isotherm ($\phi_s = A\ (\Theta)\ \phi_f$), then a complete analytical solution is possible for each half-cycle.

Under these assumptions parapump equations are

$$\alpha f(t) \frac{\partial \phi_f}{\partial z} + \frac{\partial \phi_f}{\partial t} + \frac{\partial \phi_s}{\partial t} = 0 \qquad [14]$$

$$\phi_s = A(\Theta)\ \phi_f \qquad [15]$$

We have assumed that equilibrium will prevail between adjacent phases; therefore if we know one composition, say ϕ_f, we also know the other, ϕ_s. In Equation 14 we therefore have only one dependent variable -- either ϕ_s or ϕ_f, whichever we choose. Eliminating ϕ_s with the following substitution

$$\frac{\partial \phi_s}{\partial t} = \frac{\partial \phi_s}{\partial \phi_f} \frac{\partial \phi_f}{\partial t} + \frac{\partial \phi_s}{\partial \Theta} \frac{\partial \Theta}{\partial t} \qquad [16]$$

and inserting into Equation (14) we obtain

$$\alpha f(t) \frac{\partial \phi_f}{\partial z} + \frac{\partial \phi_f}{\partial t} \left(1 + \frac{\partial \phi_s}{\partial \phi_f}\right) + \frac{\partial \phi_s}{\partial \Theta} \frac{\partial \Theta}{\partial t} = 0 \qquad [17]$$

With Θ restricted to square wave variation, $\partial\Theta/\partial t$ is zero at all but two instants in the cycle. Thus for almost all time the parametric pumping equation is simply

$$\alpha f(t) \frac{\partial \phi_f}{\partial z} + \left(1 + \frac{\partial \phi_s}{\partial \phi_f}\right) \frac{\partial \phi_f}{\partial t} = 0 \qquad [18]$$

Note that $\partial\phi_s/\partial\phi_f = A(\Theta)$.

Let the left side of Equation 18 represent a total derivative

$$\frac{d\phi_f}{dt} = \frac{\partial \phi_f}{\partial t} + \frac{\partial \phi_f}{\partial z}\frac{dz}{dt} \qquad [19]$$

Thus we have the ODE

$$\frac{d\phi_f}{dt} = 0 \qquad [20A]$$

which is valid along characteristic lines described by

$$\frac{dz}{dt} = \frac{\alpha\ f\ (t)}{1 + A(\Theta)} \qquad [20B]$$

Equation 20b is the velocity at which waves of constant concentration move through the bed.

The solution of Equation 20a indicates that ϕ_f is constant along the characteristic line. That is

$$\phi_f\ (z_2,\ t_2) = \phi_f\ (z_1,\ t_1) \qquad [21A)$$

where the solution of (20b) is

$$z_2 = z_1 + \int_{t_1}^{t_2} \frac{\alpha f(t)}{1 + A(\Theta)}\ dt \qquad [21B]$$

Equation 21b describes the trajectory in the z-t domain of constant concentrations. For $\alpha f(t)$ a square wave with magnitude α.

$$z_2 = z_1 + \frac{\alpha}{1 + A(\Theta)}\ (t_2 - t_1) \qquad [22]$$

This indicates that the constant composition lines are straight in the z-t domain, and that their slope depends on the velocity and on the temperature of the bed.

Equation 22 describes the movement of concentration waves through a column with constant temperature. It does not account for what occurs when the temperature switches from cold to hot or hot to cold every half cycle. At these times solute is redistributed between the phases. The ratio of fluid phase concentrations before and after the temperature change is

$$\frac{\phi_{f\ hot}}{\phi_{f\ cold}} = \frac{1 + A(\Theta_{cold})}{1 + A(\Theta_{hot})} \qquad [23]$$

Pigford *et al* (1969), Aris (1969), and Gregory and Sweed (1970) have extended these analytical solutions for the single half cycle to multiple cycle runs in batch and continuous systems. The development below follows that of Gregory and Sweed.

The analytical solutions are based on the movement of concentration waves within the bed and keeping track when they leave the bed and where.

Initially the fluid phase concentration is ϕ_o.

1. Batch separations

For batch parametric pumping, Gregory and Sweed presented the following equations for the reservoir concentration as a function of cycle number. The method of analysis is analogous to the one described below for continuous systems. The flow and temperature cycles are in phase ($\varepsilon = 0$). They included the possibility that the reservoir volume exceeded the volume flowed by V_T or V_B, referring to top or bottom, respectively.

On the nth cycle, for $n > k$, the top reservoir concentration is

$$\phi_{T,n} = \phi_o\ [1 + K' + (k + \beta - 1)(1-K)\ K'-K'\ (1 - \beta + \beta K^{-1})K^{n-k}] \qquad [24]$$

The bottom reservoir concentration for $n \geq 0$ is

$$\phi_{B,n} = K^n \qquad [25]$$

In these expressions

$$K = \frac{\alpha}{\alpha + V_B}\ \frac{1 + A_H}{1 + A_C} + \frac{V_B}{\alpha + V_B} \qquad [26A]$$

$$K' = \frac{\alpha}{\alpha + V_T} \qquad [26B]$$

Also

$$k + \beta = 1 + \frac{2(1 + A_H)(1 + A_c) - (1 + A_H)\,\alpha}{\alpha\,(A_c - A_H)} \qquad [27]$$

where k is the greatest integer contained in the right hand side and β is the remaining fraction.

In the limit of very many cycles, i.e., as $n \to \infty$,

$$\phi_{T,\infty} = \phi_o\,[1 + K' + (k + \beta - 1)(1 - K)\,K' \qquad [28]$$

and

$$\phi_{B,\infty} = 0. \qquad [29]$$

These equations contain a vast amount of information. Most striking is Equation 29 which shows that there is no limit to the separation in batch parapumping. The concentration of the bottom reservoir drops as K^n, an exponential decline. This is very much akin to the effect of increasing the number of stages in distillation. Thus batch parapumping separations appear to be staged, not in position, but in time. Recall that Figure 3 showed an exponential rise in separation for the toluene-heptane system.

To explain the exponential decrease in bottom reservoir concentration on physical grounds one need only consider the velocity at which concentration waves move in the bed. From Equation 20b, this velocity is slower than the bulk fluid velocity by the factor $1/[1+A(\Theta)]$. Since $A(\Theta_{COLD}) > A(\Theta_{HOT})$ in general, the velocity cold is less than the velocity hot. That is

$$\frac{\alpha}{1 + A_c} < \frac{\alpha}{1 + A_H}$$

Thus molecules entering the bottom of the column when it is hot move upward faster, hence further, than they move downward on the next cold half cycle. In effect, some molecules which get into the column act as though they have passed through a trap door, never to return. This process continues every cycle. Each cycle the same fraction, K, of the remaining molecules in the bottom reservoir is removed.

The analytical solutions also show the effects of temperature through A_H and A_C. The greater the difference between the slopes the greater will be separation. Clearly, it is not the temperature spread itself which determines separation; it is the effect of temperature on the slope of the isotherms, i.e., on the microstep.

The effect of dimensionless displacement, α, is such that, in the absence of excess reservoir volume (i.e., $V_T = V_B = 0$), the bottom concentration is independent of it. It depends only on the number of times the temperatures switch.

The excess reservoir volume has two effects. At the bottom end, this volume slows the rate of separation by increasing the capacity of the system. It has no effect on the ultimate concentration at this end, which is still zero. Therefore, a batch parapump, no matter how small, can remove any amount of solute from any size reservoir if allowed to operate a sufficient number of cycles, at equilibrium with linear adsorption isotherms.

At the top, excess volume not only slows the rate of separation but also affects the final or maximum concentration. If $V_B > 0$, there is more solute in the system initially and this eventually appears in the top reservoir, helping it to increase its concentration. With $V_T > 0$, the greater volume in the top dilutes the maximum separation.

2. Continuous separations

Analytical solutions for batch and continuous parametric pumping are based on the repeated application of Equation 22 for each cycle portion. One must also keep track of the location of concentrations in the bed and when and where they leave the bed as product or reflux. In this section we show in some detail how a few open system solutions have been developed. The method is generally applicable to any parapump system for which Equation 22 is valid.

Let us first consider the non-symmetric system. Figure 16 is a plot of some representative characteristics of Equation 20b for each portion of the nth cycle of this non-symmetric open system.

In general we first determine where the characteristics which separate regions of different concentrations intersect either the boundaries between cycle portions ($t_1 \ldots t_5$), or the boundaries at the top and bottom of the column ($z = 1.0$ and $z = 0$). These characteristics emanate from points in the column where the concentrations are discontinuous, or from the reservoir-column or feed-column intersections. The mixed product composition is computed by averaging the concentrations of those regions which intersect the end of the column during the product collection portion of the cycle. The concentrations in each region depend on conditions at the beginning of the cycle portion via Equation 21a. Repeating this procedure cycle after cycle from initial conditions, equations are obtained for product concentrations in terms of the initial conditions, equilibrium parameters, reflux ratios, and the cycle number, n. Limiting conditions can be calculated by allowing n to increase without limit.

For each combination of reflux ratios and equilibrium parameters, the origin of concentration regions which make up the product stream is different. Therefore, no single mathematical solution can cover the entire range of these parameters. Instead, several solutions are required, each one valid only for certain limited ranges of the parameters.

For the non-symmetric system the reflux ratios can be calculated in terms of the relative times for the five cycle portions,

$$R_{BOT} = \frac{t_{II} + t_{III}}{t_I} \qquad [30A]$$

$$R_{TOP} = \frac{t_V}{t_{IV}} \qquad [30B]$$

where the t's are defined in Figure 16.

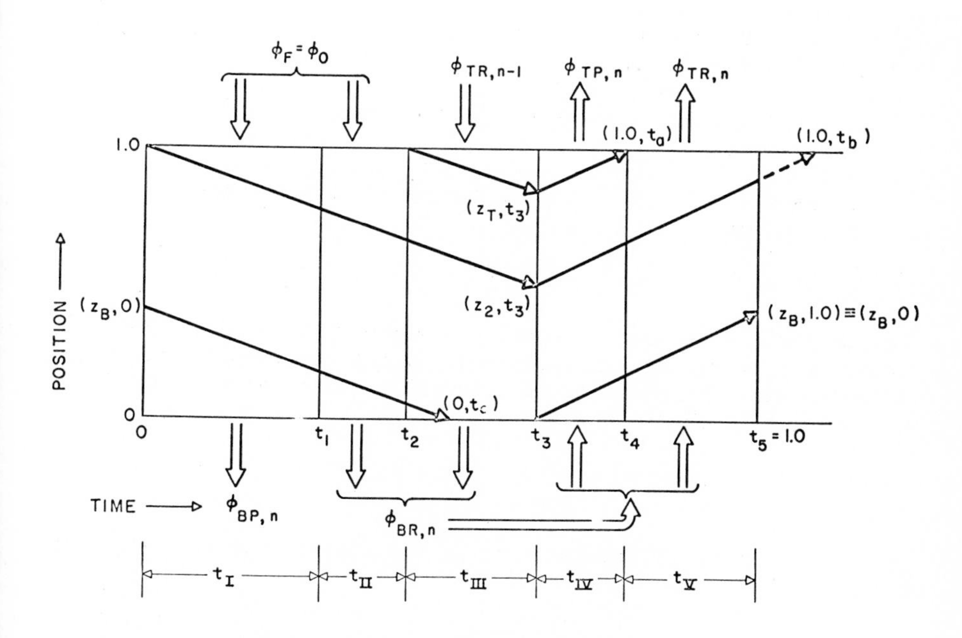

Figure 16. Some representative characteristic lines (from Equation 20b) for the nth cycle of the non-symmetric open system. [Reprinted with kind permission from the Chemical Engineering Journal, July 1970].

(The Roman subscripts refer to the duration of a cycle portion; the Arabic subscripts refer to the elapsed time since the cycle began.) To maintain constant reservoir volumes from cycle to cycle

$$t_{II} = t_{IV}$$

[30C]

and

$$t_{III} = t_{V}$$

[30D]

Finally,

$$t_I + t_{II} + t_{III} + t_{IV} + t_V = 1.0$$

[30E]

Using Equations 30 (a-e), all of the t's can be written in terms of the two reflux ratios, R_{TOP} and R_{BOT}.

On the cold downstroke the characteristic originating at $(z_B,0)$ is of special importance. z_B is the distance which solute, initially at $z = 0$, travelled on the previous hot upstroke. This characteristic can intersect the boundaries ($z = 0$ or $t = t_3$) in three different places.

1. At $z = 0$ for $0 < t < t_1$, during the bottom product collection,

2. At $z = 0$ for $t_1 < t < t_3$ during the bottom reservoir filling, or

3. At $t = t_3$ for $z \geq 0$.

Each of these three cases requires a separate solution. The constraints developed below determine which case applies to each situation based on equilibrium and reflux ratio parameters. For all cases, from Equation 22.

$$z_B = \frac{\alpha}{1 + A_H}(t_5 - t_3) \qquad [31]$$

If t_c is the time of intersection of the characteristic from z_B with $z = 0$, then

$$t_c = \frac{1 + A_c}{\alpha} z_B \qquad [32]$$

Thus

$$t_c = t_I \, R_{BOT} \frac{1 + A_C}{1 + A_H} \tag{33}$$

For the first case to exist, i.e., for t_c to lie in the range between 0 and t_1, then the bottom reflux ratio must lie within certain bounds. That is

$$0 \leq R_{BOT} \leq \frac{1 + A_H}{1 + A_C} \tag{34}$$

For the nth cycle, a mass balance under this constraint (see Figure 16) gives

$$\phi_{BP,n} = \frac{t_c}{t_I} \frac{1 + A_H}{1 + A_C} \phi_{BR,n-1} + \frac{t_I - t_c}{t_I} \phi_o \tag{35}$$

(BP refers to bottom product, BR to the bottom reservoir.) The first term is the fluid which was in the bottom reservoir on the previous half-cycle and has undergone one temperature change; and the second term is fluid, originally in the feed which has undergone an even number of temperature changes. However

$$\phi_{BR,n-1} = \phi_o \tag{36}$$

since under the above constraint, the flow from the column to the bottom reservoir is always fluid, originally of feed concentration, which has undergone an even number of temperature changes. Thus

$$\Phi_{BP,n} = \phi_o \left[R_{BOT} \frac{(A_H - A_C)}{1 + A_H} + 1 \right] \tag{37}$$

Under the constraint of Equation 34, the bottom product concentration immediately drops to the final value (37) and is independent of n.

The other cases are handled similarly, the constraints and equations for the bottom product are presented in Table I.

Since the open system may be run indefinitely, let us consider separations at limiting conditions ($n \to \infty$) where a cyclic steady-state develops. As n increases, the bottom product concentration under cases 1 and 2 approaches a limiting value

$$\phi_{BP,\infty} = \phi_o \left[R_{BOT} \frac{(A_H - A_C)}{1 + A_H} + 1\right] \tag{38}$$

TABLE I

Non-Symmetric Open System:
Bottom Product Concentration

Case	Point Intersection	Resulting Constraint	$\phi_{BP,n}$	ϕ_{BP} @ $n \to \infty$
I	$z = 0;\ 0 \leq t_c \leq t_1$	$0 \leq R_{BOT} \leq \frac{1 + A_H}{1 + A_C}$	$\left[R_{BOT} + 1 - R_{BOT}\frac{1+A_C}{1+A_H}\right]\phi_o$	$\left[R_{BOT} + 1 - R_{BOT}\frac{1+A_C}{1+A_H}\right]\phi_o$
II	$z = 0;\ t_1 \leq t_c \leq t_3$	$\frac{1+A_H}{1+A_C} \leq R_{BOT} \leq \frac{1+A_H}{A_C-A_H}$	$\phi_o\frac{1+A_H}{1+A_C}[K_1^{n-1} + K_2(1+K_1+\ldots+K_1^{n-2})]$ $K_1 = 1 - \frac{1}{R_{BOT}}\frac{1+A_H}{1+A_C}$ $K_2 = \frac{1}{R_{BOT}} + 1 - \frac{1+A_C}{1+A_H}$	
III	$t_c = t_3;\ z \geq 0$	$R_{BOT} \geq \frac{1 + A_H}{A_C - A_H}$	$\left[\frac{1 + A_H}{1 + A_C}\right]^n \phi_o$	0

[Reprinted with kind permission from the Chemical Engineering Journal, July 1970].

which decreases as the bottom reflux ratio increases, up to a critical reflux ratio

$$R_{BOT} = \frac{1 + A_H}{A_C - A_H} \quad [39]$$

(In case I this limiting concentration is reached on the first cycle.) Above this reflux ratio, the bottom product concentration approaches zero.

Similarly, for the top product concentration there are two important characteristics (shown in Figure 16) which lead to five distinct cases within the constraint that the characteristic from (1.0, 0) does not intersect z = 0 for $t < t_3$. (If z_2 is the intersection of this characteristic with $t \geq t_3$, then for $z_2 \geq 0$

$$t_3 < 1 + A_C$$

from Equation 22; but since $t_3 < 1$ and $A_C > 0$, this constraint is always true.)

Each of the five cases is analyzed separately to develop the range of validity and the equation for the top product concentration. These results are shown in Table II.

Note that at limiting conditions the top product concentration increases with increasing top reflux ratio. When the bottom product is zero the steady state material balance requires that all solute in the feed leave with the top product, no matter what the equilibrium functions. Therefore, if the bottom reflux ratio exceeds the critical value given in Equation 39, the top product is

$$\phi_{TP,\infty} = \left(\frac{1 + R_{TOP} + R_{BOT}}{R_{BOT}} \right) \phi_o \quad [40]$$

Table I and Table II give the complete transient and limiting condition solutions for the bottom and top concentrations of the non-symmetric equilibrium system. These results cover the full range of reflux ratios and equilibrium parameters. Thus, the separation which can be obtained from a non-symmetric open parapump operating at equilibrium with linear isotherms can be calculated algebraically from these tables without solving the partial differential model equations. As we will see below, these separations are the best possible with the given isotherms and system configurations. Axial dispersion and non-instantaneous interphase transfer reduce separations below this ideal case. Thus a rapid method is available to test whether this open parapump system will be suitable for a particular separation.

A grid of characteristics for a symmetric flow, open parapump

TABLE II

Non-Symmetric Open System: Top Product Concentration

Case	Point of Intersection	Resulting Constraint
I	$z = 1.0;\ t_a \leq t_4$	$R_{TOP} \leq \frac{1 + A_C}{1 + A_H}$
	$z = 1.0;\ t_b \leq t_4$	$\frac{t_4}{t_3} \geq 1 + \frac{1 + A_H}{1 + A_C}$
II	$z = 1.0;\ t_a \leq t_4$	$R_{TOP} \leq \frac{1 + A_C}{1 + A_H}$
	$z = 1.0;\ t_4 \leq t_b \leq 1$	$\frac{t_4}{t_3} \geq 1 + \frac{1 + A_H}{1 + A_C}$; $R_{BOT} \geq \frac{1 + A_H}{A_C - A_H}$
III	$z = 1.0;\ t_a \leq t_4$	$R_{TOP} \leq \frac{1 + A_C}{1 + A_H}$
	$z \leq 1.0;\ t_b = 1.0$	$R_{BOT} \geq \frac{1 + A_H}{A_C - A_H}$
IV	$z = 1.0;\ t_4 \leq t_a \leq 1.0$	$R_{TOP} \geq \frac{1 + A_C}{1 + A_H}$
	$z = 1.0;\ t_4 \leq t_b \leq 1.0$	$\frac{t_4}{t_3} \geq 1 + \frac{1 + A_H}{1 + A_C}$; $R_{BOT} \geq \frac{1 + A_H}{A_C - A_H}$
V	$z = 1.0;\ t_4 \leq t_a \leq 1.0$	$R_{TOP} \geq \frac{1 + A_C}{1 + A_H}$
	$z \leq 1.0;\ t_b = 1.0$	$R_{BOT} \leq \frac{1 + A_H}{A_C - A_H}$

<table>
<tr><th>$\phi_{TP,n}$</th><th>ϕ_{TP} @ $n \to \infty$</th></tr>
<tr><td>$\phi_o \frac{1 + A_C}{1 + A_H} [R_{TOP} + 1 - R_{TOP} \frac{1 + A_H}{1 + A_C}]$</td><td rowspan="2">$\phi_o \frac{1 + A_C}{1 + A_H} [1 + R_{TOP} - R_{TOP} \frac{1 + A_H}{1 + A_C}]$

(If $\phi_{BP,\infty} = 0$, then use Equation 40 for $\phi_{TP,\infty}$)</td></tr>
<tr><td>$\phi_o \frac{1 + A_C}{1 + A_H} [K_1^{n} + K_2(1 + K_1 + K_1^{2} + \ldots + K_1^{n-1})]$
$K_1 = 1 - \frac{1}{R_{TOP}} \frac{1 + A_C}{1 + A_H}$
$K_2 = 1 + \frac{1}{R_{TOP}} \frac{1 + A_C}{1 + A_H} - 1$</td></tr>
</table>

[Reprinted with kind permission from the Chemical Engineering Journal, July 1970].

system is shown in Figure 17.

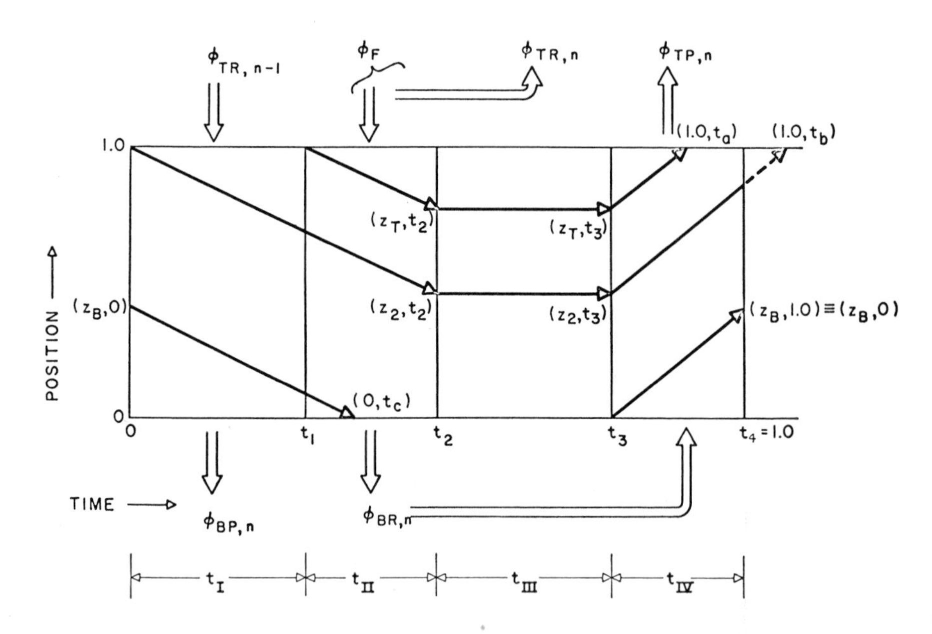

Figure 17. Some representative characteristic lines (from Equation 20b) for the nth cycle of the symmetric open system.

Again the same type of analysis is used. The reflux ratios are

$$R_{TOP} = \frac{t_{II}}{t_I} \qquad [41]$$

and

$$R_{BOT} = \frac{t_{III}}{t_{IV}} \qquad [42]$$

Also,

$$t_{II} = t_{IV} \qquad [43]$$

and

$$t_I = t_{III} \qquad [44]$$

to maintain constant reservoir volumes, where the t's are defined in Figure 17. From these relations it is obvious that

$$R_{BOT} = \frac{1}{R_{TOP}} \qquad [45]$$

The initial concentration is the same as the feed. Equations and constraints for the bottom product concentration are summarized in Table III. At limiting conditions the bottom product concentration approaches a constant value depending on the bottom reflux ratio. For $R_{BOT} \geq \frac{1 + A_H}{A_C - A_H}$, this concentration is zero.

This particular symmetric flow operation gives a simple expression for the top product concentration

$$\phi_{TP,n} = \left(\frac{1 + A_C}{1 + A_H} \right) \phi_o \qquad [46]$$

for non-zero bottom concentration because the feed is introduced directly into the top reservoir and undergoes only one temperature change before leaving as product. As discussed above, if the bottom concentration is zero, an overall material balance gives the top concentration

$$\phi_{TP,\infty} = (1 + R_{TOP})\, \phi_o$$

It is most interesting that the continuous parametric pump can produce a product close to pure solvent, even though this product is being removed at a finite rate. In distillation this cannot be done even at total reflux where no product is removed, unless the column has a very large number of stages. Thus, parametric pumping uses an increased number of cycles in the same way that staged operations use more stages. Therefore parametric pumping is staged in time rather than in space.

B. Recuperative mode

Analytical solutions of recuperative mode thermal parametric pumping are not available in the literature. However, we can speculate on how such solutions might be arrived at.

If the additional assumption is made that interphase heat transfer is instantaneous (i.e., $\gamma = \infty$), then the thermal waves in the packed column will move in the same way that concentration waves do. In effect, the heat capacity of the adsorbent slows the progress of a thermal wave [by a factor $1/(1+\beta)$] in the same way that adsorption slows the concentration wave [by a factor $1/(1+A(\Theta))$]. Thus the

TABLE III
Symmetric Open System:
Bottom Product Concentration

Case	Point of Intersection	Resulting Constraint	$\phi_{BP,n}$	ϕ_{BP} @ $n \to \infty$
I	$z = 0;\ 0 \leq t \leq t_1$	$0 \leq R_{BOT} \leq \frac{1 + A_H}{1 + A_C}$	$\phi_o[(R_{BOT} + 1)\frac{1 + A_H}{1 + A_C} - R_{BOT}]$	
II	$z = 0;\ t_1 \leq t \leq t_2$	$\frac{1+A_H}{1+A_C} \leq R_{BOT} \leq \frac{1 + A_H}{A_C - A_H}$	$\frac{1+A_H}{1+A_C}\phi_o[K_1^{n-1} + K_2(1+K_1 + \ldots + K_1^{n-2})]$ $K_1 = 1 - \frac{1}{R_{BOT}}\frac{1 + A_H}{1 + A_C}$ $K_2 = \frac{1+A_H}{1+A_C}(\frac{1}{R_{BOT}} + 1) - 1$	$\phi_o[(R_{BOT}+1)\frac{1+A_H}{1+A_C} - R_{BOT}]$
III	$z \geq 0;\ t = t_2$	$R_{BOT} \geq \frac{1 + A_H}{A_C - A_H}$	$(\frac{1 + A_H}{1 + A_C})^n \phi_o$	0

[Reprinted with kind permission from the Chemical Engineering Journal, July 1970].

analytical solution of recuperative parapumping will depend on whether the thermal waves travel more slowly, at the same speed, or faster than the concentration waves. Add to this the complication that the thermal wave velocity is independent of temperature while the concentration velocity is not, and we can see that these analytical solutions no doubt will contain many more constraints than do the direct mode solutions.

VII. NUMERICAL SOLUTIONS OF MODEL EQUATIONS

The analytical solutions above are extremely useful for calculating separations for the special case of linear isotherms, square velocity and temperature waves, no axial dispersion, and equilibrium operation. Unfortunately, these conditions are rarely encountered in real systems. However, the analytical equations do represent an upper bound on separation; axial or interphase dispersion mechanisms serve only to reduce separation.

The linearity of isotherms is often a very good assumption for low concentration streams. But, if parapumping is very successful, one end of the column will experience concentrated solutions, and the linearity becomes questionable. For us to predict behavior in real parametric pumping equipment, we need to solve the complete set of partial differential equations which describe the model.

The model equations apply within the column only, not to the boundary conditions. Thus, any solution can be readily adapted to batch, semi-batch, or continuous systems.

A. Finite difference method

One method for solving the complete set of PDE'S is to replace the derivatives with appropriate finite difference approximations, and then solve these (Lapidus, 1962).

If lowest order backward differences are used in place of the time derivatives a marching solution in time is possible starting from the initial conditions.

By choosing a particular difference scheme for the position derivatives, (lowest order central differences for the second derivative and the lowest order backward difference of the first derivative), the difference equations form a tridiagonal matrix equation which is solved readily using the Thomas algorithm (Lapidus, 1962). Rolke (1967) was successful in simulating recuperative mode separations of $NaCl-H_2O$ using this method.

If another finite difference scheme were used for the position derivatives, i.e., lowest order central differences for both derivatives, then the resulting equations would have the form of the equations for a series of continuous stirred tank reactors. This model of a packed bed was examined in depth by Deans and Lapidus (1960).

Using this model Wilhelm, <u>et al</u> (1968) calculated the separation factor response, Figure 18, for the $NaCl-H_2O$-ion exchange resin system. The figure shows the separation factor after start-up transients have died out as a function of the heat transfer and mass transfer coefficients, γ and λ respectively. The velocity function used in these computations was sinusoidal; the system was at total reflux. When γ or λ is small there is little interaction between the fluid and the adsorbent and so there is little separation. When these rate constants are large the bed approaches equilibrium operation. Note that the best separation does not coincide with equilibrium operation.

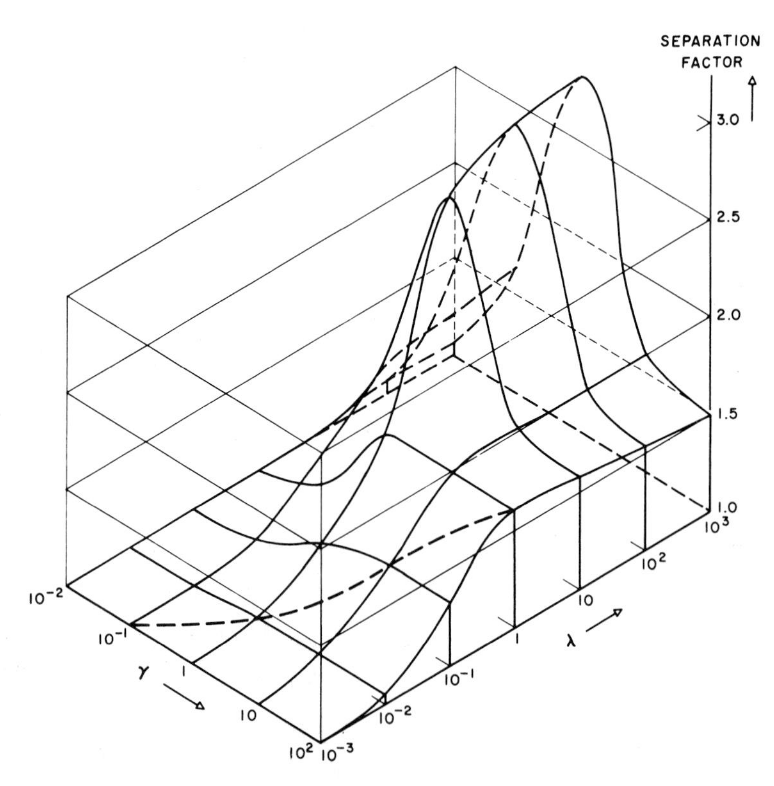

Figure 18. Calculated separation factors for a total reflux, recuperative thermal parametric pumping separation as a function of the mass and heat transfer constants, λ and γ, respectively. The velocity was sinusoidal; the system used was $NaCl-H_2O$-ion exchange resin. [Reprinted with kind permission of the American Chemical Society from IEC Fundamentals 7, 345 (1968)].

The reason for this is that the temperature distribution is more effective for finite γ than for the thermal equilibrium case.

Finite difference solutions have one important drawback however. Finite differencing in the axial direction introduces an error which shows up as additional -- and unwanted -- diffusion. If η is large, or if large concentration gradients do not exist, then the additional diffusivity is unnoticed. However, if experimental separations are found to be large (as they have been) then the gradients must also be large and the additional axial diffusivity is significant. A computational method is needed to reduce the axial dispersion due to finite differencing.

In many situations, the dispersion axially contributes less to the overall dispersion than does the noninstantaneous interphase transfer rate, the other dispersion mechanism. We may therefore lump all dispersive mechanisms into λ and eliminate η from Equation 5.

In dropping the second derivative we altered the mathematical form of the equation. It is now of the hyperbolic type rather than the parabolic. This change is important for it allows us to use numerical methods specifically applicable to hyperbolic systems. In particular, we may use the Method of Characteristics (Acrivos, 1956) or its recent modification, the STOP-GO algorithm (Sweed and Wilhelm, 1969).

B. STOP-GO algorithm

To reduce axial dispersion in calculation, Sweed and Wilhelm (1969) developed the STOP-GO method. This algorithm is a simplified version of the method of characteristics which provides a very clear physical picture of what the calculations involve. Qualitatively, the algorithm separates the simultaneous actions of flow and interphase transport into sequential steps for the purpose of computation. During the GO step, fluid is displaced axially a distance ΔZ, without axial mixing or dispersion, or interphase transfer. When the GO step is complete, the flow is STOPPED and interphase transfer begins. The time allowed for mass transfer is the same as the time for flow. In fact it is the identical time step. Thus the displacement ΔZ per time step Δt, is

$$\Delta z = \alpha \Delta t$$

The STOP-GO algorithm is especially valuable where the interphase mass transfer rate expression is very complicated, either because of nonlinearities or because many interacting components are present. This rate expression, if in lumped parameter form, consists of one or more initial value ordinary differential equations, the solutions of which can be calculated in a straightforward manner.

The STOP-GO method is shown schematically in Figure 19. Note that i refers to a position increment and j to a time increment. During the GO step fluid is displaced relative to the solid. No calculations are required because fluid $\phi_{f_{i,j}}$ which was initially adjacent to solid section 1, $\phi_{s_{i,j}}$, is now next to solid section i + 1, $\phi_{s_{i+1,j}}$. Each packet of fluid is moved exactly one step ahead. Thus, the GO part requires only a renumbering of ϕ_f subscripts.

In the STOP portion, each axial section j is a closed vessel. Whatever solute the solid phase gains must come from the fluid. The new solid concentration is calculated from the differential rate equation; the fluid is calculated by an algebraic material balance. The calculations of STOP-GO continue until the desired volume of flow is completed. This model is used in all the simulations that follow.

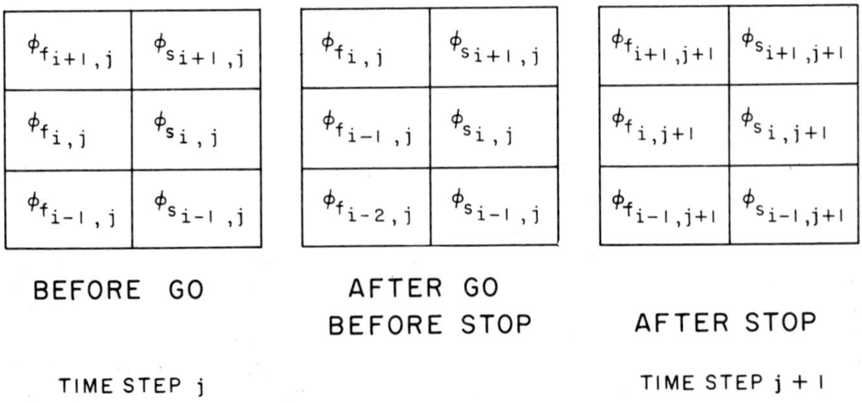

$\phi_{f_{i+1,j}}$	$\phi_{s_{i+1,j}}$
$\phi_{f_{i,j}}$	$\phi_{s_{i,j}}$
$\phi_{f_{i-1,j}}$	$\phi_{s_{i-1,j}}$

BEFORE GO

$\phi_{f_{i,j}}$	$\phi_{s_{i+1,j}}$
$\phi_{f_{i-1,j}}$	$\phi_{s_{i,j}}$
$\phi_{f_{i-2,j}}$	$\phi_{s_{i-1,j}}$

AFTER GO
BEFORE STOP

$\phi_{f_{i+1,j+1}}$	$\phi_{s_{i+1,j+1}}$
$\phi_{f_{i,j+1}}$	$\phi_{s_{i,j+1}}$
$\phi_{f_{i-1,j+1}}$	$\phi_{s_{i-1,j+1}}$

AFTER STOP

TIME STEP j

TIME STEP j + 1

Figure 19. Schematic of STOP-GO computational algorithm. [Reprinted with kind permission of the American Chemical Society from IEC Fundamentals 8, 224 (1969)]

VIII. ESTIMATION OF MODEL PARAMETERS

To this point we have completed a mathematical model of parametric pumping and the STOP-GO algorithm to solve it. However, to apply the model to a specific physical system, say the NaCl-H_2O-ion retardation resin system of Sweed and Gregory, we still need additional information.

Some of this is easy to obtain because it is part of the specification of an experimental run. Certainly, the experimenter fixes, and so obviously knows, the cycle time, τ, the phase angle between heating and flow cycles, ε, the operating temperatures, the volumetric flow rate in the bed (with the void fraction this gives the displacement, α), and the reflux ratios at each end.

There are still two quantities which cannot be determined *a priori*. They are the mass transfer coefficient λ with its dependence on cycle time, flow rate, temperature, and composition; and the equilibrium function with its dependence on concentration and temperature.

For the direct mode, the equilibrium distribution function is needed at the two operating temperatures only. In the recuperative mode the full temperature dependence is required.

It must be pointed out that the equilibrium in question here is that between the particles of adsorbent and the interstitial fluid. All solute within the particle -- whether actually adsorbed or merely in the pores -- is considered to be associated with the particle and so with ϕ_s. This equilibrium function is not the one usually reported in the literature as adsorption equilibrium. This latter is concerned with the solute actually adsorbed on the solid and disregards solute in the pores.

The reason for using the entire particle to determine ϕ_s is that the mathematical model allows only two places for solute to exist: in the fluid which flows with velocity $\alpha f(t)$, or not in this fluid. Since pore fluid does not flow with the bulk, it is lumped with the solid phase.

For their NaCl-H_2O-resin system Sweed and Gregory determined the isotherms shown in Figure 20.

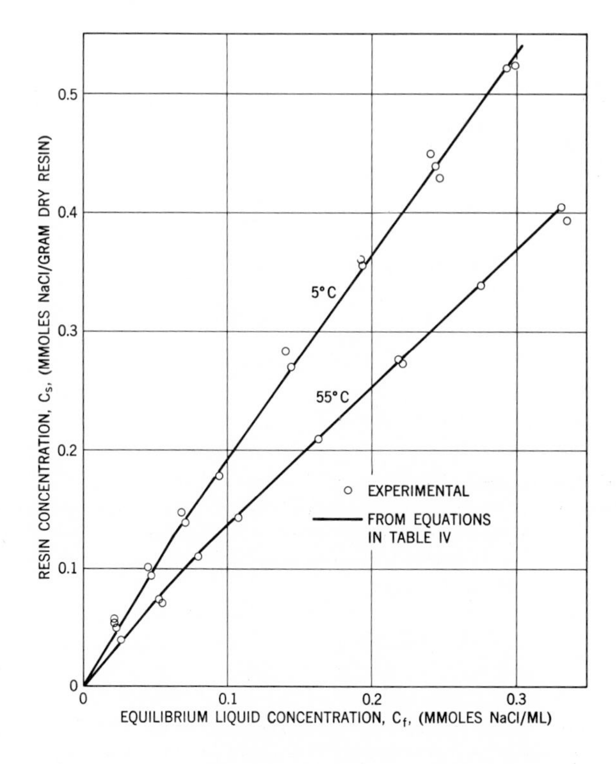

Figure 20. Adsorption isotherms for an aqueous NaCl solution on Bio-Rad AG11A8 ion retardation resin at 5° C and 55° C. [Reprinted with kind permission from the A.I.Ch.E. Journal 17, 171 (1971)].

At each temperature, 5° C and 55° C, a known amount of NaCl in solution was contacted with a known quantity of NaCl-free resin. After equilibrium was established, fluid concentration was measured and resin concentration calculated by material balance.

For use in the model, each isotherm was fitted to two straight line segments as shown in Table IV.

TABLE IV

Equations of Equilibrium Isotherms

For an NaCl-H_2O Solution

On Bio-Rad AG11A8 Ion Retardation Resin

$c_s = dc_f^*$ $\qquad$ $C_f^* \leq 0.06$ N NaCl

$c_s = ac_f^* + b$ $\qquad$ $C_f^* > 0.06$ N NaCl

	5° C	55° C
a	1.711	1.169
b	0.0225	0.0212
c	2.086	1.523

All that the model lacks now is the function for λ, the dimensionless mass transfer coefficient. λ is the product of a dimensional mass transfer coefficient, h_m, the cycle time τ, and a constant B which is fixed in each system. Thus

$$\lambda = Bh_m\tau \qquad [47]$$

If resistance to mass transfer is primarily in the boundary layer, then h_m will depend on the Reynolds number in the bed. Conversely, if flow rate does not affect h_m, then the rate determining step is not in the boundary layer and the lumped parameter rate expression (Equation 6) is only an approximation to the actual rate. In the present case, flow rate does alter the mass transfer rate.

Mass transfer coefficients in packed beds are often correlated in terms of a j-factor (Bird, et al, 1960) which in turn is a function of the Reynolds number raised to an exponent. The Reynolds number is a function of the flow rate in the bed. Thus

$$h_m \approx j\ v \qquad [48]$$

$$j \approx Re^{-a} \approx v^{-a} \qquad [49]$$

Thus

$$h_m \approx B'\ v^{1-a} \qquad [50]$$

where B′ depends on fluid properties, concentration, and temperature. Then

$$\lambda = B'\ v^{1-a}\ \tau$$

At a fixed temperature, it follows that

$$\lambda\ (v_2,\ \tau_2) = \lambda\ (v_1,\ \tau_1)\ \left(\frac{v_2}{v_1}\right)^{1-a} \left(\frac{\tau_2}{\tau_1}\right) \qquad [51]$$

A rate experiment is now required to determine whether the simple lumped parameter expression (6) is valid, and to evaluate "a" in Equation 51.

Three types of experiments can be used for taking rate data in adsorbent beds. The first method uses shallow, differential beds; the second uses deep beds which produce breakthrough curves; and the third method uses parametric pumping experiments directly. (Use of stirred tanks is not usually acceptable because the fluid flow here cannot be correlated with that in the bed.)

Although the differential bed method does give good rate information, breakthrough curves are more like the parapump process. Indeed, the model equations for a half cycle of parametric pumping are exactly those which describe a breakthrough curve experiment. Only the initial conditions are different. Thus if the breakthrough curves can be simulated accurately using the STOP-GO algorithm, we can be confident that simulation of the complete parapumping model will be successful.

The parameters in each of the breakthrough curves in Figure 21 are completely specified in the model except for the numerical value of λ. By scanning this parameter on the computer and comparing calculated and experimental breakthrough curves, one can obtain the

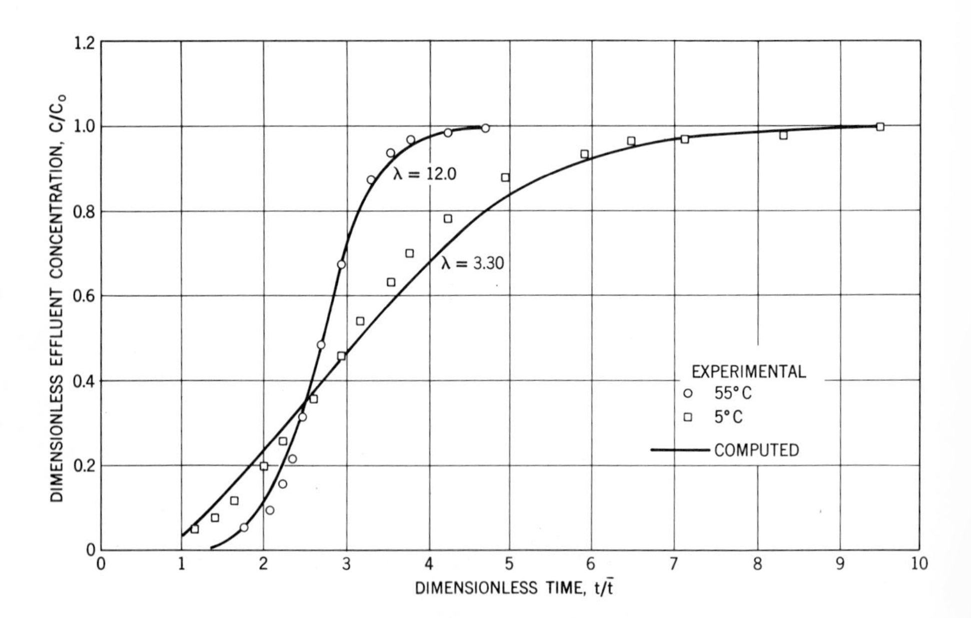

Figure 21. Breakthrough curves, experimental and calculated, for the $NaCl-H_2O$-ion retardation resin system. Flow rate corresponds to $\alpha = 0.8$ and $\tau = 20$ min. in a batch parapump.

value of λ in each case which provides the best fit. These values of λ and the corresponding STOP-GO simulations are also shown in Figure 21.

At 55° C., $\lambda = 12.0$; at 5° C., $\lambda = 3.3$. These values are for a flow rate equivalent to $\alpha = 0.8$. $\tau = 20.$ minutes. Using other flow rates in additional breakthrough curves, the effect of Reynold's number on λ could likewise be established. However, this data can also be derived from batch parametric pumping runs.

Since fluid velocity in the closed system is inversely proportional to cycle time, τ for constant displacement, α, the velocity dependence of can be calculated from batch experiments at several cycle times. The batch runs described above (Figure 6) were used for this purpose. It was found that a = 0.7.

For the NaCl-H_2O-resin system, Sweed and Gregory obtained the following functions for in batch parapumps:

$$\lambda = 12.0 \left(\frac{\alpha}{0.8}\right)^{0.3} \left(\frac{\tau}{20}\right)^{0.7} \qquad T = 55°C \qquad [52]$$

$$\lambda = 3.3 \left(\frac{\alpha}{0.8}\right)^{0.3} \left(\frac{\tau}{20}\right)^{0.7} \qquad T = 5°C \qquad [53]$$

If the system is non-batch, then the λ functions are still valid, but must be put into a form using fluid velocity rather than cycle time. Thus

$$\lambda = 12.0 \left(\frac{v}{0.24}\right)^{-0.7} \left(\frac{\alpha}{0.8}\right) \qquad T = 55°C \qquad [54]$$

$$\lambda = 3.3 \left(\frac{v}{0.24}\right)^{-0.7} \left(\frac{\alpha}{0.8}\right) \qquad T = 5°C \qquad [55]$$

where v is velocity of the fluid in the bed, cm/min.

Figures 5 and 6 show calculated and experimental concentrations for a number of different batch runs. The agreement is quite good.

The model is now complete. We may use it to design parametric pumps, to optimize them, to investigate them for whatever reason.

To get a feeling for the difference in separation between assuming instantaneous interphase transfer and assuming finite rates, one need only compare the curves on Figures 12, 13, and 14. In each case the analytical solution (i.e., $\lambda = \infty$) is better than the numerical solution (i.e., λ is finite). Thus the analytical solutions given in Tables I, II, and III are useful in that they provide an upper bound on separation; actual separations may be far smaller. This points up the importance of using the complete model for precise work. For any other solute-solvent-adsorbent system, we can similarly determine the rate and equilibrium functions and then completely simulate its behavior.

IX. OPTIMIZATION

With the complete model we are now in a position to optimize system operating conditions relative to some stated objective.

For the NaCl-H_2O-retardation resin system we can choose any one of several objectives on which to optimize. For example, which process should we use and how should we operate it to produce a particular high concentration salt solution at the maximum rate; or to produce the largest volume of potable water? These two objectives lead to different operating systems. Because of general interest in the problem of desalination we have chosen the latter objective to illustrate the optimization.

If the assumptions of equilibrium operation apply, then the analytical solutions of Tables I, II, III and Equation 46 can be used immediately to determine the optimum for the non-symmetric and symmetric systems and then to compare one system with the other. Had we developed analytical expressions for another parapump system we could similarly compare it too. However, if the rate of inter-phase mass transfer is not extremely large, as in most real separations, then the numerical solutions must be used. This requires considerable computational effort but it is essential if a valid optimization is desired.

A. Analytical

For each open system there are a number of different solutions for the bottom product concentration, depending on the bottom reflux ratio. We must investigate each solution separately and then compare results.

Since we have assumed constant velocity for both systems we can maximize product volume by maximizing the times of the cycle portions during which product is collected. For the non-symmetric system this time is $t_{I_{NS}}$.

From the definition of the reflux ratio

$$t_{I_{NS}} = \frac{1}{1 + 2\ R_{BOT}} \qquad [56]$$

If

$$R_{BOT} < \frac{1 + A_H}{A_C - A_H} \qquad [57]$$

the bottom product at limiting conditions is

$$\phi_{BP,\ \infty} = \phi_o\ [1 + R_{BOT}\ (\ \frac{A_H - A_C}{1 + A_H}\ )\ ] \qquad [58]$$

Thus

$$t_{I_{NS}} = \frac{A_H - A_C}{2(\phi_{BP,\infty}/\phi_o)(1 + A_H) - 2 - A_H - A_C} \qquad [59]$$

If

$$R_{BOT} \geq \frac{1 + A_H}{A_C - A_H} \qquad [60]$$

then

$$\phi_{BP,\infty} = 0 \qquad [61]$$

Since $t_{I_{NS}}$ decreases as R_{BOT} increases, the maximum flow rate for zero concentration product occurs when

$$R_{BOT} = \frac{1 + A_H}{A_C - A_H} \qquad [62]$$

For the symmetric system by similar arguments,

$$t_{I_S} = \frac{A_H - A_C}{2(\phi_{BP,\infty}/\phi_o - 1)(1 + A_C)} \qquad [63]$$

The corresponding concentration is found in Table III. For zero concentration product, the maximum flow rate occurs when

$$R_{BOT} = \frac{1 + A_H}{A_C - A_H} \qquad [64]$$

The ratio of $t_{I_{NS}} : t_{I_S}$ compares the rates of product formation for the two open systems.

$$\frac{t_{I_{NS}}}{t_{I_S}} = \frac{2(1 + A_C)(\phi_{BP,\infty}/\phi_o - 1)}{2(\phi_{BP,\infty}/\phi_o)(1 + A_H) - 2 - A_H - A_C} \qquad [65]$$

If this ratio is less than one the symmetric system is preferred; if greater than one, the non-symmetric gives more product. The ratio clearly depends on the fraction of solute originally in the feed which remains in the product. The ratio of times in Equation 65 is equal to 1.0 when $\phi_{BP,\infty}/\phi_o = 0.5$. If $\phi_{BP,\infty}/\phi_o < 0.5$ the non-symmetric system produces more bottom product.

For other objectives a similar analysis will determine the optimum. However, recall the restrictions under which these solutions were derived; especially the one dealing with interphase equilibration. For finite rates of mass transfer the numerical solutions must be used.

B. Numerical

Gregory (1970) has optimized each open system by fixing the feed rate and bottom product rate and then determining the best values of displacement, cycle time, and bottom reflux ratio to maximize cycle time, and bottom reflux ratio to minimize product concentration. Table V summarizes these parameters for a dimensionless feed rate of 0.52 and product rate of 0.12. These rates are fractions of the average volume flowed per half cycle.

TABLE V

Optimum Operating Parameters
To Minimize Bottom Product Concentration
Based on STOP-GO Simulation

Dimensionless feed rate = 0.52
Dimensionless bottom product rate = 0.12
Dimensionless feed concentration, $\phi_o = 1.0$

	α	R_{BOT}	τ, min.	$\phi_{BP,\infty}$
Non-symmetric	0.8	5.2	24	0.432
Symmetric	1.56	3.33	50	0.458

For this particular case the non-symmetric system produces a lower concentration product. Comparing this realistic simulation to the ideal, equilibrium case, we find for the latter that both systems would give zero concentration product. Clearly we must be careful concerning how we use the ideal solutions.

A complete optimization must of course include the economics of operating each system. Since the heat duty is a direct function of the number of cycles, it may be that the symmetric system is actually preferable. In the final analysis it is the lowest cost which determines the true optimum.

X. MULTICOMPONENT DIRECT THERMAL SEPARATIONS

Thermal

All of the separations discussed to this point were for binary mixtures, i.e., one solute and one solvent. What if several solutes are present? How can parametric pumping separate them?

A complete answer to these questions is not yet available. However, there are some situations where the binary mixture theory does apply to multicomponent mixtures. In particular, if the solutes are dilute enough so that the adsorption of one component does not affect the adsorption of any others, then each solute behaves independently. If this happens, a multicomponent solution behaves like a set of binary solutions, many independent separations occurring simultaneously in the same column. Clearly, if the solutes have sufficiently different adsorption characteristics, then each one will separate at its own rate and a relative separation of components occurs.

For example, in a batch system filled uniformly with a multicomponent mixture, the separation factor is 1.0 for all species. After a very large number of cycles, all the solute (of all species) will have migrated to one reservoir. The separation factor for each component is infinite, yet the relative concentrations among the various species is the same as it was in the initial mixture.

However, had this batch run been halted at an intermediate cycle number, some species would already be in high concentration in the reservoir, others would not be. At such a point in time, a relative separation of species would exist, and could be utilized.

The more interesting situation where solutes compete for adsorption sites has not yet been solved. However, Butts (1970) experimentally has shown for the competitive ion-exchange between H^+ and K^+ on Dowex 50, that one species can be made to accumulate in one reservoir, the other species in the opposite reservoir. This separation does not vanish after many cycles.

It may be possible to select operating conditions so that in any multicomponent system certain desired species are made to migrate in one direction, away from the undesirable ones. Certainly this is the goal of all separation processes.

XI. OTHER FORMS OF PARAMETRIC PUMPING

The PPP does not restrict the form which the energy of separation must take. Let us examine briefly a parametric pumping system which makes use of chemical energy in the form of a periodic pH gradient.

pH alternations can cause the equilibrium distribution of solutes to vary in certain materials, such as ion exchange resins.

If pH alternations can be imposed on a column of resin, then there will be a periodic interphase flux of solute. This flux, when coupled with an alternating axial solution flow, results in separation.

A. pH parametric pumping

The methods of imposing a periodic pH on a column are not so simple as those for causing temperature alternation. Heat can travel across solid boundaries, such as a tube wall; chemical species usually cannot. Thus, unless the column wall is specifically designed to be permeable to the driving species, hydrogen ion, direct chemical (in analogy with direct thermal) parametric pumping is not possible. However, if each reservoir is maintained at a different pH, then, as in the recuperative thermal case, H+ is carried into and out of the bed with the reciprocating flow. The reciprocating flow alternately causes the ion exchange resin in the bed to contact low pH solution from the acidic end, then high pH solution from the alkaline end. Periodic oscillations in pH develop within the bed, producing a recuperative pH parapump.

Unlike the other species present, H+ concentration must be maintained in each reservoir by some external means. If the reservoir concentrations were not so maintained, neutralization would occur and there would be no long term separation; both reservoirs would eventually mix. In effect, this pH maintenance is the energy input to the system.

Sabadell and Sweed (1970) used a semi-batch, recuperative pH parametric pump to separate Na+ and K+ (Figure 22). Experimental operation is quite similar to the semi-batch, direct thermal separations described above except that, instead of temperature alternations, this system uses an automatic buret to infuse hydrochloric acid into the well-mixed closed end for pH control. To minimize dilution at the closed end, the concentration of the HCl solution was kept high. The feed solution at the other end was alkaline with constant pH. In their experiments the ion exchanger was 30-50 mesh particles of IRC-84, a carboxylic polymethacrylic cation exchange resin (Rohm & Haas).

With the total reflux end held at pH ≐ 3.05 ± 0.05, and the feed at pH = 11.75 ± 0.05, with $\alpha = 1.4$ and $\tau = 16$ minutes, the separations shown in Figure 23 were typical.

In each run the ratio of Na+ to K+ in the feed was 1.0; the concentration of each ion was 0.1, 0.05, and 0.02 moles/liter for runs 1, 2, and 3 respectively.

In runs 1 and 2 the increase in cation concentration (Na+ + K+) in the acid reservoir was respectively, 1.16 and 1.31 times the feed concentration. These concentrations include the dilution from the HCl solution. If a more concentrated acid had been used the separations would have been higher. In these runs there was no separation of Na+ relative to K+. Run 3 showed no separation.

These separations are encouraging. Although not as large as direct thermal mode separations, they are of the same magnitude as the recuperative thermal separations presented above. It should be pointed out that no attempt at optimization was made. Rather, these experiments were undertaken to demonstrate that chemical energy can be used to drive a parametric pump.

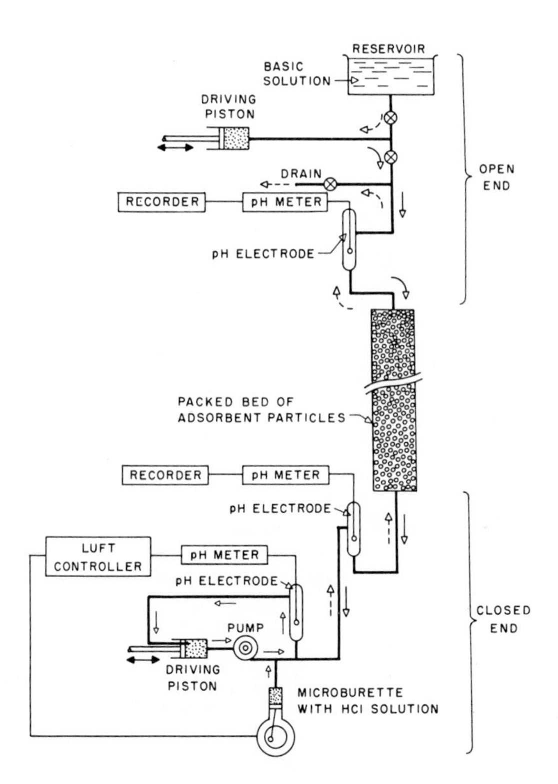

Figure 22. Apparatus arrangement for recuperative pH parametric pumping. [Reprinted with kind permission of Marcel Dekker, Inc., from Separation Science 5, 177 (1970)]

The mathematical model of pH parametric pumping is significantly more complicated than for the thermal case. In addition to the multicomponent nature of the equilibrium function and the requirement for additional fluid phase material balances like Equation 5, the pH system also has a neutralization reaction occurring within the bed. An additional term must be added to Equation 5 to account for the apparent loss of H+ and OH- in the bed.

We should realize that any chemical species -- not just hydrogen ion -- can be used to effect separation. For example, a periodic variation in Na+ could be used to separate H+, the reverse of Sabadell and Sweed's work. Also, biological materials might be separated by parapumping using variable ionic strength.

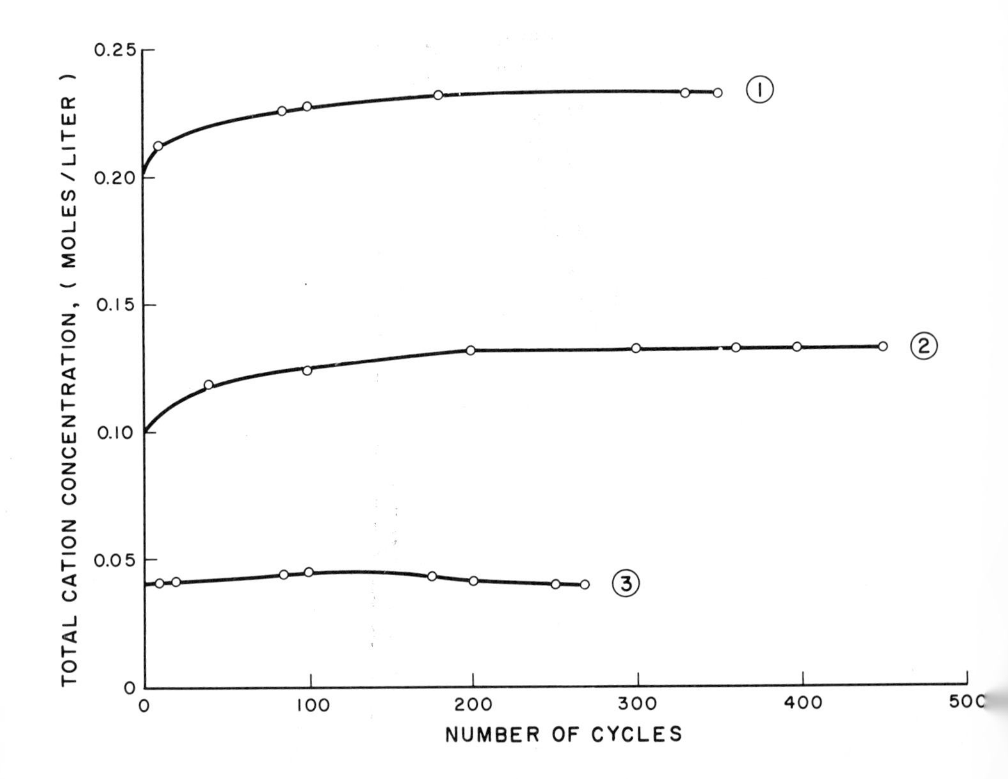

Figure 23. Total cation concentration of the fluid phase measured at the total reflux (acidic) end. [Adapted with kind permission of Marcel Dekker, Inc., from Separation Science 5, 180 (1970].

Other driving forces have been used. Horn and Lin (1969) have used electrical energy to drive the periodic microstep and so parapump electrophoresis. Skarstrom (1959) and Alexis (1967) have used pressure as the periodic variable to separate gases. In the next few years we will no doubt see additional examples of parapumping processes.

XII. SUMMARY

This chapter has reviewed the Principle of Parametric Pumping and several of its present implementations. These have shown that parametric pumping processes can produce large separations in the unusual framework of a periodic system. For the direct thermal mode, for example, the equilibrium theory predicts no limit to the separation factor. Already experimental separation factors in excess of 10^5 have been achieved. For the open direct thermal systems, the equilibrium theory predicts that above a critical reflux ratio at one end of the column, one product will contain all of the solute in the feed while the other product will be pure solvent. Thus for a finite production rate it is possible to get complete separation.

For both recuperative and direct thermal systems, model equations were developed which allow simulation and optimization of non-equilibrium parapump processes with a minimum of experimentation. The only experiments required are those to evaluate the model parameters.

The field of parametric pumping separations is still wide open to investigation. The behavior of multicomponent systems is one of the most important of these. Others include a detailed analysis of chemically driven systems, of systems driven by more than one thermodynamic variable, and parametric pumping implementations in other than the fixed bed arrangement. Any system that obeys the two-step Principle of Parametric Pumping is acceptable. The only limitation is one's own imagination.

ACKNOWLEDGEMENT

The author wishes to express his appreciation to Dr. R. Alan Gregory for his assistance in preparing this chapter. The author is indebted to the National Science Foundation for its support of much of the research presented herein, through Grant GK-1427, and to the late Professor Richard H. Wilhelm who introduced the concept of parametric pumping and encouraged the author to first investigate it.

XIII. NOMENCLATURE

All quantities are dimensionless unless otherwise indicated.

A	Slope of a linear adsorption isotherm
h_m	Dimensional mass transfer coefficient (see Appendix)
j	'$_j$'- factor for mass transfer in a packed bed
K, K'	Constants defined by Equation 26
k	The greatest integer in Equation 27
Pe	Peclet number
q	Equal to the right-hand side of Equation 1
R_{TOP}, R_{BOT}	Reflux ratio at a column end
Re	Reynolds number
t	Time
V	Alternating velocity in the Tinker Toy model
V_T, V_B	Excess reservoir volume in the top and bottom respectively
z	Distance measured from the bottom of the column
α	Fluid displacement
β	Ratio of heat capacities
β	The non-integer part of Equation 27
γ	Interphase heat transfer coefficient
ε	Phase angle between flow and heating cycles
ξ	Axial thermal diffusivity
η	Axial mass diffusivity
Θ	Temperature
λ	Interphase mass transfer coefficient

τ	Cycle time, minutes
ϕ_{BP}	Concentration of bottom product
ϕ_{BR}	Concentration of bottom reservoir
ϕ_{f_i}	Concentration of component i in the fluid
ϕ_o	Fluid concentration in the feed and/or initially in the bed
ϕ_{s_i}	Concentration of component i in the solid phase
ϕ_{TP}	Concentration of top product
ϕ_{TR}	Concentration of top reservoir
ω	Cycle frequency

XIV. REFERENCES

Acrivos, A., Ind. Eng. Chem. 48, 703 (1956).

Alexis, R. W., Chem. Eng. Progr. Symp. Ser. 63, No. 74, 51 (1967).

Aris, R., Ind. Eng. Chem. Fundamentals 8, 603 (1969).

Bird, R. B., Stewart, W. E., Lightfoot, E. N., "Transport Phenomena", Wiley, New York, 1960.

Bringhurst, F. R., B. S. E. thesis, Princeton University, 1970.

Butts, T. J., personal communication, (1970).

Coulombe, J., Favreau, I., Clin. Chem. 9, 102 (1963).

Deans, H. A., Lapidus, I., A. I. Ch. E. J. 6, 656 (1960).

Goldstein, R. J., B. S. E. thesis, Princeton University, 1969.

Gregory, R. A., Ph. D. dissertation, Princeton University, 1970.

Gregory, R. A., Sweed, N. H., Chem. Eng. J. 1, 207, July 1970.

Horn, F. J. M., Lin, C. H., Berichte der Bunsengessellschaft fur Physikalische Chemie 73, 575 (1969).

Jenczewski, T. J., Meyers, A. I., A. I. Che. E. J. 14, 509 (1968).

Lapidus, I., "Digital Computation for Chemical Engineers", McGraw-Hill, New York, 1962.

McAndrew, M.A., Ph. D. dissertation, Princeton University, 1967.

Pigford, R. I., Baker, E. III, Blum, D. E., Ind. Eng. Chem. Fundamentals 8, 144 (1969).

Rolke, R. W., Ph. D. dissertation, Princeton University, 1967.

Rolke, R. W., Wilhelm, R. H., Ind. Eng. Chem. Fundamentals 8, 235 (1969).

Sabadell, J. E., Sweed, N. H., Separation Science 5, 171 (1970).

Skarstrom, C. W., Ann. N. Y. Acad. Sci. 72, 751 (1959).

Sweed, N. H., Gregory, B. A., A. I. Ch. E. J. 17, 171 (1971).

Sweed, N. H., Wilhelm, R. H., Ind. Eng. Chem. Fundamentals 8, 221 (1969).

Wakao, N., Matsumoto, H., Suzuki, K., Kawahara, A., Kagaku Kogaku 32, 169 (1968).

Wehner, J., F., Wilhelm, R. H., Chem. Eng. Sci. 6, 89-93 (1956).

Wilhelm, R. H., Rice, A. W., Bendelius, A. R., Ind. Eng. Chem. Fundamentals 5, 141 (1966).

Wilhelm, R. H., Rice, A. W., Rolke, R. W., Sweed, N. E., Ind. Eng. Chem. Fundamentals 7, 337 (1968).

Wilhelm, R. H., Sweed, N. H., Science 159, 522 (1968).

USE OF LIQUID ANION-EXCHANGERS IN REVERSED-PHASE EXTRACTION CHROMATOGRAPHY

U. A. Th. Brinkman
Department of Analytical Chemistry
Free Reformed University
Amsterdam, The Netherlands

I. INTRODUCTION

A. Liquid Anion-Exchange

In 1948, Smith and Page (129) introduced high-molecular-weight amines as extractants of inorganic and organic acids from aqueous solution and they compared this extraction process with sorption by anion-exchange resins. The long-chain aliphatic and aromatic amines and related substituted quaternary ammonium salts were little exploited until 1955, but since that time extraction of acids and of metal ions from acid solution with this class of extractants has received considerable attention. Over the years, these processes have become known as liquid anion-exchange, which can be defined for most purposes as follows (110): liquid anion-exchange refers to liquid - liquid extraction systems, that operate, at least formally, by interchange of anions at the interface between an aqueous solution and an immiscible solvent (the diluent), with negligible distribution of the extractant to the aqueous phase. Interchange of anions is generally assumed in preference to the addition of neutral species. Although these alternatives may sound drastically different, they represent only an arbitrary choice of description for equilibrium extractions, as the alternatives are thermodynamically equivalent, and cannot be distinguished by any measurements made at equilibrium. As an example, metal extraction may be represented by:

$$\underline{n}\ RR'_3N^+X^-_{org.} + MX_p{}^{n-}_{aq.} \rightleftharpoons (RR'_3N^+)_n MX_p{}^{n-}_{org.} + \underline{n}\ X^-_{aq.} \quad [1a]$$

or

$$n\ RR'_3N^+X^-_{org.} + MX_{p-n\ aq.} \rightleftharpoons (RR'_3N^+)_n MX_p{}^{n-}_{org.} \quad [1b]$$

(R, alkyl; R', alkyl or H).

Extensive investigation on a wide variety of liquid anion-exchangers has shown that for practical application, certain basic requirements must be fulfilled, such as compatibility with a diluent, low aqueous solubility and sufficient chemical stability. Although exceptions are known, usually only monofunctional saturated nitrogen compounds fulfill these requirements of a useful extractant.

Intensive development led to commercial application in 1957, with rapid expansion since then in the production of uranium and related elements from ores, and in nuclear fuel reprocessing. In the early stages, most work was confined to the development of analytical methods and separation techniques; in recent years, descriptive chemistry and physicochemical aspects of liquid anion-exchange have been increasingly studied. In summary, solvent extraction by liquid exchangers is presently a very popular technique, as may be ascertained from review articles and books, such as those by Coleman and coworkers (109,110), Green (39), Marcus and Kertes (122,124), Moore (125), and Olenovich _et al_. (126).

B. Chromatography

In the late fifties, chemists thought of coupling the favorable features of liquid exchangers with a chromatographic process. In the field of anion-exchangers, Testa (92), in 1961, was the first to realize the potentialities of this technique. He showed that paper treated with Tri-n-octyl amine (TnOA) behaves like an anionic resin film, and that consequently many separations of cations that form anionic complexes can be carried out by this new procedure. Shortly after, the use of TnOA-loaded supports in column chromatography was also described (19,40). Chromatography on impregnated thin layers, introduced in 1963 (55), has become increasingly popular since 1965, when a small-scale technique was elaborated (6).

Presently, about 100 papers have been published on the use of liquid anion-exchangers in what is often called "reversed-phase" or "reversed-phase partition" chromatography. Actually, this term carries little meaning except within its historical background. Therefore, the term "extraction chromatography" introduced by Hulet (47) may be preferred as a brief and more descriptive expression. "Reversed-phase extraction chromatography" has also been recommended (17). This term pictures the intimate connection between chromatography and liquid - liquid extraction while it also indicates that it does not comprise systems where the extractant is the mobile phase. Therefore, it has been selected for the title of the present Chapter

A number of review articles on the subject has already appeared (3,4,13-15,17,26,27,39,51,56,82,86,87,94,99). Those by Cerrai (13-15) and Testa (94) mainly describe the work of their own groups; the present author's dissertation (4) summarizes the work published in refs. 6,8-12 and 90. All papers published up to the end of 1967 are covered in an excellent review by Cerrai and Ghersini (17). The bibliographies by Eschrich and Drent (26,27) also may be recommended since both the relevant titles and abstracts of the publications are given.

In the present Chapter, we have tried to cover all publications on the application of liquid anion-exchangers to inorganic analysis, to early 1970.

C. Abbreviations; Terms

Liquid anion-exchangers. The term "liquid anion-exchangers" is used for substituted quaternary ammonium salts, and for high-molecular-weight amines, both in the free-base and salt form.

Abbreviations used for the exchangers are summarized in Table I

Elements. Unless otherwise mentioned, the elements appear in the oxidation states as shown: Mn(II), Fe(III), Cu(II), Se(IV), Te(IV), Re(VII), Pt(IV), Au(III), Hg(II), and U(VI).

II. MATERIALS

A. Liquid Anion-Exchangers (see Table I)

Some 30 - 35 liquid anion-exchangers have been used for revers[e]-phase extraction chromatography, but only ten of these have become more or less popular.

In addition to the data in Table I, the following may be report[ed]

TABLE I

Characteristics and availability of some liquid anion-exchangers.

Exchanger	Abbreviation used in review	Class	Approx. molecular weight	Manufacturer*
Aliquat 336	Aliquat	quaternary	475	General Mills
Adogen 464	-	quaternary	431	Archer-Daniels-Midland
Tridodecylamine	TLA	tertiary	522	e.g., General Mills, Rhône-Poulenc
Tri-n-octylamine	TnOA	tertiary	353	e.g., K & K, Fluka
Tri-i-octylamine	TiOA	tertiary	353	e.g., K & K
Alamine 336	Alamine	tertiary	392	General Mills
Tri-n-butylamine	TBuA	tertiary	185	e.g., B.D.H.
Amberlite LA-1	LA-1	secondary	372	Rohm-Haas
Amberlite LA-2	LA-2	secondary	374	Rohm-Haas
Primene JM-T	Primene	primary	311	Rohm-Haas

*General Mills, Kankakee, Illinois
Archer-Daniels-Midland, Minneapolis, Minnesota
Rhône-Poulenc, Lyon, France
K & K Laboratories, Plainview, New York
Fluka, Buchs, Switzerland
B.D.H., Poole, England
Rohm & Haas, Philadelphia, Pennsylvania

Aliquat 336, obtainable as the chloride, is a methyltri-n-alkylammonium salt ("methyltricaprylylammonium chloride") with 27 - 33 carbon atoms on an average. The alkyl groups mainly consist of octyl and decyl chains. The commercial product is stated to have a minimum purity of 88%.

Adogen 464 is a methyl-tri-n-alkylammonium chloride with a minimum percentage of quaternary ammonium salt of 92%.

Tri-n-octylamine is by far the most widely studied liquid anion-exchanger. It is obtainable from many firms and is mostly used as a practical-grade product.

Tri-i-octylamine is a branched-chain amine, which principally contains 3,5-, 4,5- and 3,4-dimethylhexyl chains.

Alamine 336, also called tricaprylylamine, has three straight-chain alkyls, mainly octyls and decyls. The number of carbon atoms totals 26 - 32, and the minimum and typical tertiary amine content are 90% and 95%, respectively.

Tridodecylamine (trilaurylamine) is a viscous liquid with a melting point close to 16°C. The commercial product has a purity of approximately 99% (67).

Tri-n-butylamine has successfully been used in several studies, in spite of the high solubility of some of its salts in water.

Amberlite LA-1, a viscous yellow liquid (viscosity 72 cp at 25°C), is an N-dodecenyltrialkylmethylamine. The unsaturated amine contains 24 - 27 carbon atoms.

Amberlite LA-2 is a saturated secondary amine with a structure analogous to that of Amberlite LA-1. It has a much lower viscosity, and is slightly more basic.

Primene JM-T, a trialkylmethylamine with 18-24 carbon atoms, is the primary amine almost exclusively used.

The liquid exchangers generally have a specific gravity of approximately 0.80 at room temperature. A complete list of all exchangers not mentioned in Table I - together with the pertinent references - is given in Table II.

The undiluted liquid anion-exchangers must be handled with care (39): it is important to provide adequate ventilation to prevent inhalation of the vapors; moreover, contact with the eyes and prolonged contact with the skin should be avoided. As a rule, the technical-grade exchangers can be used without further purification. In a few instances, however, purification procedures have been described. Hulet (47) redistills TiOA at approximately 1.5 mm Hg pressure and selects the fraction that boils between 132°C and 142°C. Murray and Passarelli (59) report that LA-2 tends to become cloudy upon standing; either distillation at reduced pressure or simple filtration gives a satisfactory product with no discernible difference in behavior. TLA is purified (67) by treatment with an adsorbent, which preferentially removes the secondary and primary amines present. Cerrai and Ghersini (16) strip out most of the iron impurity contained in Aliquat by shaking a solution in benzene three times with an equal volume of 0.1 N HCl. Primene JM-T consists of a relatively large number of amines with large differences in molecular weight and boiling point. For this material, therefore, a rough purification by repreated vacuum distillation may be recommended: a nearly colorless and fairly homogeneous end-product remains (120). Alamine 336 may be substituted by Alamine 336-S, which contains over 99% of tertiary amine.

B. Supports

Table III lists the supports used in reversed-phase extraction chromatography; their trade names and manufacturers are given (when known). In thin-layer chromatography, cellulose MN 300 HR and silica gel DO are most widely used. Supports containing $CaSO_4$ ("G") are employed more rarely. Mild leaching of this binder by acid eluants produces no difficulties (59), but may be expected to have an effect on some detection procedures.

No general agreement exists concerning the support to be

TABLE II

Supplementary list of liquid anion-exchangers.

Exchanger	Reference
Quaternary	
Tetrahexylammonium iodide	8
Tridodecylmethylammoniumchloride	62
Hyamine 1622	8
Hyamine 10X	8
Hyamine 2389	8
Tertiary	
Adogen 364	8
Adogen 368	8, 9
Methyldioctylamine	48
Amberlite XE-204	48
Tribenzylamine	8
Cyclohexyldi-n-octylamine	48
Cyclohexyldidodecylamine	48
Tri-2-ethylhexylamine	48
Dodecylbenzyloctylamine	48
Tri-n-pentylamine	63, 65
Tri-n-hexylamine	63, 65
Tri-n-heptylamine	63, 65
Tri-n-nonylamine	63, 65
Cyclohexyldi-2-ethylhexylamine	48
Secondary	
Didodecylamine	4, 80
Di-2-ethylhexylamine	4
Primary	
Hexadecylamine	80

TABLE III
Supports used in reversed-phase extraction chromatography.

Support	Specification	Manufacturer
Thin layer		
Cellulose powder	MN 300 Hr	Macherey-Nagel
	MN 300 G	Macherey-Nagel
Silica gel	DO	Fluka
	MN-N-HR	Macherey-Nagel
	SG-41, Whatman	Balston
	G	Merck
Polymers	Corvic D55/3 (PVC)*	I.C.I.
	Corvic R51/83 (PVC-PVA copolymer)*	I.C.I.
Paper		
Paper	Whatman No. 1 (CRL/1)	Balston
	Whatman No. 4	Balston
	S & S 2043	Schleicher-Schüll
Column		
Cellulose powder	Whatman No. 1	Balston
Silica gel	chrom. grade	Schuchardt
Diatomaceous silica (kieselguhr)	Celite, Hyflo Super Cel	Johns-Manville
Activated carbon	Carbopol-H extra	-
Glass powder	Pyrex-G2	Pyrex
Polytetrafluoroethylene	Algoflon	Montecatini
	Fluoroplast-4	-
Polytrifluoromonochloro-ethylene	Kel-F	Minnesota Mining, Applied Sci., Analabs
	Hostaflon C2	Hoechst
	Voltalef 300 CHR	Rhône-Poulenc
	Plaskon CTFE2300	Allied Chem.
PVC-PVA*	Corvic	I.C.I.
PVC*	Vipla	Montecatini
Polyethylene	Microtene 710	US Ind. Chem.
	capillary	-
Polyisoprene	Moplen	Montecatini
Polystyrene-divinylbenzene	beads	Dow Chem.

*Abbreviations: PVC, polyvinylchloride; PVA, polyvinylacetate.

preferred for thin-layer work. Most authors (33, 59, 68) agree that with silica gel as a substrate small initial spots of solute are more easily formed, and slightly more discrete spots are obtained after development. They emphasize, however, that with cellulose powder and Corvic more robust layers are obtained, which have less tendency to flake off the plates. Silica gel surpasses many substrates when the dipping technique described by Brinkman *et al.* (10) is employed. Moreover, silica gel is not attacked by 12 N HCl, whereas with paper, cellulose powder and the resin-loaded Amberlite SB-2 papers the use of 8 - 10 N HCl represents the upper limit.

A large variety of products has been recommended for use in column chromatography, as is evident from the enumeration in Table III. The more popular ones are several polytrifluoromonochloroethylenes, notably Kel-F, and various types of kieselguhr. Since the products used by different authors often have been purchased from different manufacturers, vary in mesh size, and have been subjected to diverse pretreatments, only little can be said about their relative merits. A few observations are worth mentioning, however. Some authors (1, 54) prefer chromatographic-grade silica gel to activated carbon, cellulose or Kel-F powder. Cellulose is partially destroyed by 8N and stronger HCl, and the exchanger is removed by the acid solution and appears in the eluate.

Kel-F powder shows a higher retention for the exchanger, but the preparation of the column bed requires great care and the flow rate is rather low. In this connection, we may remark that Colvin (22) adds 40 vol. % of asbestos to Plaskon, another polytrifluoromonochloroethylene, to give porosity to the column and thus increases flow rate.

A comparison of 6 polymer support materials has been made by Testa (94), again because of various problems--high cost, but also poorly reproducible results--encountered with Kel-F. Preliminary data indicate, for example, that Algoflon and Moplen constitute good substituents for Kel-F. However, the latter product still is superior as far as retention of the exchanger and chemical stability are concerned. The other materials tested, Microtene, Corvic and Vipla, show poorer column performance, tailing being fairly excessive.

III. METHODS

A large number of methods has been described to produce a properly conditioned support impregnated with the liquid anion-exchanger, especially in the case of column chromatography. In the following sections, some typical and frequently applied techniques and experimental procedures are described.

A. Paper and Thin Layers

As an example, the use of exchanger-HCl systems will be discussed. Amines in solution are converted to their HCl salts by equilibration with an aqueous solution containing an excess of HCl. The organic solution is separated, filtered and/or dried. Quaternary ammonium chlorides are often equilibrated in the same way, in order to convert any free amine present into its HCl salt. Conversion to the salt prior to impregnation of the support is advantageous, as that causes a more regular development and a better defined solvent front. Amines are most commonly diluted with chloroform or benzene but other solvents may be chosen at will. This may be concluded from data by Ishimori et al. (48) who, using 12 different diluents, found the R_F values of Fe, Cu and Co not deviating more than 0.02 R_F units from the average value. Still, a volatile and "inert" diluent is preferable.

Thin layer. Chromatoplates are mainly prepared according to one of the two following techniques, illustrated with silica gel (10, 97) and cellulose (33) as the support, respectively.

(a) The organic solution of the impregnant is mixed thoroughly with silica gel ($SiO_2/CHCl_3$ 1.2/2;v/v), stored overnight and agitated again before use. Microscope slides are dipped into the resultant suspension. After evaporation of the diluent, superfluous material is wiped off the back of the slide, and a small margin is made along the edges. The plates are put on a frame and a 3.0-cm wide ruler is placed over them. A groove is made on the plate by moving a pencil along the upper edge of the ruler. Spots are applied by moving a pointed filter paper strip impregnated with the sample solution, vertically along the lower edge. The distance between groove and starting point is then exactly 3.0 cm. After development, drying of the chromatoplates and detection of the ions, the R_F values are measured with an appropriate R_F meter.

(b) A cellulose suspension is made by slurrying 15 g of cellulose with 70 ml of the organic solution of the impregnant. Chromatoplates are prepared at a fixed layer thickness using a Shandon apparatus; solvent (chloroform) is evaporated by air-drying. The plates are spotted and again air-dried and development is then carried out in a double saturation chamber. This consists of a saturation chamber inside, a tightly sealed polyethylene bag; a wick of glasswool is employed to ensure rapid saturation of the small space inside the inner chamber.

Paper (18). Paper strips are dipped into or slowly pulled through a solution of the anion-exchanger in benzene. After soaking the paper, the excess liquid is allowed to drip off the paper, and the remaining diluent is removed with hot air. The chromatographic procedure is simple and follows the usual practice. Apparatus for simultaneous ascending chromatography of 12 samples is described in ref. 13.

Some quantitative data have been reported (71, 74, 77) on the uptake of TnOA salts by Whatman No. 1 paper from 0.2 M solution in benzene. Typical results (in mg/cm^2) are: Cl^-, 0.60; Br^-, 0.73; I^-, 0.76; SCN^-, 0.84. These data correspond with an uptake of 1.2 - 2.0 $\mu mol/cm^2$. For the thiocyanate system, a similar result has been obtained with TBuA and TiOA (71).

B. Performance

Temperature. Paper and thin-layer chromatography are generally carried out at or close to ambient temperature, but this still implies a temperature range from 15°C to 30°C (65, 80). Using radial chromatography, Ishimori et al. (48) have studied the influence of the temperature on the chromatographic process. With increasing temperature, developing time decreases, but the effect is only small above approximately 20°C. The R_F values of 4 ions investigated also change with the temperature, the magnitude and even the direction of the effect, depending upon both the ion tested and the normality of the acid eluant. It is noteworthy that often relatively large variations occur between 10°C and 20°C.

Development. The development time for a fixed distance is strongly dependent upon the composition of the eluant and the amount of exchanger present on the support. In paper chromatography, 2 to 4 hours may represent an average time for acid eluants when the length of run is 10 cm to 15 cm and 0.05M to 0.20M impregnating solutions are used. Development has proved difficult only with certain Hyamines and tribenzylamine (8) (and hexadecylamine (80)). The irregular solvent fronts, and the elongated and badly formed spots encountered with these materials are probably due to the crystallization of the exchanger on the paper.

As far as speed is concerned, thin-layer chromatography is only slightly superior to paper chromatography: 1 to 2 hours are required for 12 cm to 15 cm travelling distance at not too high impregnating molarities (33, 38). There are definite advantages, however, for thin-layer work when the scale of the chromatogram is reduced to about the size of a microscope slide. Here, development times generally do not exceed 15 min; when using a radial technique, 2 to 3 min are even sufficient (8, 10).

The migration rate generally decreases with increasing normality of the acid eluant. For instance, in the 0.3 M Primene-cellulose system (33), the development time for a 14-cm run amounts to 1, 1.5, 2.5 and 5 hours when using 1, 3, 6 and 9 N HCl, respectively. Increasing the molarity of the impregnating solution also reduces the migration rate, but the effect is not too inconvenient up to 0.25 M to 0.30 M (8, 33). A particularly large increase of development time is observed when concentrated salt solutions are employed instead of acid eluants: with concentrated $LiNO_3$ solutions, for example, development may take up to 12 hours for a 15-cm run on paper (53).

Solvent systems containing organic components--rather frequently used nowadays in column chromatography with resin exchangers--cannot be employed with the liquid exchangers because they are stripped from the support by the organic component of the eluant (21).

Extractant loading. The R_F values of most metal ions, i.e., all those for which sorption is due to an anion-exchange mechanism (cf. section IV), may be significantly altered by varying the molarity of the impregnating solution, and thus, the extractant loading. Sorption increases with increasing molarity, as has been amply demonstrated for the Cl^- (8, 33, 59), Br^- (77), SCN^- (37) and NO_3^- (18, 93) system. As the only exception chromatography on paper treated with 0.1 M (16, 73) and 0.2 M (101) TnOA in benzene yields nearly identical R_F spectra; no explanation can as yet be offered for this result. In practice, the use of a high exchanger concentration is not recommended, in view of experimental inconveniences such as low migration rates (see above), diffuse spots and ragged solvent fronts (37). It is preferable to select a more strongly sorbing exchanger that can be employed at a relatively low loading. (However, see also ref. 34.)

For a discussion of the nature of the exchanger present on the loaded support, refer to section VII-C.

Reproducibility. Good reproducibility of the R_F values is generally claimed but pertinent data have only been collected by Graham and coworkers (33, 37). From results obtained for both cellulose-HCl (1-9 N) and cellulose-NH_4SCN (1-7 M) systems, it follows that in all cases the R_F values for the ions tested are reproducible to $\pm$ 0.02 of an R_F unit. In order to obtain such highly reproducible values, it is necessary to use a double saturation chamber (cf. section III-A), and to apply a suitable standard solution on every plate. It may be added here that other workers (8, 10, 90, 91) use simple closed jars such as Hellendahl staining jars, and omit chamber saturation. This will somewhat decrease the reproducibility, but on the other hand it increases both ease and speed of working.

Detection. The ions are detected using radioactive techniques or, more generally, by conventional chromogenic reagents. Useful summaries of visualization procedures may be found in refs. 10, 38, 90, 91 and 101.

C. Columns and Column Performance

As an example of column preparation, the technique described in ref. 43 has been chosen. The same paper gives details for the determination of column characteristics such as capacity and free-column volume, and the calculation of separation factors, distribution coefficients and HETP. A description of some conventional elution techniques and their application in separation analysis may be found in ref. 17.

Support material of the desired mesh size (dried and/or treated with dimethyldichlorosilane if required) is slurried with a solution of the exchanger, here a quaternary ammonium salt, in a suitable diluent, such as acetone, benzene or chloroform. The diluent is allowed to evaporate, and the column is filled by gently tamping the dried impregnated support into layers several millimeters thick, forming a loosely packed bed. After preconditioning with some bed volumes of eluant, the column is ready for use.

When amines are employed as exchangers, conversion of the free base to the appropriate salt form is done by equilibrating the amine with an aqueous acid solution prior to (19) or after (58), and even during (28) the introduction of the support into the column. With the latter method, a suspension consisting of support material, organic amine solution and aqueous acid solution is poured into the column. (For further details see the quoted papers.)

It is well known that column preparation is a critical step; an exhaustive discussion of this topic is given in ref. 17. Nevertheless, many authors mention the successful performance of exchanger-impregnated columns. The column packing material can be prepared in a reproducible form, resulting in a uniform bed density free from pressure zones and trapped air bubbles (46). The columns show excellent stability towards acid eluants, and are sometimes used for as many as 10 to 15 elutions with no apparent loss in exchange capacity (28, 44, 46).

Amine salt columns show slightly better stability than the free base columns (69), although the difference over a period of a few weeks is small. Watanabe (102) reports that TiOA-kieselguhr, stored in water for 2 years, is nearly equal to freshly prepared material in its ability for sorbing metal ions.

According to Ramaley and Holcombe (79), Aliquat - Kel-F introduced into the column as a dried powder, in the manner described above, tends to bleed and lose capacity with time, whereas this does not occur when the column is filled with an Aliquat - Kel-F slurry in chloroform, and the diluent is washed out with the desired eluant. A similar observation has been made for the Aliquat-kieselguhr system (62). On the other hand, Horwitz (43-45) and Smułek and Zelenay (85) successfully used dry-filled columns for the separation of adjacent pairs of transplutonium elements, and of Ni and Co, respectively.

The column material is also reasonably stable to radiation, and is, for instance, not substantially damaged after absorption of an accumulated dose of about 0.5 Wh (43). However, due to alpha-radiation, color changes and noticeable gas pockets occur which cause an increase in tailing.

Well defined elution curves, narrow peaks, and great selectivity range are among the advantages of the exchanger-impregnated columns. Exchange rates, moreover, generally are higher than those achieved with resins (19, 46). This may probably be explained by the fact that with the liquid exchangers, the active groups are fixed on the surface of an inert carrier and are thus more readily accessible for an exchange reaction with the aqueous phase. The more efficient utilization of the available exchange sites may also be the reason for the appreciably higher distribution coefficients obtained in reversed-phase extraction chromatography than are noted with resin exchangers in the rare earth-thiocyanate system (46).

A rather extensive study of the influence of several parameters on column performance has been made by Horwitz and his group (43, 45). Table IV shows the effect of increasing the capacity of the Aliquat-diatom material on HETP: the plate height decreases, i.e., column performance increases, with decreasing capacity, although the effect is small below 0.43 mM/g of column material. A similar effect has been reported for the lanthanide-HDEHP (hydrogen di(2-ethylhexyl) phosphoric acid, a liquid cation-exchanger) system; here, however, HETP remains essentially constant below 0.3 mM/g.

TABLE IV

The effect of column capacity on HETP for Am(III)$^{+}$

Capacity of column material ($mM\ NO_3^-$ /g material)	HETP (mm/plate)
0.11	0.40
0.27	0.45
0.31	0.49
0.43	0.53
0.58	0.96

$^{+}$Column: Aliquat nitrate-diatom earth; 0.61 ml bed vol.; 100 mm high; 100-300 mesh. Flow rate, 1 ml·cm^{-2}·min^{-1}. Room temperature (43). mM, millimoles.

The influence of temperature and mesh size on HETP and the separation factor $S.F._{Es/Cf}$ is given in Table V. The effect of mesh size again parallels that reported for the lanthanide-HDEHP system. The data show that the column performance of the coarse material is approximately the same at high temperature as that for the fine mesh size material at room temperature. However, from the standpoint of a practical Es/Cf separation, the decrease in HETP at 80°C is nullified by the decrease in separation factor. The Aliquat-diatom earth system is relatively insensitive to flow rate. Especially below 1.5 $ml \cdot cm^{-2} \cdot min^{-1}$ (quite contrary to the results known for liquid cation-exchanger systems), the flow rate does not influence the number of free column volumes to peak maximum (45).

The HETP values of approximately 4 mm, which may be calculated from the data of Sastri et al. (81), and which are higher than usual (0.5 - 1.5 mm), are probably also due to the use of a relatively high exchanger capacity and large grain size of the column material.

IV. THIN-LAYER AND PAPER CHROMATOGRAPHIC RESULTS

In order to evaluate the various aspects of thin-layer and paper chromatography, the literature data will be discussed with special regard to the following points: (a) the sorption sequence of the liquid anion-exchangers; (b) the plots of R_F values versus the molarity of the eluants, called R_F spectra, of the elements in selected exchanger-eluant systems; (c) the influence of variations in the experimental conditions on the R_F spectra.

A. The Sorption Sequence

From a systematic investigation of the sorption sequence for some 16 liquid anion-exchangers in the HCl system, two conclusions have been drawn (4, 8): 1.) Each of the ions investigated shows approximately the same behavior with all exchangers and, although obviously either very strong or very low sorption may lead to quite different R_F spectra, for most ions the analogy is indeed very pronounced. This conclusion is to be expected when the same mechanism governs the sorption process for a given ion for all exchangers. It has been amply confirmed for both HCl and other eluant systems (see, for example, refs. 9, 16 and 71). Particularly illustrative examples may be found in Figures 1-16 of ref. 8 and Figures 1-4 of ref. 9. 2.) The sorption sequence of the exchangers in HCl medium is: quaternary $\simeq$ tertiary > secondary > primary. To illustrate the appreciable difference in sorption strength between primary and secondary amines, it may be mentioned that the R_F values in the 0.3M Primene-cellulose system (34) still are distinctly higher than those for a 0.1 M LA-1 system (10).

The sequence given above also holds when using solutions of HNO_3, HF HBr, HI, HSCN, and some of their salts, as eluants. This is evident from a vast amount of data reported in the literature (9, 19, 48, 59, 72, 73, 80). Only two exceptions have been noted: methyldioctylamine falls in the secondary amine class (48) (probably due to the relatively high water-solubility of its salts) and hexadecylamine sometimes shows a remarkable and unexplained high sorption (80).

In summary, for all monobasic acids investigated, the sorption

TABLE V

The effect of temperature and mesh size on HETP and separation factor+

Temperature (°C)	Mesh size	HETP (mm/plate)	$S.F._{Es/Cf}$
24	100 - 300	0.63	1.46
41	100 - 300	0.48	1.41
80	100 - 300	0.35	1.29
24	200 - 400	0.37	1.46

+Column: Aliquat nitrate-diatom earth; 100 mm x 2.8 mm; capacity, 0.15 millimole NO_3^- /ml of bed. Eluant, 4.8 M $LiNO_3$ - 0.05 N HNO_3. Flow rate, 1 ml · cm^{-2} · min^{-1} (45). $S.F._{Es/Cf}$: separation factor ratio Einsteinium/Californium.

strength of the liquid anion-exchangers increases in the same order. Deviations from this sequence occur only seldomly, but a more subtle division cannot easily be introduced. TiOA, for instance, surpasses Aliquat in the HCl system, but the latter exchanger shows higher sorption strength with SCN^- and LiCl (8, 9, 59).

Little, if any, information is available regarding the relationship between the chain length and structure of the exchangers and their sorption strength. In studies on the nitrate and thiocyanate system, Pang and Liang (63, 65) do not find a distinct relationship between the number of C-atoms in the amines and the R_F values of six test ions, when using tri-n-alkylamines with 12 to 27 C-atoms. Their choice of ions, however, is somewhat unfortunate, since, for example, in the nitrate system four ions have $R_F \geqslant 0.88$ for five out of the six exchangers studied. Only a single conclusion may be drawn from their data, namely that the sorption strength increases in the order TBuA $<$ $(n\text{-}C_5H_{11})_3N$ to $(n\text{-}C_8H_{17})_3N \leqslant (n\text{-}C_9H_{19})_3N$.

Most surprisingly, Waksmundzki and Przeszlakowski (71, 100) report for SCN^- systems that TBuA has a somewhat higher sorption strength than both TnOA and TiOA!, notwithstanding the strong similarity between these amines. This sequence has been corroborated by work in our laboratory (98). The solubility of TBuA salts in aqueous solution probably plays a role here. In that case, the contradictory results obtained with TBuA-thiocyanate systems (63, 71) may be due to the presence of HNO_3 and HCl in the equilibration solution and eluant, respectively, employed by Pang and Liang. Since TBuA is a weak exchanger in the nitrate system (see above), and since the solubility of TBuA-HCl in aqueous solution is so high as to make reversed-phase extraction chromatography impossible (98), the added mineral acids will certainly decrease the sorption strength of the TBuA-SCN^- system.

Next to monobasic acids, only a single dibasic acid has been systematically investigated, namely H_2SO_4. With this acid, it has occasionally been reported that the extraction efficiency of anion-exchangers decreases in the order primary > secondary > tertiary. This result has been confirmed for various ions in chromatography (4,7; cf. section IV-E). Primene distinctly shows stronger sorption than LA-1, Alamine, Adogen 368, and the quaternary ammonium salt Aliquat. However, the sorption of these four exchangers is rather low and does not permit us to make a reliable further division.

It has been suggested (4, 106) that the reversal of the sequence is due to the fact that the metal-sulfate complexes present in the organic phase will, on the average, be more highly charged than, for example, the metal-chloro anions. This necessitates the attachment of a larger number of amine H^+ cations in the case of the sulfate complexes, thus favoring the sorption by means of primary amines over that involving more bulky secondary and tertiary exchangers.

B. Chloride Systems

In paper and thin-layer chromatography most attention has been devoted to the chloride system. A summary of all published data in this field is given in Table VI.

Thin-layer chromatography has been carried out (10) for some 50 to 60 elements, using silica gel as the support. Alamine, LA-1 and Primene were selected to represent strongly, moderately and weakly sorbing exchangers respectively. The data are compared both mutually and with R_F spectra for chromatography on silica gel treated with chloroform containing no anion-exchanger. The results are interpreted (4) by classifying the ions into three groups, which are characterized by the following predominant phenomena: A, no anion-exchange; B, anion-exchange; C, effects such as hydrolysis and precipitation (see Table VII). For all ions of groups B and C, the R_F spectra are shown in Figs. 1 and 2.

With group A, the formation of negative metal-chloro complexes is negligible and $R_F > 0.9$ is found either in the presence or the absence of an anion-exchanger. With the ions in group B, negative complexes are extracted by the exchangers at least in part of the HCl concentration range (cf. Equation 1, section I-A), which results in $0.0 \leqslant R_F < 0.9$. Group C includes the ions for which the R_F spectra found in the presence and absence of anion-exchangers show a close resemblance; however, in contrast with group A, $R_F < 0.9$ is found. This indicates that the role of anion-exchange is negligible: precipitation, hydrolysis and/or adsorption determine the form of the R_F curves, at least, to a considerable extent. Some of these phenomena are fairly sensitive to small differences in experimental conditions. This may explain why the literature data for the group-C ions show a distinctly inferior agreement, as compared with the data for groups A and B.

The striking similarity of the R_F spectra of metal ions obtained with different liquid anion-exchangers, has already been stressed in the previous section. A further point of interest relates to the effect of variation in the experimental conditions on the R_F spectra. Fortunately, data of manifold origin are known for Aliquat, TiOA, LA-1, Primene, and, especially, TnOA. Some illustrative examples concerning the latter exchanger (8, 16, 23, 73, 101) are given in Fig. 3. The good agreement observed here, holds irrespective of the ions

TABLE VI. Summary of paper and thin-layer chromatographic data for liquid anion-exchanger - Cl^- systems.[+,++]

Exchangers[+++] quaternary	tertiary	secondary	primary	various	Number of ions	Tlc /pc	Ref.	Remarks[++++]
		LA-1			12	t,p	6	
Ali	TnOA TiOA Ala	LA-1,-2	Pri	many,untr	13	t,p	8	
	Ala	LA-1	Pri	untr	55	t	10, 4	
Ado 464	Ala	LA-1			8	t,p	12, 4	LiCl
Ali	TnOA				6	p	16	
	TnOA				6	t	23	
			Pri		3	t	33	
			Pri	untr	58	t	34	
			Pri		.	t	35,36	densitom.determin.
	TnOA TLA	LA-1,-2	Pri	many	4	p	48	
	TiOA				.	t	55	val.
Ali	TiOA	-2			3	t	59	LiCl
	TiOA				3	t	68	val.
	TiOA				4	t	70	val.
	TnOA	-2			40	p	73	
	TnOA	-2			7	p	74	
Ali	TLA	DLA	$C_{16}NH_2$		38	p	80	val.
		LA-1			20	t	90	val.
		LA-1			15	t	91	val.
	TnOA				14	p	92	

TABLE VI (cont.)

TnOA	33	p 101	NaCl

+Solutions of HCl are employed as eluants in all cases. The use of salt solutions is indicated in the last column.

++The paper by Chang (108) does not record new experimental data, as is suggested in CA, 60 (1964) 8609d, but summarizes the results of ref. 92.

+++Abbreviations: Ali, Aliquat; Ado, Adogen; Ala, Alamine; DLA, dilaurylamine; Pri, Primene; untr, untreated.

++++"val" indicates that only single R_F values-- not R_F spectra --are given.

TABLE VII. Classification of ions according to their behavior in anion-exchanger - Cl^- systems (4, 10).

A		B				C
no anion-exchange		anion-exchange				precipitation, hydrolysis, etc.
Be	Mg	V(V)	Cr(VI)	Mn	Fe(II)	Ti
Al	p	Fe(III)	Co	Cu	Zn	Ge
Sc	V(IV)	Ga	As(III)	Se	Br^-	
Cr(III)	Ni	Mo	Ru	Rh	Pd	Zr
As(V)	Y	Ag	Cd	In	Sn(II)	Nb
La	Er	Sn(IV)	Sb(III)	Sb(V)	Te	Ba
Tl(I)	Th	I^-	W	Re	Ir	Hf
		Pt	Au	Hg(I)	Hg(II)	
		Tl(III)	Pb	Bi	U	

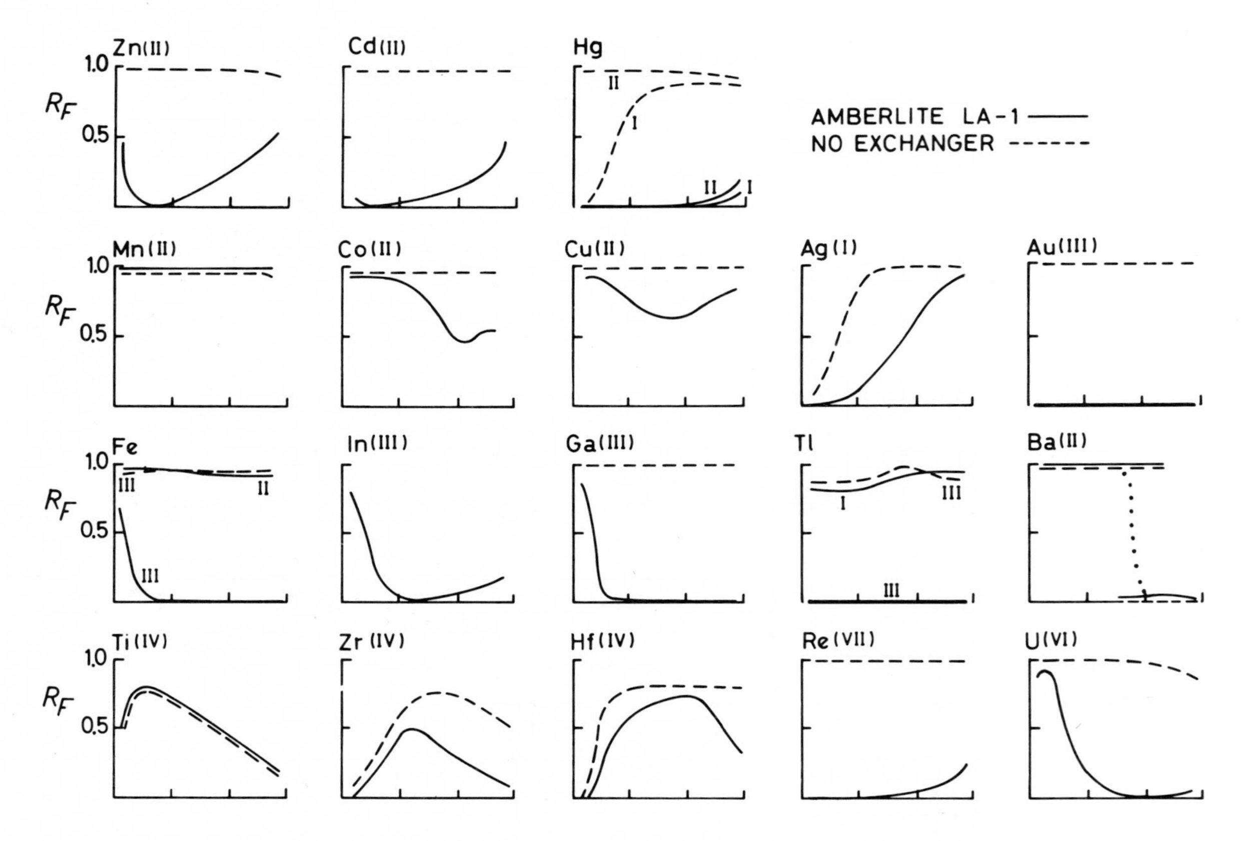

Figure 1. R_F vs. N HCl spectra for group-B and -C ions, using silica gel impregnated with 0.10 M LA-1 (——) and untreated silica gel (- - - -) (10).

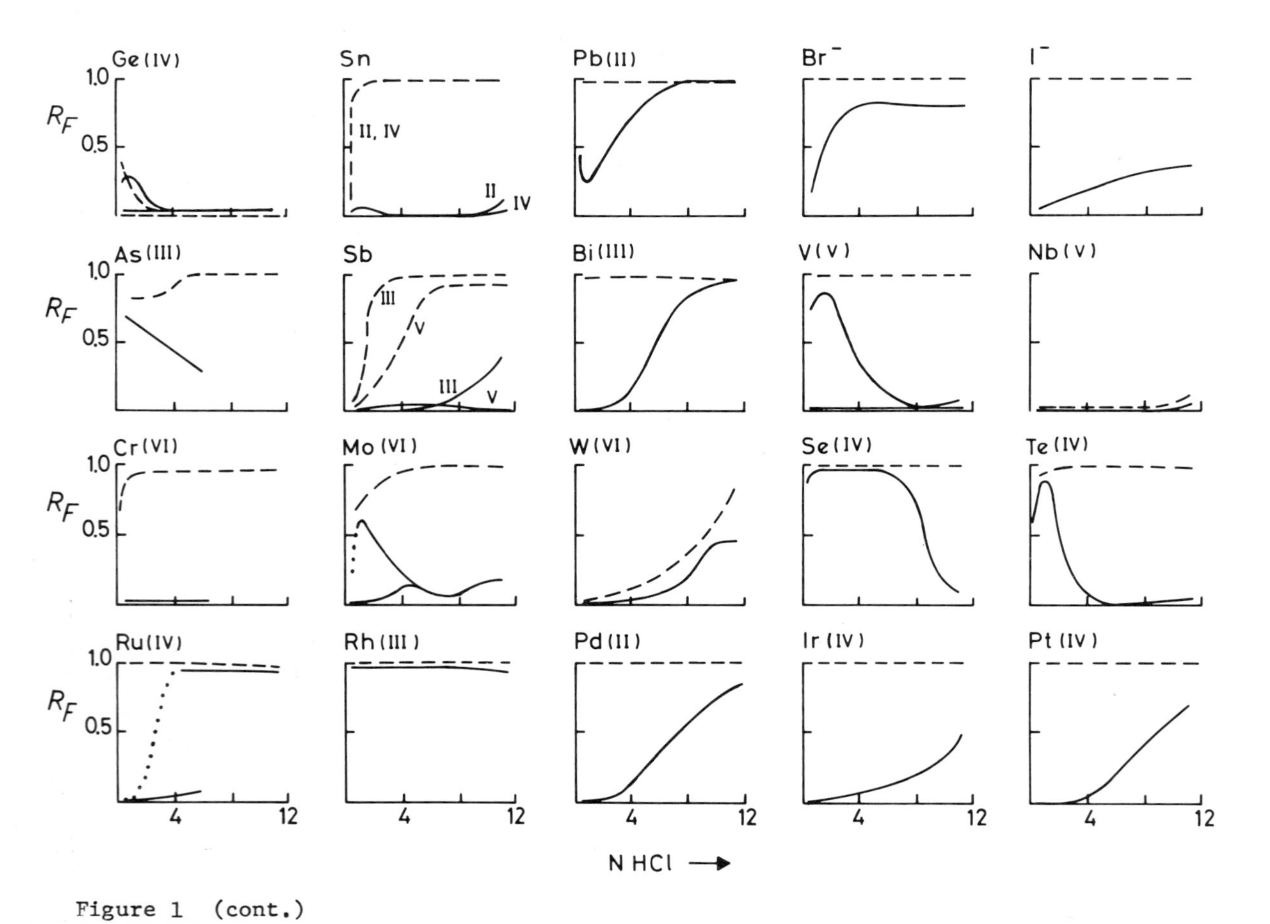

Figure 1 (cont.)

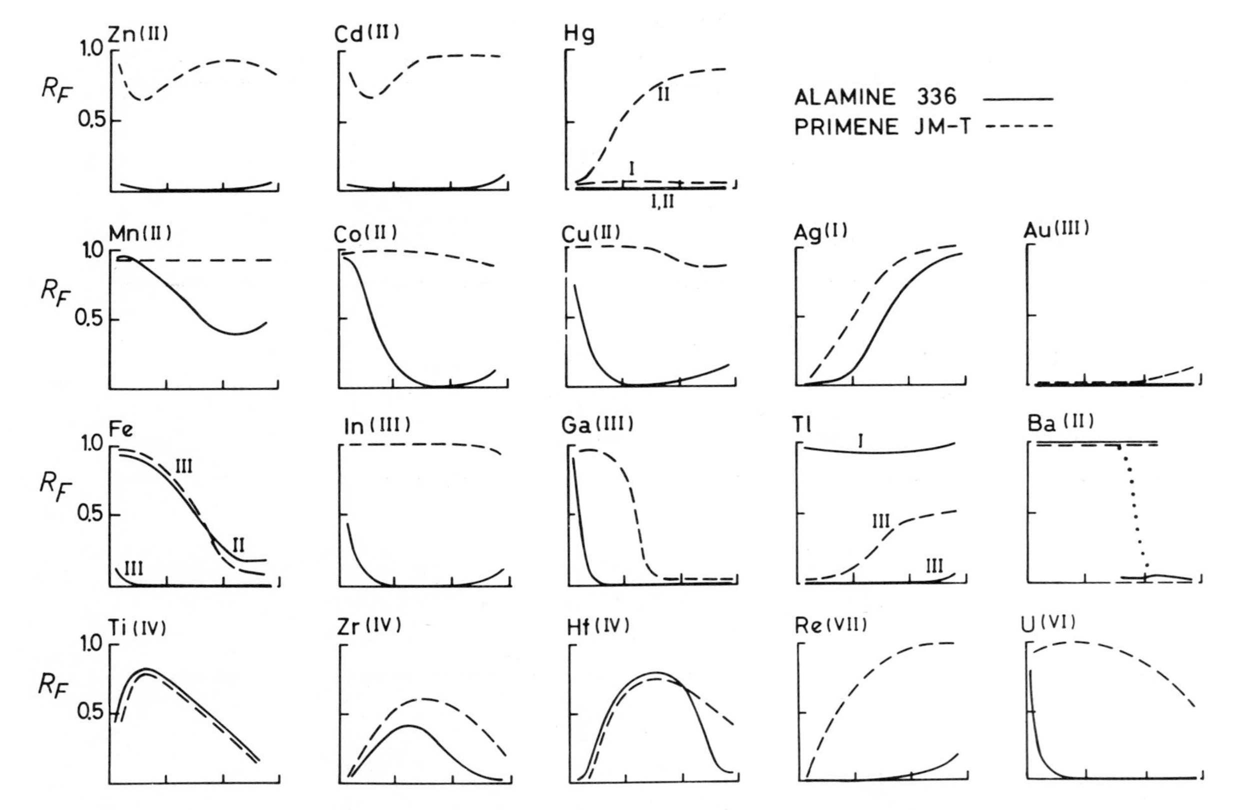

Figure 2. R_F vs. N HCl spectra for group-B and -C ions, using silica gel impregnated with 0.10 M Alamine (———) and 0.10 M Primene (- - - -) (10).

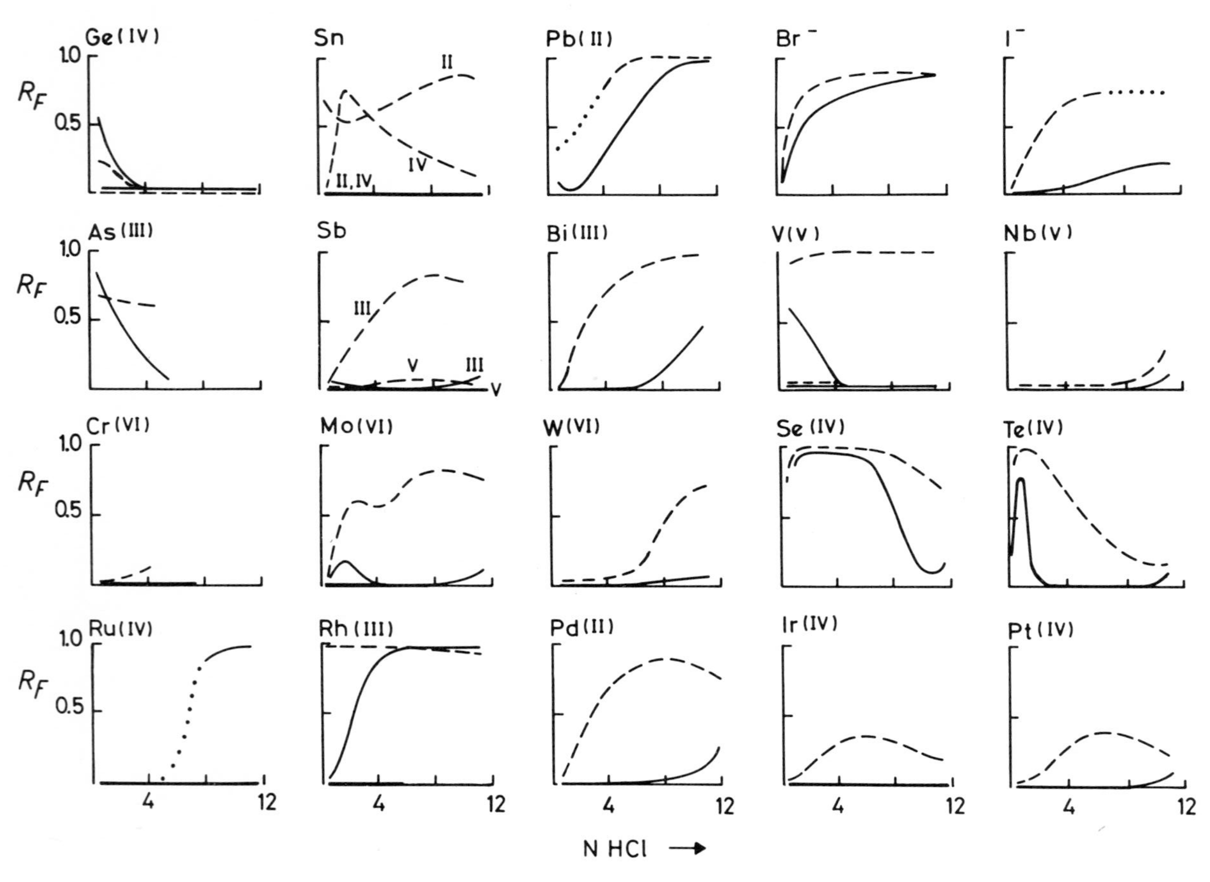

Figure 2 (cont.)

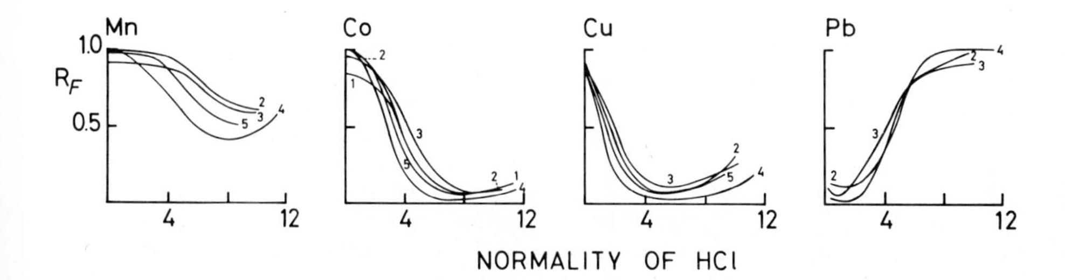

Curve Number	Molarity of TnOA	Diluent	Support	Reference
1	0.10	benzene	paper	16
2	0.20	benzene	paper	101
3	0.10	benzene	paper	73
4	0.10	chloroform	silica gel	8
5	0.10	benzene	cellulose	23

Figure 3. Comparison of literature data on the R_F spectra for ions in the TnOA-HCl system.

and/or exchanger selected for the comparison.

Two aspects are worthy of mention. When comparing data on Primene-impregnated and untreated silica gel (10) with those collected for Primene-impregnated and untreated cellulose (34), the agreement--though still being fair--is somewhat less satisfactory than with the other systems cited. Discrepancies occur especially above 5 to 6 N HCl. These are at least partially due to the fact that a large number of metal ions is significantly sorbed on untreated cellulose, whereas they move with the solvent front when employing untreated silica gel. Moreover, ions of such diverging character as Na and K, and Co, Mn and Cu (none of which are sorbed on untreated cellulose with 1 to 9 N HCl) surprisingly show very similar R_F spectra on Primene-cellulose for 5 to 9 N HCl. Their sorption increases approximately linearly with increasing acidity. It may be added that the behavior of untreated cellulose closely parallels that of Whatman no. 1 paper (121). Here a distinct difference between silica gel and cellulose-based support media turns up, which becomes manifest when sorption due to the exchanger is relatively small. It certainly must be borne in mind when interpreting the results for impregnated layers.

The support material also influences the behavior of the exchanger with which it is impregnated. Sorption is stronger for impregnated silica gel layers than for the corresponding paper system (8). On the other hand, the R_F values for silica gel are invariably higher than those obtained for a cellulose system (33), as has also been found when employing tri-n-butylphosphate (TBP) as the extractant (105). Two possible reasons have been suggested (37) to explain this phenomenon. First, a higher weight-to-volume ratio of support to impregnant exists in the case of silica gel compared with cellulose; this effectively reduces the loading of the former. Second, TBP has a greater tendency to associate with cellulose, probably by hydrogen bonding in the amorphous regions of the cellulose, than with silica gel; this results in a less effective uptake of the exchanger from the slurrying solvent by the silica gel during the preparation of the plates. These two effects, which are superimposable, possibly also explain the behavior observed for the liquid anion-exchangers.

LiCl; NaCl. Development with slightly acidified solutions of LiCl or NaCl instead of HCl has been carried out on both paper (12, 101) and thin layers (12, 59). Substitution of HCl by NaCl solutions (101) does not lead to significant changes in the R_F spectra, except for Sb(III). (The different curves found for this ion both here and in Br^- and I^- systems may well be due to hydrolysis effects occurring when employing the slightly acidic salt solutions as eluants). On the other hand, distinctly stronger sorption occurs when LiCl is used instead of HCl, the effect being more pronounced at high Cl^- concentrations (Figure 4). For the interpretation of the effect in terms of competitive sorption of HCl_2^-, the reader is referred to section VII-C. Concentrated LiCl solutions are only seldom used as eluants, since the ions are sorbed too strongly and development takes a rather long time (cf. section II-B).

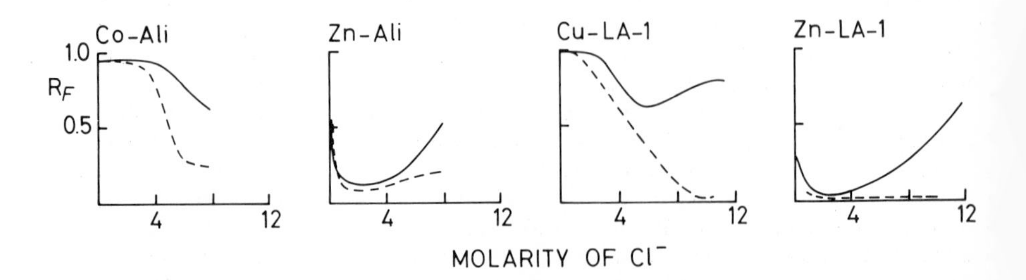

Figure 4. Comparison of R_F spectra for some ions, using HCl (———) and LiCl (- - - -) solutions as eluants. Systems: Ali, 2% Aliquat-cellulose (59); LA-1, 0.10 M LA-1 - silica gel (10).

C. Other Halogenide Systems

Bromide and Iodide systems. Only a few workers have devoted attention to the use of eluants containing Br^- and I^-. Paper chromatography has been carried out for approximately 40 ions using up to 6 to 7 M HBr, NaBr (pH=2), and NaI (pH=2) (73, 77). Two amines were selected, TnOA and LA-2. Secondary (LA-1), tertiary (TiOA, Adogen 368), and quaternary (Aliquat, Adogen 464) exchangers have been employed in thin-layer chromatography on silica gel (9). R_F spectra for some 10 ions are presented, using 0.5 to 8 N HBr and 0.5 to 7.5 N HI as eluants.

For both bromide and iodide systems, the agreement between thin-layer and paper chromatography is excellent, when comparing R_F spectra obtained on supports loaded with TiOA and TnOA, or LA-1 and LA-2 (Fig. 5). Again the R_F values on impregnated paper are higher than those on the corresponding thin layers. As is the case with chloride systems, the substitution of HBr by NaBr does not lead to significantly altered R_F spectra. Even though the experiments are carried out with up to 6 M NaBr, this result is not contrary to expectation, since it is known from liquid-liquid extraction data (107) that the uptake of HBr_2^- by liquid anion-exchangers occurs less readily than that of HCl_2^-. Comparative chromatographic data for iodide systems are lacking, but here the effect may be expected to be even smaller.

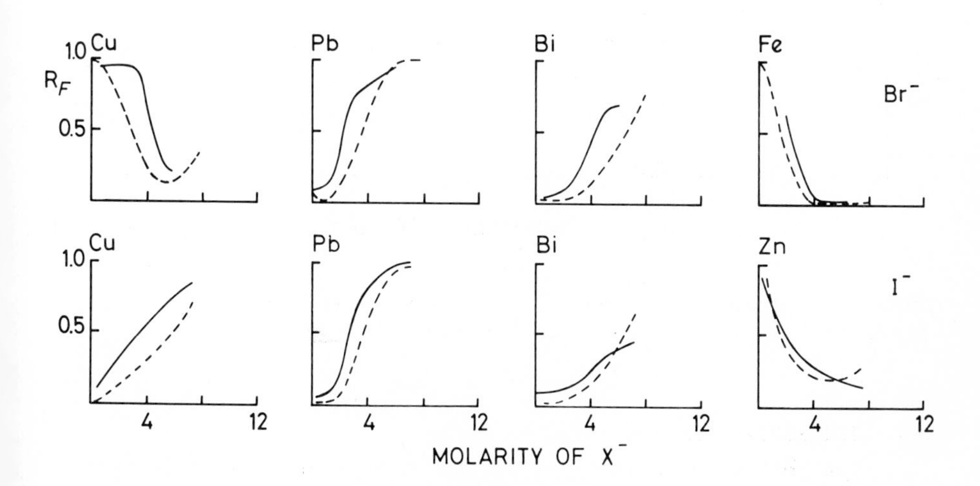

Figure 5. Comparison of literature data on R_F spectra for Br^-- and I^--containing eluants. ———, 0.10 M LA-2 - paper system (73); - - - -, 0.10 M LA-1 - silica gel system (9).

Although the amine-HI and aqueous HI solutions darken owing to oxidation of I^- to I_2, the impregnated thin layers are only slightly colored. Moreover, in detection procedures involving H_2S and NH_3, for example, a disappearance of the interfering color occurs due to these reductants. The I_2 - I^- complexes present in the eluant are held tightly by the impregnated support, forming a small zone at the lower end of the thin-layer plate. If, therefore, the spots are applied a few millimeters higher than usual, no further precautions are necessary with respect to the presence of I_2 in HI (9).

The R_F curves obtained for Cu and Fe in iodide systems, may be attributed to Cu(I) and Fe(II), the reductions Cu(II)→ Cu(I) and Fe(III)→ (Fe(II)) being fairly complete (9).

HF. Employing paper impregnated with TnOA and LA-2, Przeszlakowski (72) has carried out chromatography for 39 ions, using solutions of 0.012 to 13.4 N HF as eluants. Only 4 ions, Au, Pt, Ta and Re, stay at or close to the point of application over the whole HF concentration range. The remaining ions migrate with the solvent front or show an increase--never a decrease--in R_F with increasing acid concentration (Fig. 6).

Polyvinylchloride sheets coated with amine HF-loaded cellulose have been employed for thin-layer work (5). Alamine and LA-2, and some 25 ions have been studied. The results are closely analogous to those reported for the experiments on paper.

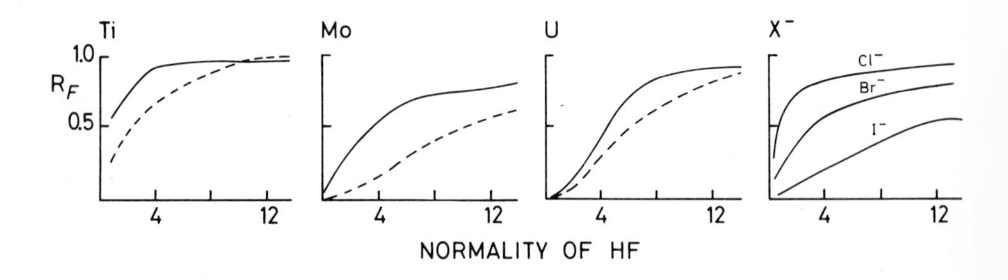

Figure 6. R_F vs. N HF spectra for 6 ions. Systems: 0.10 M TnOA - paper (- - - -), and 0.10 M LA-2 - paper (———) (72).

D. Thiocyanate Systems

Systematic investigations of the thiocyanate system have been repeatedly carried out. This may well be due to the fact that, whereas HBr and HI generally give results quite analogous to those obtained with HCl, more diverse characteristics are manifested by HSCN. Rather much attention has been devoted to the choice of suitable experimental conditions. The use of silica gel as a support leads to the occurrence of fairly irregular solvent fronts (9, 37). On cellulose layers, however, the solvent front is uniform except at very high molarities of the mobile phase. As another disadvantage of the silica gel, the presence of traces of iron causes the Fe(III) to be sorbed onto the thin layers as a red thiocyanate complex. The color interferes slightly with the detection of some metal ions. In paper chromatography, no special difficulties have been encountered.

Graham and Carr (37) have compared two methods used to convert amines to their thiocyanate salt form: (a) treatment of amine hydrochlorides with an aqueous solution of NH_4SCN, and (b) equilibration of the amine with a pure aqueous HSCN solution. Numerous data show that the R_F values, within the limits of experimental error, are largely independent of the method of preparation of the amine salt. Therefore, the simpler and more convenient method (a) may be preferred. On the basis of the good mutual analogy between the R_F spectra discussed below, it may be assumed that the use of thiocyanate solutions made by acidifying KSCN solutions with $HClO_4$ (9) or by mixing NH_4SCN and H_2SO_4 (71), yields useful alternatives.

R_F spectra have been collected for 6 to 40 ions on paper impregnated with LA-1, TiOA, Aliquat and the Adogens 368 and 464 (9), LA-2, TnOA, TiOA and TBuA (71, 73, 100) and tri-n-alkylamines having carbon chains from 4 to 9 C-atoms (63). Thin-layer chromatography has been carried out on cellulose--and incidentally silica gel--treated with LA-2 (37, 38) and LA-1 (9). Sorption again increases when substituting silica gel by cellulose (37). It is remarkable, though, that the effect is strongly dependent upon the metal ion tested.

The agreement between the data for the various exchangers is excellent (9, 71), and this also holds when comparing the results obtained for different exchanger-support systems (Fig. 7). Strictly speaking, this is somewhat surprising. Among the eluants employed are solutions of NH_4SCN, solutions of NH_4SCN acidified to pH=2 with H_2SO_4, and KSCN - $HClO_4$ mixtures of various composition. It is known (46, 71, 100) that the final acidity of the eluant has a pronounced effect on the R_F values of most metal ions (Table VIII). Evidently, the acidity does influence the R_F values, but the form of the curves remains essentially the same.

Graham and Carr have collected a vast amount of data on liquid anion-exchangers, representing primary (Primene), secondary (LA-2) and tertiary (Alamine) amines, and quaternary ammonium salts (Aliquat) (32). Unfortunately, at present only their results for LA-2 are available (38). By comparing the latter data with those in references 9, 71 and 73, and by slightly modifying the basis of classification proposed by Graham and Carr, have we been able to draw up the classification presented in Table IX. Ions for which the data are conflicting (Ta and Sb(III)), or as yet are too scarce, have been left out. Group A comprises the ions moving with the solvent front. Here, absence of anionic-complex formation may generally be assumed. However, it is known from the literature that Cr(III) forms a series of well defined anionic thiocyanate complexes. Presumably (38), $Cr(H_2O)_6^{3+}$ is, under the conditions of the system studied, kinetically inert, so that replacement of the water ligand with the thiocyanate ligand does not take place. For all ions summarized in group B, anion-exchange may be indicated as the predominant

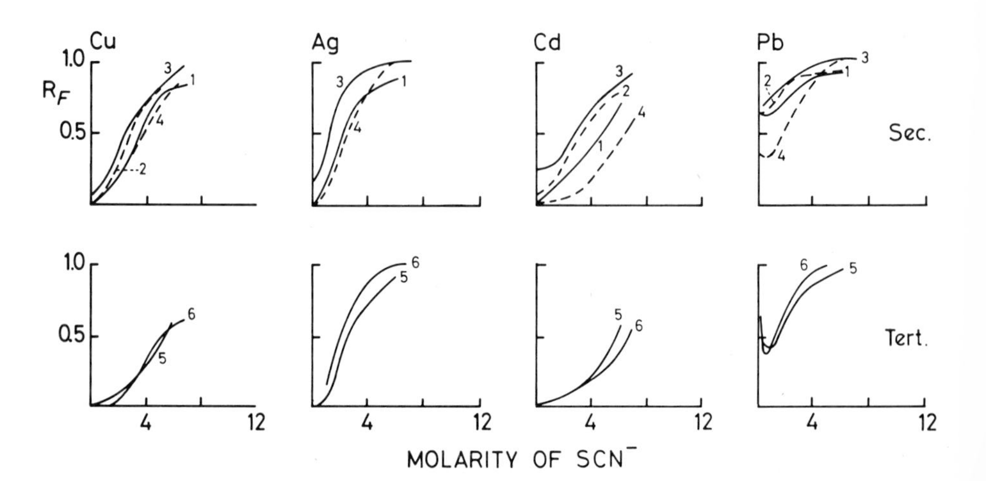

Figure 7. R_F spectra for 4 ions in the SCN^- system. Secondary amines: 1,———, 0.15 M LA-1 - paper (9); 2, - - - -, 0.10 M LA-1 - silica gel (9); 3,———, 0.10 M LA-2 - paper (73); 4, - - - -, 0.10 M LA-2 - cellulose (38). Tertiary amines: 5,———, 0.15 M TiOA - paper (9); 6,———, 0.20 M T(n or i)OA - paper (71).

TABLE VIII. Influence of the acidity of the eluant on R_F values in the SCN^- system.+

HCl (N)	R_F Ga	Cd	Hg	Bi	U
0.03	0.04	0.33	0.07	0.09	0.00
0.17	0.07	0.54	0.19	0.24	0.02
0.33	0.13	0.68	0.25	0.40	0.07
0.50	0.23	0.78	0.27	0.44	0.11
0.65	0.33	0.80	0.27	0.51	0.18

+Paper treated with 0.2 M TBuA in benzene; eluant: 6 M NH_4SCN acidified with HCl to final acidity as indicated (71, 100).

TABLE IX. Classification of ions according to their chromatographic behavior in anion-exchanger-thiocyanate systems (38).

A			B			
no anion-exchange			anion-exchange			
Mg	Ca	Be	Sc	Ti	V(IV)	Mn
Cr(III)	Ge	Fe	Co	Ni	Cu	Zn
As(III)	As(V)	Ga	Zr	Mo	Pd	Ag
Se	Sr	Cd	In	Sn(II)	Te	Hf
Y	Ba	W	Re	Au	Hg	Tl(I)
alkali's	rare earths	Pb	Bi	Th	U	

mechanism. Precipitation of insoluble thiocyanates and/or adsorption to the support--which lead to R_F=0.0 at all SCN^- concentrations--in most cases may be ruled out, as becomes obvious when comparing the data in ref. 38 with those in ref. 73. In the latter case, sorption appears to have been weaker, and almost all the ions show some migration at high SCN^- molarities. More definite proof no doubt can be obtained by collecting data for the weaker exchanger, Primene, and/or for cellulose not treated with an anion-exchanger.

E. Nitrate Systems

With HNO_3-containing eluants, high R_F values ($>$ 0.7) are consistently found for the majority of the elements studied, both in paper (80) and thin layer (4, 9, 98) chromatography. Some more interesting R_F spectra are given in Fig. 8, which shows the analogy between paper and thin-layer data to be good. This holds (98) for nearly all ions mentioned in ref. 80. However, in a few cases, namely Sb(III), Zr and Hf, high R_F values in paper chromatography contrast with low ones in thin-layer work. It has been suggested (9) that a specific interaction between these ions and the silica gel is responsible for the sorption.

As is the case in the chloride system, the substitution of HNO_3 by one of its salts often brings about an increase of the sorption. This has first been demonstrated by Cerrai and Testa (18) for La, U and Th. The use of $LiNO_3$, $NaNO_3$, NH_4NO_3, $Ca(NO_3)_2$ and $Al(NO_3)_3$ instead of HNO_3 appreciably lowers the R_F values of the metal ions; besides, an increase in the R_F values above approximately 5 M NO_3^-, which occurs with HNO_3 for U and Th, is never observed. Similar results have been obtained by Knoch *et al.* (52, 53) for several actinides and their fission products. A few illustrative data are presented in Table X.

No definite conclusions can be given regarding the magnitude of the salt effect. For example (cf. Table X), the substitution of $LiNO_3$ by $Al(NO_3)_3$ significantly increases the sorption of La in one paper (18), whereas it does not have a distinct effect in another (53).

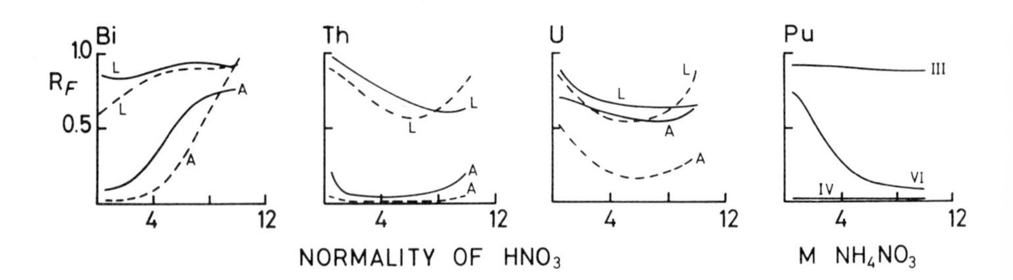

Figure 8. R_F spectra for some ions, using HNO_3 (U, Th and Bi) and NH_4NO_3 (Pu) as eluants. U, Th and Bi: A, ———, 0.05 M Aliquat-paper (80); A, - - - -, 0.10 M Aliquat-silica gel (9); L, ———, 0.15 M LA-1 - paper (9); L, - - - -, 0.10 M LA-1 - silica gel (9). Pu (in different oxidation states): 0.10 M TnOA - paper (53).

Secondly, the substitution of $LiNO_3$ by NH_4NO_3 increases the R_F values for U, but decreases those for La (18).

Interesting results have been obtained for Np and Pu (53). The various oxidation states of these two actinides--Np(IV-V-VI) and Pu(III-IV-VI)--are identified by their R_F spectra in the TnOA-NH_4NO_3 system (Fig. 8) and good separation procedures are outlined. The authors emphasize that with Np the eluant should not contain more than 1 N HNO_3 to avoid oxidation and disproportionation. This implies that R_F spectra for Np for 0 to 10 N HNO_3--as also given in their paper--must be regarded with some caution. A similar remark may be made regarding Pa(V), where R_F values of 1.0 and 0.0 are reported for the same eluant; this is probably due to the presence of fluoride ions in the Pa solution in the former case.

Employing $LiNO_3$ solutions, Testa (93) has studied the behavior of various rare earths and related elements on paper treated with TnOA. The R_F values increase from the lighter to the heavier rare earths, in accordance with the behavior on resin exchangers and in liquid-liquid extraction; the mean separation factor, with a value of 1.45 for two adjacent elements, is very similar to that found for the resin Dowex-1. The R_F values for the rare earths decrease appreciably with increasing molarity of the $LiNO_3$ solution (Table XI). Y and Sc behave as the heavier lanthanides; Th has R_F=0.0 at all $LiNO_3$ concentrations. Similar results have been obtained by Pang and Liang (64, 65).

A strict comparison of the data cannot be made, since different impregnant molarities and development techniques have been used.

TABLE X

Effect of nature and concentration of the eluant on R_F values in the NO_3^- system+

Eluant	Ref.	R_F values at an NO_3^- molarity of:				
		1	2	4	6	8
Uranium						
HNO_3	18	0.76	0.69	0.58	0.57	0.62
	53	0.80	0.68	0.54	0.55	0.60
$LiNO_3$	18	0.26	0.23	0.18	0.12	-
	53	0.54	0.33	0.10	0.08	0.02
NH_4NO_3	18	0.59	0.51	0.35	0.24	0.17
	53	-	-	-	-	-
$Al(NO_3)_3$	18	0.67	0.39	0.16	0.08	-
	53	0.63	0.42	0.20	0.09	-
Lanthanum						
HNO_3	18/53	higher than 0.85 at all concentrations				
$LiNO_3$	18	0.90	0.89	0.88	0.87	-
	53	0.93	0.84	0.80	0.73	0.63
NH_4NO_3	18	0.91	0.86	0.74	0.66	0.36
	53	-	-	-	-	-
$Al(NO_3)_3$	18	0.88	0.68	0.36	0.25	-
	53	0.83	0.77	0.72	0.70	-

+Ref. 18: Whatman No. 1 paper treated with 0.1 M TnOA in benzene; 0.005 N HNO_3 in eluants; radial development. Ref. 53: S & S 2043b paper treated with 0.1 M TnOA in xylene; 0.5 N HNO_3 in eluants; ascending development.

TABLE XI. R_F values of some rare earths as a function of $LiNO_3$ molarity+

Rare earth	R_F value at $LiNO_3$ molarity of: 2	4	6	10
La	0.33	0.05	0.02	0.00
Sm	0.86	0.42	0.06	0.00
Dy	0.91	0.83	0.13	0.00
Yb	0.91	0.91	0.61	0.03

+Paper treated with 0.2 M TnOA in benzene; Whatman CRL/1 paper; ascending development (93).

Moreover, the final acidity of the eluant--which has an effect on the R_F values similar to that observed for the SCN^- system (18, 64, 65, 93)--is not the same in all cases. According to Pang and Liang, the acidity of the impregnant solution also influences the R_F values. For instance, with 0.2 M TnOA, equilibration with at least 0.5 N HNO_3 is required; with lower concentrations, the spots of the elements investigated become elongated and even show tailing. Equilibration with 4 N instead of 0.5 N HNO_3 causes an increase in the R_F of 0.06 to 0.12. Since amines easily take up nitric acid above an amine: HNO_3=1:1 mole ratio (107), the increasing R_F values may well be due to the transport of part of the excess acid from the sorbed amine-HNO_3 to the acid-deficient mobile phase.

F. Miscellaneous

$HClO_4$. In reversed-phase extraction chromatography with perchloric acid 0 to 10 N on thin layers, or 0 to 8 N on paper (since 10 N acid clogs the paper), the majority of the elements investigated (9) consistently give R_F values of 0.85 to 1.0. Sorption and tailing occur with Sb(III), Pb and Hg. The R_F spectra for these three ions hardly change if, instead of paper impregnated with an anion-exchanger, untreated paper is used. Owing to the absence of anionic-complex formation, other factors, e.g., hydrolysis and adsorption, may determine the R_F picture, giving rise to much the same curves, whether exchangers are present or absent (9). This conclusion parallels that drawn for group-C elements in the HCl system (cf. section IV-B).

H_2SO_4. Polybasic acids have not been employed as yet, with the exception of H_2SO_4. However, even for this acid only a few data (55, 92, 103) have been reported, despite the rather frequent use of sulfuric acid in liquid-liquid extractions. This lack of data may be partly due to experimental inconveniences, such as troublesome detection of some ions and streaking of the spots (7) or difficulties in handling paper in contact with concentrated aqueous H_2SO_4 solutions.

However, the results recently obtained in thin-layer chromatography (7) certainly are of interest, since for many ions the normal sorption sequence is reversed. The phenomenon is particularly pronounced with Sc, Fe, Zr, Ce(III) and CE(IV), Hf, Bi, Th and U(IV).

However, no reversal of the sorption sequence occurs with V(V), U(VI) and Mo. These data, some of which are pictured in Fig. 9, confirm the results obtained with liquid-liquid extraction (111, 117). For the hypothesis forwarded to explain the results, one is referred to section IV-A.

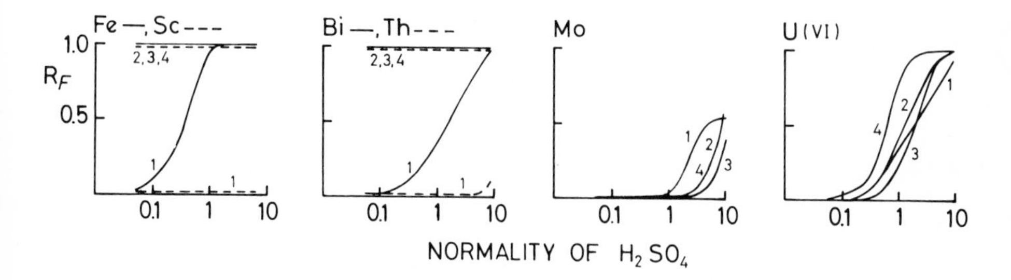

Figure 9. R_F vs. N H_2SO_4 spectra for 6 ions, using silica gel impregnated with 0.10 M exchanger solutions. 1, Primene; 2, LA-1; 3, Alamine; 4, Aliquat (7).

<u>Organic acids</u>. The use of organic acids as eluants has so far only been described by Przeszlakowski *et al*. (78). Papers are impregnated with TnOA or LA-2, and aqueous solutions of 0.05 to 6 N HCOOH, CH_3COOH, and $CH_2ClCOOH$ serve as mobile phases. R_F spectra for 4 ions are pictured in Fig. 10. Out of the 33 elements investigated, only U, Se, and (not drawn here) As(V) show minima in their

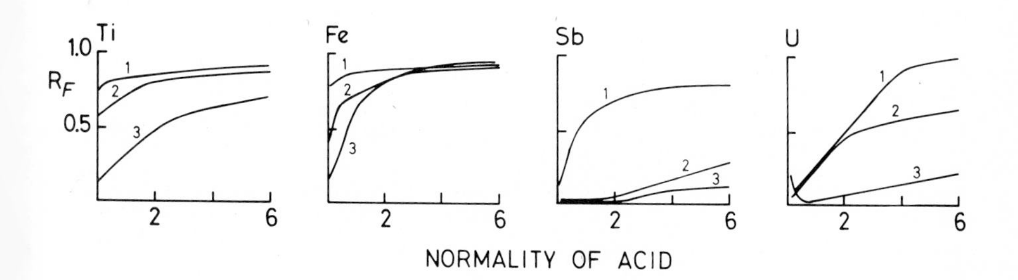

Figure 10. R_F spectra for 4 ions, using paper impregnated with 0.10 M TnOA. Eluants: 1, $CH_2ClCOOH$; 2, HCOOH; 3, CH_3COOH (78).

R_F spectra, which moreover merely occur with acetic acid as the eluant. The R_F values for all ions decrease in the order $CH_2ClCOOH >$ $HCOOH > CH_3COOH$, which is also the order of decreasing ion-exchange affinity of these acids towards anion-exchange resins. The substitution of TnOA by LA-2, investigated for the $CH_2ClCOOH$ system, causes small and insignificant changes in the R_F values only. Further research will be necessary before definite conclusions can be drawn regarding the prevailing mechanism in the liquid anion-exchanger-organic acids systems.

V. SEPARATIONS ON PAPER AND THIN LAYERS

A. Qualitative Analysis

Approximately 30 papers concern themselves with paper or thin-layer chromatographic separations. The exchangers mostly employed are TnOA, TiOA, LA-1 and LA-2 and Primene. Obviously, especially with the two tertiary amines mentioned, a large number of good substitutes is available. Solutions of chlorides and nitrates are by far the most popular eluants.

A concise survey of all papers recording separations is given in Table XII. As a rule, the feasibility of a separation can be read directly from R_F spectra such as have been given in section IV. Therefore, only a few selected results are described below. Some examples of chromatograms are presented in Fig. 11.

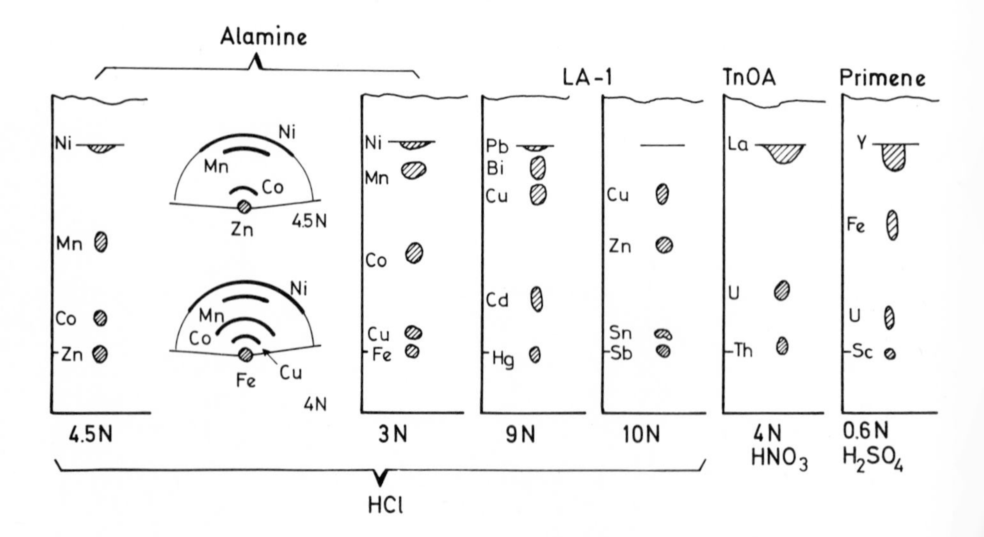

Figure 11. Examples of separations obtained on silica gel impregnated with liquid anion-exchangers. Systems as indicated (4, 7, 10).

TABLE XII

Survey of paper and thin-layer chromatographic separations.

Separations	Exchangers[+]			Paper/ thin l.	References
Cl^-					
Ni-Co-Cu-Fe; Ni-Cu-Zn; Al-Zn-Fe; (Pb, Cd)-Bi-Hg		LA-1		t	6
17 sepns. of 23 elements; e.g. Al-U-Ti-Fe; Ni-Co-Mn-Au-Fe; Hg-Cd-Cu-Bi-Pb; As-Bi-Sb	Ala	LA-1	Pri	t	10.4
Zr-Hf; Ni-Co-Fe	TnOA			p	16
8 sepns.; rare earths + related	TnOA	(+HDEHP)		p	20
Zn-Cu-Ni-Co-Mn; Zn-Fe	TnOA			t	23
15 sepns. of 25 elements; e.g. Zn-Cd-Hg; Pb-Bi-Sn-Cd; Au-Ag-Cu; Ba-Sr-Ca-Mg			Pri	t	34
Fe-Co-Cu	TnOA			p	48
6 2- and 3-component sepns.; 9 elements	TiOA			t	55
Zn-Co-Mn	TiOA			t	68
Zn-Co; Cd-In	TiOA			t	70
9 sepns. of 25 elements; e.g. Tl-In-Ga; Ir-Pt-Pd-Ru; Te-Re-Mo-Se; Fe-In-Zn-Al		LA-2		p	73
30 sulfidic ores		LA-1		t	90.4
23 non-sulfidic ores		LA-1		t	91
10 sepns. of 14 elements; e.g. U-Zr-Th; Zn-Co-Mn-Ni; Zr-Hf	TnOA			p	92
15 sepns. of 32 elements; e.g. Te-Se; Hg-Cd-Cu-Bi-Ag-Pb; Zn-In-Ga	TnOA			p	101

TABLE XII (cont.)

Separation	Extractant		Type	Ref.
NO_3^-				
U-Th-La	TnOA		t	4
U-Th-La	TnOA		p	18
U-Th-Yb-La	TnOA (+HDEHP)		p	20
6 sepns.; actinides + related	TnOA		p	52,53
Fe-Mo-U	TiOA		t	55
U-La-(rare earth, Y)	TnOA		p	65
10 sepns.; rare earths + Related	TnOA		p	93
SCN^-				
Am-Tm-Eu	Ali		p	2
Zr-(U, Th); (Zr, U)-Th-(rare earth, Y)	TnOA, TBuA		p	63
18 sepns. of 26 elements; e.g. As-Sb-Sn; Fe-Al-Mn; Sc-Th-La; Sn-Bi-Cd-As	TnOA, TBuA		p	71
SO_4^{2-}				
6 sepns.; e.g. Th-U-Ce; Re-Mo-V; Sc-U-Ce-Y		Pri	t	5
Mo-(Fe,U)-Cu-Ni; U-Fe	TiOA		t	55
(U,Mo)-Fe-Cu-Ni	TnOA		p	92
Zr-Hf	TnOA		p	103
Miscellaneous				
11 sepns. of 16 elements; e.g. Re-W-Mo; I^--Br^--Cl^-; Mo-U-V-Re	TnOA	(F^-)	p	72
16 sepns. of 20 elements; e.g. Zn-Co-Mn; Ag-Pb-Hg-Sb; Ge-As-Pb-(Sb, Sn)	TnOA	(Br^-, I^-)	p	77
6 sepns.; e.g. Hg-Bi-Fe-Co; Mo-U-La; W-V-Ga	TnOA	(HCOOH)	p	78

+Abbreviations: Ala, Alamine; Ali, Aliquat; Pri, Primene.

H_2S separation scheme. Ag, Pb and Tl(I) may be separated on LA-1-treated supports, using 0.5 to 1 N HCl (98). A separation of Ag, Pb, Hg and Sb(III) has been achieved in the system TnOA - 7M NaI (77).

The As-subgroup (As, Sb, Sn) yields to separation in the TBuA - 7 M NH_4SCN system (71), while separation of Ge, As and Pb from each other and from Sb or Sn occurs on TnOA-impregnated paper with 4.5 or 5 N HBr as the eluant (77). The analysis of the Cu-subgroup (Hg, Cu, Cd, Pb, Bi) may be carried out on LA-1-impregnated silica gel, using approximately 9 N HCl (10). A separation of Ag and the five elements of the Cu-subgroup has also been reported (101).

No complete separation has as yet been elaborated for the NH_3 group. However, successful analyses of mixtures of Al - U - Ti - Fe, Al - U - Fe - Zr (HCl;10), and Fe - Al - Mn (NH_4SCN;71) have been reported.

The $(NH_4)_2S$ group has been subjected to many investigations. A very sharp separation of Ni, Co, Mn and Zn occurs with a strongly sorbing exchanger, employing 3 to 5 N HCl as the eluant (10, 23, 92). The analysis of the well known mixture Ni - Co - Mn - Cu - Fe is successful in the Alamine - 3 N HCl system, both with ascending and radial methods (10).

The ions of the carbonate group yield to separation on a cellulose layer in the 0.3 M Primene - 8 N HCl system. The spots of Mg and Ca are contiguous; the separation of Ca from Sr and Ba, which tails at the start, is clear-cut (34).

Lanthanides and actinides. Solutions of HNO_3 (4, 18) and NH_4SCN (63) may be used to separate U, Th and La or Ce on paper treated with TnOA; a thin-layer separation has been reported for the Primene-H_2SO_4 system (7). Chromatography with $LiNO_3$ solutions on TnOA-loaded paper has been employed for the separation of Cm(III), Am(III) and Pu(III) of Pu(IV), Th, U, Np(V) and Am(III) (52) and of various 3- and 4-component mixtures such as Th - Ce - Nd - Gd, Th - U - La - Sc, Pr - Sm - Dy and Ce - Sm - Ho (93). Knoch and coworkers (53) have obtained sharp separations of the various oxidation states of Np (IV-V-VI) and of Pu(III-IV-VI), when using acidified NH_4NO_3 solutions as eluants (cf. Fig. 8). NH_4NO_3 has also been used to separate U, La and rare earths (65). An ingenious separation technique has been worked out by Cerrai and Testa (20), who treat their paper for a small part with TnOA, and impregnate the remaining surface with the cation-exchanger HDEHP. Using HCl solutions of varying normality, they separate U - Sc - Yb - Y - La and U - Th - Yb - Gd.

Minerals and alloys. A systematic investigation of the analysis of both sulfidic and non-sulfidic ores has been carried out by Sijperda and De Vries (90, 91: cf. 4). Chromatography is carried out on LA-1-impregnated silica gel in all cases. Three eluants are employed, 2, 6 and 10 N HCl. This choice offers a favorable combination of separations for the elements for the 40 minerals concerned. It is interesting to note that the R_F values found by the authors do not deviate significantly from those predicted from the R_F spectra, in spite of the presence of HNO_3, HNO_2 , suspended sulfur particles and possibly H_2SO_4. For mixtures containing elements such as Mo, U, V, W, Re and Fe, various separations have been elaborated using solutions of HF (72) and occasionally H_2SO_4 (7) as eluants. Interesting separations of Ir, Pt, Pd and Ru (LA-2 - HCl on paper; 73) and of Y, Fe, U and Sc (Primene-H_2SO_4 on SiO_2; 7) have also been reported.

Pb, Bi, Sn(IV) and Cd, the components of Wood's metal, have been separated on Primene-treated cellulose powder with 5 to 8 N HCl (34). Au, Ag and Cu, the main constituents of plated work, can easily be separated with HCl-containing eluants (10, 34). An excellent separation of Cu, Ni and Zn, present in various kinds of German silver, has been carried out with LA-1 - 6 N HCl (6).

Miscellaneous. Zr and Hf have been separated on TnOA-loaded paper, both with 8 N HCl + 5% conc. HNO_3 (92), 9 N HCl (16) and 2 N H_2SO_4 (103). In the latter paper, optimum separation conditions have been elaborated in regard to the molarity of TnOA in the impregnant solution.

Mixtures of anions have hardly been investigated. The only major exception is a separation of Cl^-, Br^-, and I^- with 10 N HF on TnOA-impregnated paper (72).

B. Quantitative Analysis

In an early paper, Testa (92) has briefly discussed the quantitative separations of a few μg of iron from mg quantities of nickel by filtering a sample solution acidified with HCl through a filter paper treated with TnOA. Pang and Liang (64, 65) have devised a procedure for the isolation and quantitative determination of 1 μg of Sm, Eu or Gd from 1 mg of La. After chromatography with NH_4NO_3 on TnOA-treated paper, the rare earth spot is cut out and the element determined spectrophotometrically.

Recently, a thorough investigation (35) has been made of some of the parameters involved in the direct densitometric determination of Zn on cellulose layers impregnated with Primene-HCl. The zinc spots are located by direct spraying of a chromogenic reagent onto the surface of the layers. The major source of error appears to be the unevenness of the application of the reagent. In spite of this, an accuracy of about 5% can be achieved at the 1-μg level. A spot-removal technique used in conjunction with absorptiometric analysis is comparable in accuracy with the direct densitometric technique (36). The latter, however, permits a larger range of metal ions to be investigated than does the former; it is also quicker. No loss of accuracy occurs when mixtures are investigated, as has been demonstrated for Zn - Cd (1 μg - 1 μg) and Fe - Co (99 μg - 1 μg).

VI. COLUMN CHROMATOGRAPHY

Some 40 papers have been published on reversed-phase extraction column chromatography. Part of these publications deals mainly with the separations of actinides and lanthanides, while the remainder is dedicated to light transition metals. Therefore, it seems appropriate to deal with the two topics in separate sections.

Distribution coefficients have been calculated from column chromatographic data and batch experiments for the following systems TiOA - HCl (30 elements; 69), Aliquat nitrate (rare earths, actinides; 43, 45, 46) and chloride (Fe, Co; 59), and TnOA - HCl (9 elements; 25, cf. 24). The data are used to predict the feasibility of separations (cf. below) and to investigate aspects such as the analogy between column chromatography and liquid-liquid extraction (cf. section VII-A).

A. Transition Metals

In one of the first papers on column chromatography, Cerrai and Testa (19) describe the separation of Ni, Co and Fe on TnOA-treated cellulose. Ni is removed from the bed with 8 N HCl; after its collection, Co is eluted with 3 N HCl, and Fe with 0.2 N HNO_3. Analogous separations have been recorded for Aliquat-cellulose (59, 81) and TnOA-Hyflo Super Cel (83) columns. Optimal separation conditions have been evaluated (83) and applied (85) in the separation of micro amounts of ^{58}Co from a neutron-irradiated nickel target. Ni is eluted with 8 N HCl, and Co with water. Good separations are observed for a Co:Ni ratio in the range $1:10^3$ to $1:2 \cdot 10^7$. A subject of special care was the purity of the eluting Co fractions which were intended for medical use and therefore had to be Ni-free. The TnOA-Hyflo column fully satisfies this requirement, whereas Co-fractions from Dowex-1 X10 and Amberlite IRA-400 contain micro amounts of Ni.

Ni and Fe, and also Al and Fe, have been separated by Jentzsch _et al_. (49) in a polyethylene capillary column (10 m x 0.5 mm), covered on the inner surface with a thin film of LA-2. Using HCl of varying normality as the eluant, amounts of 10 μg are separated and collected within 30 min.

Mikulski and Stroński (57, 58, 89) and Sastri _et al_. (81) describe the separations of Ni from Co or Cu, Fe from Mn, Co or Zn, and Zn or Cd from Fe and from Co, Cu or Mn. In the latter separation, Co, Cu and Mn are eluted in the first free column volume with 1 N HCl, and Fe is eluted with 0.01 N HCl. The column is washed with water, and Zn and Cd are subsequently eluted with an ammonia-ammonium acetate buffer. A sharp separation of Mn, Co, Cu and Zn on TiOA-Corvic has been elaborated by Pierce and Henry (69); HCl of varying normality (5.89 to 0 N) is used as the eluant.

Using TnOA sorbed onto Hyflo Super Cel, Smułek (84) separates both micro and macro amounts of Fe and Mn: small amounts of Mn are easily separated from as much as 15 mg of Fe, while good results are also obtained in the separation of 100 mg of Mn from tracer concentrations of Fe; 4 to 8 N HCl is employed for the elution of Mn, and 0.2 N HNO_3 for that of Fe.

Español and Marafuschi (28) determine tungsten in stainless steels by activation analysis, using a preliminary separation on a Kel-F column loaded with TnOA. Cr(III) is eluted with 10 N HCl, and subsequently a 7 N HCl - 1 N HF mixture is utilized to elute W, which appears immediately and is totally recovered. Co is rapidly eluted with 3 N HCl, and Fe with 1 N HNO_3. The method is suitable for both low and medium tungsten concentrations. A separation of Cr (VI), W (VI) and Mo (VI) has been achieved on an Aliquat - Kel-F column with 8 and 2 N HCl as eluants (89).

B. Actinides and Lanthanides

Uranium and thorium have been separated from each other and/or from various other elements by many authors. First in this field were Cerrai and Testa (19), who separate milligram amounts of Th, Zr and U on a TnOA-cellulose column with HCl (10 and 6 N) and HNO_3 (0.05 N) as the eluants. A nearly identical procedure has been described for an Aliquat - Kel-F column (81). When analyzing zirconium ores, it is sometimes necessary to separate small quantities of U and Th from a large amount of Zr. In such a case, it is convenient to use a procedure in which Zr is not retained on the bed while the two trace elements are sorbed, e.g., by complexing them in 10 M NH_4NO_3. The same separation procedure has been applied to a La - U - Th mixture (19). When La and Th are present only, the former element may well be eluted with 5 N HNO_3 instead of 10 M NH_4NO_3 (81). U and Th have been separated on an Aliquat - Kel-F column, with both U - Th and Th - U as order of elution (94).

Zr and La may be separated by exploiting the fact that whereas La is slightly complexed in concentrated NH_4NO_3 , Zr is not complexed at all (19). A better separation is obtained, however, when using 10 N HCl to elute La, with subsequent removal of Zr from the column with 6 N HCl (81). Two more separations involving Zr are mentioned in the next section.

Testa (95; cf. ref. 96) determines ^{239}Pu in urine. Plutonium is isolated by coprecipitation, oxidized to Pu(IV), and selectively sorbed on a TnOA-HNO_3 - Kel-F column. After washing with 1 N HNO_3 to eliminate completely U and Th, Pu is removed by elution with an H_2SO_3 solution, which brings about a reduction to Pu(III), thereby destroying the strongly sorbed $Pu(NO_3)_6^{2-}$ complex. The decontamination factors for U, Th, Po and Ra are greater than 150. According to Testa, among the transuranium elements, only Np(IV) can seriously interfere.

A U(Pu)-Np separation has been achieved on a column prepared from Kel-F (79.5%), TLA (16.5%) and octanol-2 (5%) (29, 30). Np(IV) is sorbed from a nitric acid solution containing iron(II) sulfamate (Pu(IV)→ (III); cf. above). After the elution of U, Pu and fission products, Np is eluted with an H_2SO_4 - HNO_3 mixture. The method is very selective for Np, and has been applied to solutions in which the weight ratio U/Np is greater than 10^{10}. A simultaneous separation of Np and Pu from U has also been described (31). Tetravalent Np and Pu are sorbed from a nitrate solution containing Fe(II)-Fe(III) in order to adjust the valency states. After washing the column with HNO_3 and HCl, Np and Pu are eluted with H_2SO_4 - HNO_3 .

Hamlin and coworkers (40, 41) state that elution of U from TnOA- or TiOA - Kel-F columns is difficult. For their Pu(III) - U separations they distinctly prefer a column impregnated with TBP or LA-1: sorption from either 1 N H_2SO_4 or 9 N HCl and back-extraction with 1 N HNO_3 gives 98 to 100% recovery for U (42).

Improving a previous technique (58) for the separation of U and Th from Pa, Stroński (58) elutes Th from an Aliquat - Kel-F column with 10 N HCl; 2 N HCl is employed to remove Pa(V), and U is quantitatively eluted with 0.1 N HCl (Figure 12B).

Mo has been separated from U on TiOA-Corvic (69) and TnOA-glass powder (66) columns. In the former paper, Mo is first eluted with 1.22 N H_2SO_4 , and U with water. The reversed sequence is obtained (66) by sorbing Mo onto the column from a 0.2 N H_2SO_4 solution, collecting U in the eluate, and stripping Mo with 2 N HCl. Miscellaneous separations involving U include those of (Tb, Hf) - U - (Cd, Zn) (57, 81) and of (Cs, Sr, Cu) - (U,Fe) (67) on polyfluorochloro-

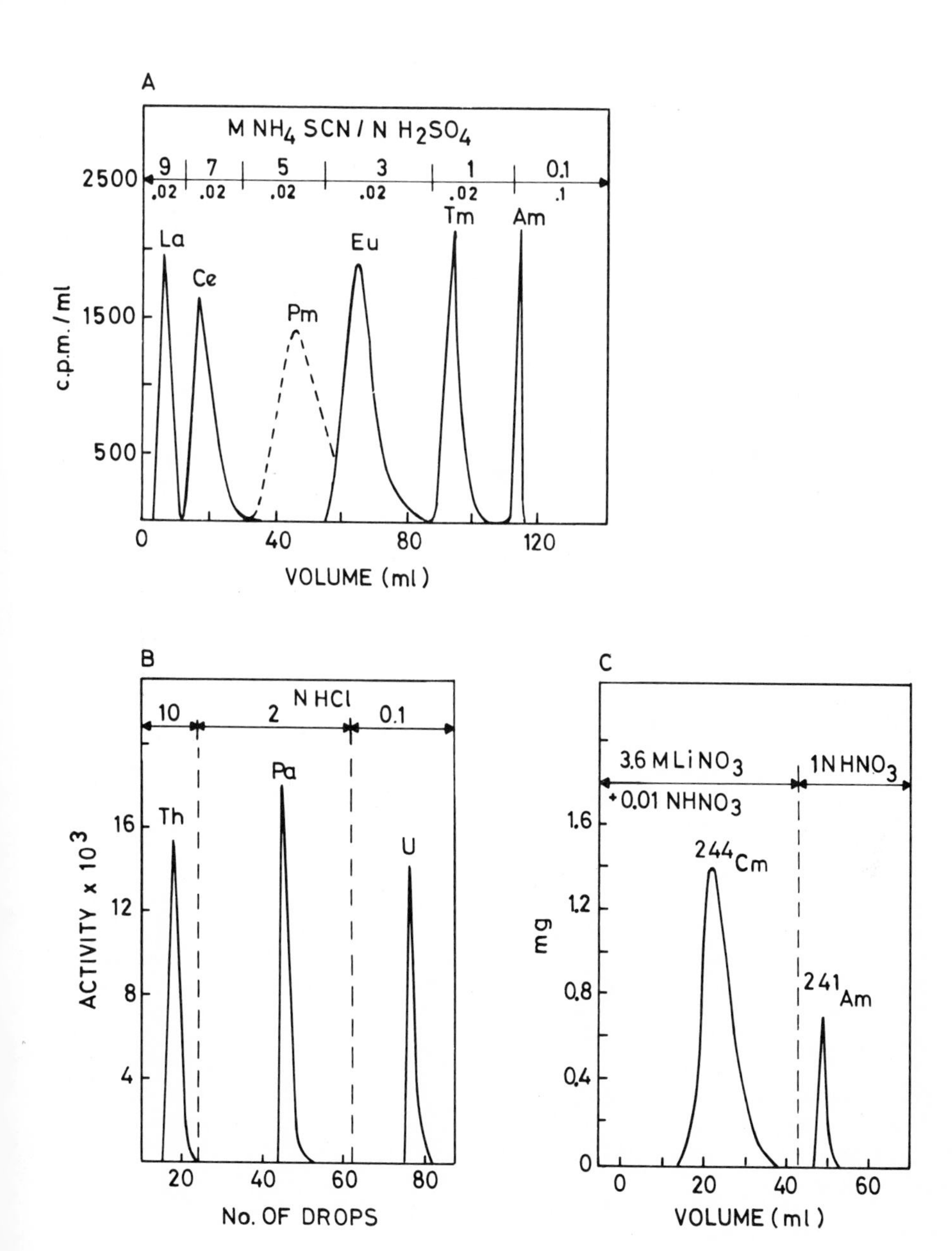

Figure 12. Examples of column chromatographic separations. A, 0.30 M Aliquat thiocyanate-Celite; bed, 100 x 5.4 mm^2; flux, 0.78 $ml \cdot cm^{-2} \cdot min^{-1}$ (2). B, 0.10 M Aliquat chloride-Kel-F; bed, 60 x 5.0 mm^2; flux, 0.25 $ml \cdot cm^{-2} \cdot min^{-1}$ (89). C, Aliquat nitrate-diatom earth; bed, 140 x 8 mm^2; capacity, 0.21 mM NO_3^-/ml bed; flux, 1 $ml \cdot cm^{-2} \cdot min^{-1}$ (43).

ethylene supports impregnated with a strongly sorbing exchanger, using HCl-containing eluants. Nanogram quantities of the rare earths as a group, together with Be, have been separated (54) from U by elution with 8 N HCl on a TnOA-SiO_2 column.

A single paper (60) has been devoted to tetravalent uranium, namely to the reextraction of U(IV)-sulphate complexes from a TnOA-treated Fluoroplast column with aqueous HF. The authors claim a stepwise formation of UF_4 from $(R_3HN^+)_4U(SO_4)_4^{4-}$.

A thorough investigation of the behavior of Am, Y and nearly all rare earths in the NH_4SCN system has been made by Huff (46). The distribution coefficients $(K_d)_c$ of the rare earths generally increase with atomic number, except for the Eu-Gd pair, which fall in the same position on the log K_d vs. log M SCN^- graph. The separation factor for adjacent elements varies between 1.3 and 1.6 for La through Er; in the region Er - Lu there is a marked increase in this factor to approximately 2. A separation factor of about 60 can be realized for the 4f and 5f analogues, Eu and Am. The addition of mineral and organic acids to the thiocyanate aqueous phase leads to a decrease in the K_d values.

NH_4SCN gradient elution systems have been elaborated (46) for the separation of the adjacent elements Tm - Yb - Lu, and of multi-component mixtures of La, Y, Eu, Ho, Er, Tm, Yb, Lu, and Am. Cross-contamination, which occurs especially in the latter case, may probably be reduced by extending the range of NH_4SCN concentrations employed in the gradient, as has been done (2) for the separation La - Ce - Pm - Eu - Tm - Am (9.0 - 0.1 M NH_4SCN; cf. Fig. 12A). A rapid separation of Am from the rare earths as a group is described in the same paper.

Chloride and nitrate systems have also been employed to separate Am and/or rare earths. On a column prepared by absorbing Alamine onto polystyrene-divinylbenzene beads (61), slow elution with slightly acidified 13 M LiCl gives a complete separation of Eu from Am. The separation is excellent, but the polymeric beads have a low capacity for the amine and throughput is limited by low equilibration rates. $LiNO_3$ solutions of varying molarity have been employed to separate Eu-Am and various pairs of adjacent rare earths (88; cf. Table XIII). Using a $LiNO_3$ gradient, Testa (93) separates fairly large amounts of Yb, Nd, La and Ba on a TnOA-cellulose column.

Among the actinides, the separation of the pair Pu-Am may easily be accomplished with HCl (57) on a TnOA-Teflon column; Am(III) is washed out with 4 N HCl in the first free volume of the column, and PuO_2(II) is eluted with 0.1 N HCl.

A rapid and efficient method for separating trivalent Am and Cm has been described by Horwitz _et al_. (43, 44) and by Van Ooyen (62). Following the procedure given in ref. 43, Cm is eluted from an Aliquat nitrate-kieselguhr column before Am with 3.6 M $LiNO_3$ - 0.01 N HNO_3. The separation factor is constant within the experimental error, between 3.5 and 4.6 M $LiNO_3$, having a value of 2.79. In actual practice, Am can be stripped from the column with a small volume of 0.15 N HNO_3, at the start of its breakthrough (Fig. 12C). Under these conditions, approximately 2 hours are required for a complete separation and recovery. More than 99% of the Cm and Am are recovered in a radiochemically pure state. The procedure has

TABLE XIII

Separation of pairs of rare earth elements on an Aliquat nitrate - Kel-F column⁺

Rare earths	$LiNO_3$ (M)	Sepn. fact.	Rare earths	$LiNO_3$ (M)	Sepn. fact.
Nd - Pr	3.2 - 2.8	-	Er - Tm	5.7	-
Sm - Pm	3.0	-	Yb - Tm	6.0	4.90
Gd - Eu	5.25	1.71	Tm - Lu	6.0	5.00
Ho - Er	5.8	1.34	(Eu - Am	4.2	2.50)

⁺Column: Kel-F (250-325 mesh) loaded with 5% (w/w) of Aliquat nitrate; 60 mm x 5 mm. Flow rate: 0.25 $ml \cdot cm^{-2} \cdot min^{-1}$ (88).

sucessfully been applied to the preparation of a multicurie quantity of high-purity ^{242}Cm in an alpha-scattering device for remote analysis of the lunar surface.

Attempts to separate trivalent Am and Cm by oxidation of the former ion to Am(VI) and sorption of anionic americyl sulfate complexes on a TiOA-diatom earth column have failed, regardless of oxidants and/or aqueous sulfate concentrations employed (47). The most likely cause is reduction of Am(VI) by the amine.

For the difficult Cf(III)-Es(III) separation, the column chromatographic system described above can be copied, introducing minor changes only (45). Using 4.8 M $LiNO_3$ - 0.05 N HNO_3 as the eluant, both tracer and microgram quantities of Cf and Es are separated, Cf eluting first. The separation factor of 1.46 is somewhat higher than that for the currently employed alpha-hydroxyisobutyrate separation system, and comparable to the factor for the EDTA system. A difference between the extraction-chromatographic separation and the other two systems is the reversal of the elution positions of Cf and Es. An interesting application of both procedures in the production of isotopically pure ^{253}Es is described in the quoted paper.

C. Miscellaneous

Stroński (89) reports difficulties encountered in the separation of Ta and Re. The anions of these elments, formed on dissolving the metals in molten KOH and KNO_3, are sorbed on an Aliquat-Kel-F column. Re can be eluted by means of NH_3 (1:1). However, the Ta(V)-anion can only be collected together with the exchanger, by washing the column with acetone. A more successful result has been described by Katykhin (50), who uses a TOA-Fluoroplast support. W, Au, Os and Ta are eluted from the column with the help of various complexing agents and Re again is extracted with ammonia.

Zr and Nb have been separated (69) on a TiOA-Corvic column with

H_2SO_4 (7.33 to 6.67 M); both peaks are sharp and show little tailing. In spite of their chemical similarity, Zr and Hf are separated fairly easily on a TnOA-cellulose column, employing 8 N HCl + 5% conc. HNO_3 (19). Hf appears in the effluent after about one column volume and is removed very rapidly. After its collection, Zr is best eluted with HNO_3.

One method for isolating Pd from nuclear waste solutions has been demonstrated by Colvin (22). The high-salt basic waste solutions are passed through an Aliquat-Plaskon column, which preferentially retains the palladium. Upon elution with three 8 N HNO_3 - 1.5 N NH_3 elution cycles, the total recovery of Pd from the column is 97%.

As, Ge and In are separated (69) on TiOA-Corvic with HCl (8.6 - 0 N), and Te and Sb(V) on TnOA-Kel-F, using 0.3 N HCl and 10% EDTA (pH=10) as eluants (57).

Finally, we may mention an interesting separation of the four halogen anions (79). Samples containing a total of between 1 and 100 μM of the halogenides are successfully separated on Aliquat - Kel-F by a gradient elution technique, using 0.1 M sodium acetate and various concentrations of $NaNO_3$ as eluants. Acetate is preferred above nitrate and sulfate, since it is the only eluant tested which completely separates the difficultly separable F^- and Cl^-.

VII. THEORETICAL ASPECTS

Apart from direct application of extraction chromatography to chemical analysis, the comparison of chromatographic data on liquid anion-exchangers and those on (a) liquid-liquid extraction using the same exchangers, and (b) anion-exchange resin chromatography, has repeatedly attracted attention. (In the literature, the, at times, remarkable analogous behavior of liquid exchangers and neutral organophosphorus compounds such as tri-n-butylphosphate and tri-n-octylphosphine oxide is also a point of discussion. However, as a rule this subject is only mentioned in order to explain the mechanism of the sorption by the neutral p-compounds by analogy with the data obtained with liquid anion-exchangers and not the reverse. This subject will therefore not be considered here.)

A. Chromatography and Liquid-Liquid Extraction

A quantitative relationship exists (13, 33) between the distribution coefficient K_d of a substance in a liquid-liquid extraction system ($K_d = C_{org}/C_{aq}$), and the R_F value obtained with the exchanger is fixed on a support and the aqueous phase is used as eluant:

$$K_d = A_L/A_S \cdot (1/R_F - 1) \qquad [2]$$

or

$$\log K_d = R_M + \log A_L - \log A_S \qquad [3]$$

where $R_M = \log (1/R_F - 1)$, and A_L and A_S are the areas of cross-section of the mobile and stationary phase, respectively.

Unfortunately, both the eluant composition and the ratio A_L/A_S often are not constant along the chromatogram in paper and thin-layer chromatography; besides, the role of the supposedly inert diluent and support, present in extraction and chromatography, respectively, cannot wholly be neglected. Various examples for HCl, HBr and HNO_3 systems--pictured in refs. 4 and 18--show that indeed a fair agreement exists between the changes in the values of R_M and of log K_d with varying normality of the aqueous acid solution. However, it is quite difficult to establish a quantitative relationship between them.

In column chromatography, a general pattern of extraction, which is similar to that found in liquid-liquid extractions, has also been observed (46, 69; cf. Fig. 13). Since column chromatography is carried out under more closely controlled conditions than are paper and thinlayer work, a quantitative comparison is more valuable in this case. Horwitz and coworkers (43, 45) have observed that the nitrate ion dependency for the extraction of trivalent Am, Cm and Cf into Aliquat nitrate in xylene and for the sorption of these ions on Aliquat nitrate-diatom earth columns shows a very close parallel between the two systems over the whole NO_3^- concentration range studied. The authors state that, once the difference in magnitude between $(K_d)_c$ and $(K_d)_1$ has been determined for a given extractant concentration and column capacity, the elution position from a column can be predicted from liquid-liquid extraction data. Such predictions, however, still must be regarded with some care, as is evident from other data in ref. 43: substitution of column chromatography by extraction brings about a slight but distinct change, not only in the magnitude but even more strongly in the nitrate ion dependency of the separation factor Am/Cm.

It is of interest to note that the range of R_M and log $(K_d)_c$ values is rather more limited than that of log $(K_d)_1$ values (17). As an example, the latter may well be measured from +4 to -4, whereas R_M values vary between +2 and -2 only, even when assuming that the limits are set by $0.01 \leqslant R_F \leqslant 0.99$. This aspect must be borne in mind when predicting extraction experiments on the basis of chromatographic data, and vice versa.

B. Liquid and Resin Exchangers

As far as the principle of the methods is concerned, the analogy between the phenomena of extraction by liquid exchangers and of sorption with anionic resins is very marked. The present author has made an extensive comparison of extraction chromatographic data and results obtained on Amberlite SB-2 paper that contains the strongly basic polystyrene resin Amberlite IRA-400. In both the HCl (10,121) and the HBr (9,104) system, the agreement between the data is good; similar results have been obtained by Przeszlakowski (73). Testa (93) reports that the average separation factor of 1.45 for two adjacent rare earths on paper loaded with $TnOA \cdot HNO_3$ is very similar to that observed for the resin Dowex-1. Another illustration of the analogy between the processes of resin and liquid anion-exchange is presented in Fig. 13.

The closely corresponding results add to the suggestion that the mechanism (and thus the species involved) are identical in the two cases. However, despite the apparent analogy, this conclusion may well be partially incorrect, since the physical aspects of the two

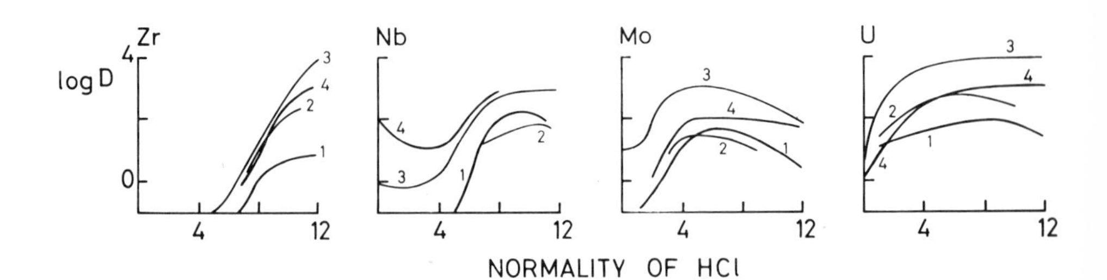

Figure 13. Comparison of literature data on various extraction techniques and/or exchangers in HCl systems. D is the volume distribution ratio D_v (69, 119), or the distribution coefficient K_d (25, 118). Liquid exchangers: 1, solvent extraction; 0.11 M TiOA in xylene (118). 2 and 3, batch extractions; 1 g TiOA-treated support - 20 ml aqueous phase (25, 69). Resin exchanger: 4, Dowex-1 x 10 (119).

processes are substantially different (13).

Two examples may be quoted as illustrations (cf. refs. 4 and 116): (1) with tetravalent actinides, the species, $M(NO_3)_6^{2-}$ and MCl_6^{2-}, are extracted by both liquid and resin exchangers. With hexavalent uranium, however, in studies with secondary-quaternary exchangers only $UO_2(NO_3)_3^-$ is present, whereas both $UO_2(NO_3)_3^-$ and $UO_2(NO_3)_4^{2-}$ are found when resins are used (2). For light transition metals such as Fe, Zn and Co, the metal-tetrachloro anions are generally proposed as the species predominantly present in the organic phase with both liquid extractants and resins. On the other hand, with Cu it is highly probably that both $CuCl_3^-$ and $CuCl_4^{2-}$ occur in the resin phase, whereas only the latter ion is observed in the case of liquid exchangers.

C. Mechanism of the Process

Sorption mechanism. Several studies have been made concerning the sorption mechanism of an exchange reaction such as

$$\underline{n}\ RR'_3N^+X^-_{org.} + MX_p{}^{n-}_{aq.} \rightleftharpoons (RR'_3N^+)_n MX_p{}^{n-}_{org.} + \underline{n}\ X^-_{aq.} \qquad [1a]$$

In the literature, two types of sorption have been described, termed "partition" and "exchange" by Cerrai and Ghersini (17). (Since anions are exchanged in both cases, the proposed terms are somewhat misleading.)

The partition mechanism implies that the anion-exchanger merely coats the support particles and completely covers them, so that a stationary phase is obtained obeying the laws of partition chromatography. For such a true partition system the distribution coefficient K_d is, by definition, a constant. Thus, assuming A_L to be constant, Equation 3 may be written as

$$R_M = \log A_S + \text{constant} . \qquad [4]$$

If the effective thickness of the stationary phase, or the amount of exchanger, is directly proportional to the concentration of this impregnant in the solution used to slurry the support, then the latter should also be linearly related to the R_M values (33, 37) and the slopes of R_M vs. log loading plots should be equal to 1. In the exchange mechanism, the liquid anion-exchanger is considered to be fixed within the support: the stationary phase is rather similar to an anion-exchange resin, or a solution of an exchanger in a diluent, the concentration of the active groups varying with the amount of loaded extractant. In that case, the distribution coefficient is not a constant, but varies with $[RR'_3N^+X^-]^n$, provided the other variables are kept constant (for a derivation of this well known result, see ref. 130). Since in the system considered, both A_L and A_S will be only slightly dependent upon the amount of exchanger loaded on the support, they may be assumed to be constant. Substitution of these data in Equation 3 yields:

$$R_M = \underline{n} \log[RR'_3N^+X^-] + \text{constant} , \qquad [5]$$

or a plot of R_M vs. log extractant loading will have a slope of $\underline{n}$. For column chromatography, the same considerations apply, when R_M is substituted for the logarithm of the true retention volume (17).

In principle, therefore, slope analysis is a powerful method to verify which of the two viewpoints on the retention mechanism by the stationary phase best fits liquid anion-exchangers. However, it must be recalled that: 1) the conclusions refer to ideal conditions and 2) slope analysis is fraught with difficulties (cf. 123). As an additional complication of slope analysis, when applied to extraction chromatography, we must mention that strictly speaking R_M values should be plotted versus the logarithm of the actual amount of exchanger present on the loaded support, instead of the logarithm of the concentration of the impregnant solution. This consideration may become important at high loadings (30). Still, it certainly is of interest to scrutinize the pertinent data in the literature.

Slopes with values of 1, 2 and 3 have repeatedly been reported by Przeszlakowski (74-76). These results seem to favor the exchange mechanism, but care must still be taken here. In the first of the three papers cited, the support is impregnated with a mixture of varying amounts of the exchanger and decalin (which is nonvolatile), the total loading of the support being kept constant. Thus, the constancy of the ratio A_L/A_S is ensured, in spite of the variable amounts of exchanger present. Unfortunately, however, this deviation from the normal reversed-phase chromatographic procedure brings about the occurrence of a stationary phase that will almost certainly behave as a "solution." Therefore, no meaningful conclusions regarding the sorption mechanism can be drawn from these data. This is even more true for the results presented in refs. 75 and 76. These relate to experiments in which solutions of amine salts in organic diluents are employed as the mobile phase, and solutions of inorganic acids or their salts are used to impregnate the support ("moist-paper technique"). Although these data are worth mentioning because of other aspects involved (see below), they do not contribute to our knowledge of the problem now under discussion.

Slopes with a value of 1 have been obtained for various metal ions in chloride (33, 59) and thiocyanate (37, 46) systems. Some of the data relate to the extraction of monovalent anions so that again no choice between partition and exchange can be made. However, the results for Zn and Pd in the chloride and for Cd and Cu in the thiocyanate system, as well as the close parallelism between the various R_M vs. log loading plots with the occurrence of sudden breaks in these parallel lines at approximately the same loading (see the next section), clearly favor the partition mechanism.

More work evidently will be necessary before a generally valid conclusion can be drawn regarding the sorption mechanism. Since much will depend upon the selection of extraction systems for which without doubt $n \neq 1$ (cf. Equations 1a and 5), the elucidation of the nature of the metal-containing species present in the organic phase will be of paramount importance. This aspect is discussed below.

Nature of the exchanger. In their studies of R_M vs. log loading plots, Graham and coworkers (33, 37) have also discussed the nature of the extractant present on the loaded support. As has been described above, the authors observe a linear relationship between R_M and log loading for various chromatographic systems. The slopes of the plots are equal to approximately 1 up to about 0.5 M for the Cl^-, and 0.3 M for the SCN^- system. At higher concentrations, a break occurs in the parallel lines for all metal species investigated, and the plots show an upward trend. The metal ion spots now become diffuse, the running time increases drastically, and the solvent fronts become ragged. According to Graham, this sudden change in the slope of the plots may well be due to a change in the nature of the impregnant from a monomeric to a polymeric species. At low concentration the amine salt molecules are essentially monomeric so that all the functional groups are available for the ion-exchange process and equilibrium is rapidly attained and sharp zones are established on the chromatogram. At higher concentrations, however, micellar aggregates are formed and transferred from the slurrying solvent to the layers. Diffusion of the complex metal anions within these polymers now becomes important, leading to excessive R_M values and slowing down the rate of attainment of equilibrium, thus contributing to the phenomenon of zone-spreading. Moreover, the monomer→polymer transition probably causes a drastic reduction in the effective capillary volume of the cellulose. This causes smaller volumes of the mobile phase to be drawn up the plate (so that A_L decreases) and even further increases the R_M values. (That a similar lowering of A_L occurs at relatively low loadings, already shown in the work of Murray and Passarelli (59), may be at least partially attributed to the rather unusual experimental conditions employed by these authors.)

This hypothesis attractively explains the experimental data. Still, it must be regarded with some caution, since: 1) the aggregation of amine salts often already occurs at very low concentrations, 2) the formation of aggregates during the evaporation of the diluent from the impregnated layer has been neglected and 3) there is no general agreement concerning the different reactivities of monomeric and polymeric species of the amine salts. Finally, the occurrence of breaks with a downward trend in chromatography on paper loaded with exchanger thiocyanate salts (71, 100) (though, admittedly, at very high loadings) adds to the confusing and complicated character of the matter at hand.

Complex metal anions. Three methods are currently applied to elucidate the nature of the metal-containing species present in the organic phase: 1) determination of the charge of the anion from the maximum-loading ratio $(exchanger/M)_{org}$, 2) determination of the slope of a log K_d (or R_M) vs. log exchanger concentration plot, to give the charge of the extracted species and 3) identification of the species from their absorption spectra. The first method yields valuable information, but the loading ratio is not always equal to the charge of the anion, due to the formation of species such as $R_3HN^+MCl_4^- \cdot R_3HN^+Cl^-$ (see below). The results so achieved must therefore be supplemented by those of methods 2 and 3. The difficulties inherent in the use of slope analysis (especially in extraction chromatography) have been amply discussed in one of the previous sections; fortunately, spectral identification may be regarded as a reliable method.

Brinkman et al. (4, 11, 12) have studied liquid-liquid extraction systems in which undiluted exchangers are used as the organic phase, thus passably simulating conditions pertinent in extraction chromatography. Solutions of HCl and LiCl are used as aqueous phases; for Co, the bromide system has also been investigated. The interpretation of ultraviolet (and infrared) spectra and the determination of maximum loading of the exchanger are employed as methods of investigation. Przeszlakowski (74) has evaluated R_M vs. log loading plots for several metal ions in the LA-2·HCl (+ added decalin; see above) system. The combined data are summarized in Table XIV.

The results are according to expectations based on solvent extraction data in nearly all cases (4). Zinc is the only distinct exception, the proposal of the mononegative $ZnCl_3^-$ (74) being less plausible than that of $ZnCl_4^{2-}$ (12). The results obtained for Te cannot easily be verified as yet. Regarding Fe(III) and In(III), we must emphasize that the amine/M(III) ratio of 2:1, observed at low sequence Cl^- concentrations, for amine·H^+ salts only (not for ammonium salts), is not due to the presence of MCl_5^{2-} instead of MCl_4^- anions. As is also known from solvent extraction, only mononegative anions occur in the organic phase, and the 2:1 ratio is best explained by the formation of the species $R_2R'HN^+MCl_4^- \cdot R_2R'HN^+Cl^-$.

R_M vs. log [TnOA·HCl] plots have been obtained by Przeszlakowski (79) using his moist-paper technique. Although the experiments cannot be classified under the heading "reversed-phase extraction chromatography," the results are important because of the intermediate

TABLE XIV

Summary of equilibria governing reversed-phase extraction chromatography for chloride systems (4, 12, 74, 75)$^{+}$

Divalent metal ions (Co, Ni, Mn, Cu, Zn, Cd, and UO_2)

$$2\ R_2R'_2N^+Cl^-_{org.} + MCl_4^{2-}{}_{aq.} \rightleftharpoons (R_2R'_2N^+)_2MCl_4^{2-}{}_{org.} + 2\ Cl^-_{aq.}$$

$$2\ R_2R'_2N^+Br^-_{org.} + CoBr_4^{2-}{}_{aq.} \rightleftharpoons (R_2R'_2N^+)_2CoBr_4^{2-}{}_{org.} + 2\ Br^-_{aq.}$$

Trivalent metal ions (Fe, Ga, In)

$$R_4N^+Cl^-_{org.} + MCl_4^-{}_{aq.} \rightleftharpoons R_4N^+MCl_4^-{}_{org.} + Cl^-_{aq.}$$

$$(\underline{n}+1)\ R_2R'HN^+Cl^-_{org.} + MCl_4^-{}_{aq.} \overset{++}{\rightleftharpoons} R_2R'HN^+MCl_4^- \cdot \underline{n}R_2R'HN^+Cl^-_{org.} + Cl^-_{aq.}$$

Miscellaneous (Re, Te)

$$R_2R'HN^+Cl^-_{org.} + ReO_4^-{}_{aq.} \rightleftharpoons R_2R'HN^+ReO_4^-{}_{org.} + Cl^-_{aq.}$$

$$R_2R'HN^+Cl^-_{org.} + TeCl_5^-{}_{aq.} \rightleftharpoons R_2R'HN^+TeCl_5^-{}_{org.} + Cl^-_{aq.}$$

$^{+}$(1) R, alkyl; R', alkyl or H. (2) Interchange of anions has been assumed in preference to the addition of neutral species. (3) The formulas for the monomeric alkylammonium salts have been consistently written, though it is known that aggregation of these entities may occur. (4) The formulas written for the metal-containing anions in the aqueous phase do not necessarily represent the species predominantly present.

$^{++}$Ga: $\underline{n}$ = 0 under all conditions; Fe and In: $\underline{n}$ = 0 or 1 dependent on the Cl^- concentration in the aqueous phase. The data on Ga pertain to liquid-liquid extractions only.

place of that technique between normal liquid-liquid extraction and reversed-phase extraction chromatography. The author finds amine/metal ratios of 2:1 for Fe, U, In and Co, and of 1:1 for Te, thereby confirming the data given in Table XIV. The linear relationship between R_M values for Fe and In, and log [HCl], parallel to the relationship between log K_d and log [HCl] reported for solvent extraction, is also worth mentioning.

Huff (46) has studied log $(K_d)_c$ vs. log loading plots at constant NH_4SCN concentration for a selected number of rare earth elements. Linear curves are obtained, having a slope of approximately 1. Unfortunately, one cannot decide whether this value merely points to a true partition mechanism, or demonstrates the presence of $R_4N^+M(SCN)_4^-$ in the organic phase. The latter interpretation, given by Huff, is consistent with solvent extraction data reported in the literature.

Using acidified solutions of NH_4SCN to impregnate the paper, and organic solutions of TBuA, tribenzylamine and LA-2 thiocyanate as mobile phases, Przeszlakowski (76) has observed linear relationships between the R_M values for various metal ions and the logarithm of the concentration of the exchanger salt. Slopes with a value of 2 are reported for Co, Cd, Zn, Hg, Pd, In, Ga, Fe, Ti and U. (With the latter ion this value is found with LA-2, but a value of 3 is observed with both tertiary amines.) The results seem to indicate the presence of dinegative metal-thiocyanate complexes in the organic phase, which conclusion is confirmed for some of the ions, e.g., Co and Zn, from solvent extraction data. For other ions, either insufficient or conflicting data are reported in the literature ($Fe(SCN)_5^{2-}$ or $Fe(SCN)_4^-$; di- and trinegative U-containing anions or the attachment of an additional molecule of tertiary amine salt to $(R_3HN^+)_2UO_2(SCN)_4^{2-}$), so that a detailed discussion at this moment is not opportune.

Acid effect. Desorption occurs in extraction chromatography for many metal ions when mobile phases containing high concentrations of acids (but not their lithium salts) are employed (12, 18). This phenomenon, which is well known from liquid-liquid extractions, has repeatedly been investigated for HCl/LiCl systems. The desorption and decreasing extractability have occasionally been attributed to the formation of undissociated complex metal acids in the aqueous phase (33). However, this hypothesis is not in accord with the very high acid strength of these complex acids, which probably are stronger than HCl. Besides, the HCl effect also occurs with Br^-, I^-, and ReO_4^- (4, 113), all of which are derived from acids doubtlessly stronger than HCl.

A more recent suggestion has been that the extraction by acid beyond the amount necessary to neutralize the amine (which increases with increasing aqueous HCl concentration) plays an important role. Infrared studies (114) have shown the excess acid to be present as the HCl-solvated Cl^- ion HCl_2^-. The extraction of complex metal-chloro anions from aqueous HCl solutions may be greatly influenced by the formation of this hydrogen dichloride anion, as it is less highly hydrated and more readily extractable than the simple Cl^- ion, and so competes much better than the latter in anion-exchange. Thus for a complex anion MCl_p^{n-}, the sorption from dilute HCl solutions will be almost identical with that from NaCl (up to 4 M only) and LiCl, but it increasingly falls below the LiCl values as the concentration rises towards the region where HCl_2^- can form and start extracting (112, 114, 4, 12, 59). For the application of the theory outlined here to chromatography on resin-loaded paper, see (4).

Desorption due to the extraction of excess acid also occurs in HNO_3, and presumably HSCN, systems (see sections IV-E and IV-D, and Table VIII). A mechanism somewhat similar to the acid effect has recently been indicated for NH_4SCN systems (115, 38). Here, the formation of species such as $R_2R'HN^+SCN^- \cdot NH_4SCN$ probably must be taken into account.

VIII. CONCLUSION

Within a decade, a vast amount of data has been compiled on the use of liquid anion-exchangers as the stationary phase in extraction chromatography. As a rule, the results reported by different authors show a close correspondence. Many interesting separations, both qualitative and quantitative, have been described, and the results presented in this review indicate, by means of examples, what can be accomplished in this field. The principal merits of such processes are their simplicity and versatility and extraction chromatography compares favorably, in many respects, with chromatography involving the use of resin exchangers.

Data on reversed-phase extraction chromatography and liquid-liquid extraction show a definite, though only semi-quantitative, correspondence. Thin-layer chromatography may therefore serve to predict the solvent extraction behavior of an element; the reversed procedure has been recommended in the case of column chromatography. In this connection, attention may be called to the "moist-paper" chromatography of Przeszlakowski, which appears to be valuable when studying aspects such as the composition of the alkylammonium metal-complex salts in the organic phase (75, 76), and the influence of the diluent on the extraction efficiency (127, 128). In spite of the manifold successes scored in the application of extraction chromatography to inorganic analysis, the fundamental aspects of the behavior of the liquid anion-exchangers still deserve much further study. Progress distinctly has been made in the evaluation of the nature of the metal-containing species present in the organic phase: for various complex metal anions, spectral identification, loading data and slope analysis have been shown to concur. Identical results have moreover been obtained with dilute and undiluted exchangers (4, 12). This strongly suggests that the metal complexes identified in amine phases in solvent extraction are also present in the stationary phase in chromatography. However, the equilibria regulating the extraction/sorption processes are not well understood as yet, partly because of interfering aggregation phenomena (33, 37). Moreover, positive conclusions regarding partition versus exchange mechanism and metal extraction through either interchange of anions or the addition of neutral species have hardly been reached as yet. The latter problem has briefly been touched upon by Przeszlakowski (74, 76), but his results are not unequivocal. Besides, complications such as the extraction of excess acid have not been taken into account, and a treatment in terms of activities rather than concentrations has not always been given. This imperfection also invalidates conclusions drawn elsewhere (71, 73) regarding affinity series of metal ions in exchanger-(pseudo) halogeno systems.

In summary, the stoichiometric composition of the alkylammonium metal-complex salts recorded in Table XIV as a rule is correct. More research will be necessary, however, before a valid quantitative description of the metal extraction equilibria can be given. Successful attempts may be expected to originate from the combined use of spectroscopic measurements and the determination of loading data in the field of liquid-liquid extraction, with reversed-phase extraction chromatography supplying confirmatory evidence.

Acknowledgments. Sincere thanks are due to Mr. G. de Vries for his continuous cooperation and interest in this work, and to Dr. G. Ghersini for many stimulating discussions.

REFERENCES

The references have been listed in two separate series. The first series records all papers dealing with reversed-phase extraction chromatography; the second one refers to papers discussing related topics.

1. K. Akerman, Z. Kozak and D. Wiater, Przemysl Chem., 42, 26 (1963).
2. P. G. Barbano and L. Rigali, J. Chromatog., 29, 309 (1967).
3. U.A. Th. Brinkman, Chem. Techn. Revue, 21, 529 (1966).
4. U.A. Th. Brinkman, Liquid Anion-Exchangers in Analytical Chemistry, Dissertation, Free University of Amsterdam, 1968.
5. U.A. Th. Brinkman, P.J. J. Sterrenburg and G. de Vries, in preparation.
6. U.A. Th. Brinkman and G. de Vries, J. Chromatog., 18, 142 (1965).
7. U.A. Th. Brinkman and G. de Vries, in preparation.
8. U.A. Th. Brinkman, G. de Vries and E. van Dalen, J. Chromatog., 22, 407 (1966).
9. U.A. Th. Brinkman, G. de Vries and E. van Dalen, J. Chromatog., 23, 287 (1966).
10. U.A. Th. Brinkman, G. de Vries and E. van Dalen, J. Chromatog., 25, 447 (1966).
11. U.A. Th. Brinkman, G. de Vries and E. van Dalen, Z. Anorg. Allgem. Chem., 351, 73 (1967).
12. U.A. Th. Brinkman, G. de Vries and E. van Dalen, J. Chromatog., 31, 182 (1967).
13. E. Cerrai, Chromatog. Reviews, 6, 129 (1964).
14. E. Cerrai, in Stationary Phase in Paper and Thin Layer Chromatography (K. Macek and I.M. Hais, eds.), Elsevier, Amsterdam, 1965, p. 180.
15. E. Cerrai, CISE Rept-103, Milan (1966).
16. E. Cerrai and G. Ghersini, Energia Nucl. (Milan), 11, 441 (1964).
17. E. Cerrai and G. Ghersini, J. Chromatog., 13, 211 (1964); 15, 236 (1964); 24, 383 (1966).
18. E. Cerrai and C. Testa, J. Chromatog., 5, 442 (1961).
19. E. Cerrai and C. Testa, J. Chromatog., 6, 443 (1961).

20. E. Cerrai and C. Testa, J. Chromatog., 8, 232 (1962).
21. E. Cerrai and C. Triulzi, J. Chromatog., 16, 365 (1964).
22. C.A. Colvin, U.S. At. Energy Comm., ARH-SA-28, (1969)
23. D. McCormick, R.J.T. Graham and L.S. Bark, Intern. Symp. IV, Chromatographie, Électrophorèse, Bruxelles, 1966, Presses Academiques Européennes, Brussels, 1968, p. 199.
24. R. Denig, N. Trautmann and G. Herrmann, Angew. Chem., 79, 247 (1967).
25. R. Denig and N. Trautmann, AED-Conf., 68-196-014 (1968).
26. H. Eschrich and W. Drent, Eurochemic Rept., ETR-211 (1967).
27. H. Eschrich and W. Drent, Eurochemic Rept., in the press.
28. C. E. Español and A. M. Marafuschi, J. Chromatog., 29, 311 (1967).
29. D. Gourisse and A. Chesné, Comm. Energie At. Rept., R-3245 (1967).
30. D. Gourisse and A. Chesné, Anal. Chim. Acta, 45, 311 (1969).
31. D. Gourisse and A. Chesné, Anal. Chim. Acta, 45, 321 (1969).
32. R.J.T. Graham, personal communication.
33. R.J.T. Graham, L.S. Bark and D.A. Tinsley, J. Chromatog., 35, 416 (1968).
34. R.J.T. Graham, L.S. Bark and D. A. Tinsley, J. Chromatog., 39, 200 (1969).
35. R.J.T. Graham, L.S. Bark and D.A. Tinsley, J. Chromatog., 39, 211 (1969). Intern. Symp. V, Chromatographie, Électrophorèse, Bruxelles, 1968, Presses Academiques Européennes, Brussels, 1969, p. 478.
36. R.J.T. Graham, L.S. Bark and D.A. Tinsley, J. Chromatog., 39, 218 (1969).
37. R.J.T. Graham and A. Carr, J. Chromatog., 46, 293 (1970).
38. R.J.T. Graham and A. Carr, J. Chromatog., 46, 301 (1970).
39. H. Green, Talanta, 11, 1561 (1964).
40. A.G. Hamlin and B.J. Roberts, British Pat. 900,113 (1962).
41. A.G. Hamlin, B.J. Roberts, W. Loughlin and S.G. Walker, Anal. Chem., 33, 1547 (1961).
42. T.J. Hayes and A.G. Hamlin, Analyst, 87, 770 (1962).
43. E.P. Horwitz, C.A.A. Bloomquist, K.A. Orlandini and D. J. Henderson, Radiochimica Acta, 8, 127 (1967).
44. E.P. Horwitz, K.A. Orlandini and C.A.A. Bloomquist, Inorg. Nucl. Chem. Letters, 2, 87 (1966).
45. E.P. Horwitz, L.J. Sauto and C.A.A. Bloomquist, J. Inorg. Nucl. Chem., 29, 2003 (1967).
46. E.A. Huff, J. Chromatog., 27, 229 (1967).
47. E.K. Hulet, J. Inorg. Nucl. Chem., 26, 1721 (1964); Norway At. Energy Inst. Rept., KR-56 (1963).
48. M. Ishibashi, H. Komaki and M. Demizu, Mitsubitshi denki giho, 39, 907 (1965).
49. D. Jentzsch, G. Oesterhelt, E. Rödel and H.G. Zimmermann, Z. Anal. Chem., 205, 237 (1964).
50. G.S. Katykhin, Programme and Abstracts of Reports, 13th Conf. on Nuclear Spectroscopy, Izd. AN SSSR, Moscow-Leningrad, 1963, p. 28.
51. G.S. Katykhin, Zh. Analit. Khim., 20, 615 (1965).
52. W. Knoch and H. Lahr, Radiochimica Acta, 4, 114 (1965).
53. W. Knoch, B. Muju and H. Lahr, J. Chromatog., 20, 122 (1965).
54. R. Krefeld, G. Rossi and Z. Hainski, Mikrochim. Acta, 1965, 133.
55. P. Markl and F. Hecht, Mikrochim. Acta, 1963, 970.

56. D.L. Massart and A.M. Massart-Leen, Ind. Chim. Belge, 34, 203 (1969).
57. J. Mikulski, Nukleonika, 11, 57 (1966); Polish Inst. Nuclear Phys. Rept., INP-412/C, Cracow, 1965.
58. J. Mikulski and I. Stroński, J. Chromatog., 17, 197 (1965).
59. R.W. Murray and R.J. Passarelli, Anal. Chem., 39, 282 (1967).
60. A.S. Nazarov and S.V. Gronov, Trans. Moscow Chemical Techn. Inst. Mendeleev, 47, 159 (1964).
61. Oak Ridge Nat. Lab. Staff, U.S. At. Energy Comm. Rept., ORNL-3314, p. 110 (1962).
62. J. van Ooyen, Solvent Extraction Chemistry (D. Dyrssen, J.O. Liljenzin and J. Rydberg, eds.) North-Holland Publ. Co., Amsterdam, 1967, p. 485.
63. S.W. Pang, C.C. Lei and S.C. Liang, Hua Hsueh Hsueh Pao, 30, 160 (1964).
64. S.W. Pang and S.C. Liang, K'o Hsueh Tung Pao, 1964 (2) 156.
65. S.W. Pang and S.C. Liang, Hua Hsueh Hsueh Pao, 30, 401 (1964).
66. D.C. Perricos and J.A. Thomassen, Norway At. Energy Inst. Rept., KR-83 (1964).
67. M. Petit-Bromet, CEA Report, R-3469 (1968).
68. T.B. Pierce and R.F. Flint, J. Chromatog., 24, 141 (1966).
69. T.B. Pierce and W.M. Henry, J. Chromatog., 23, 457 (1966).
70. T.B. Pierce and P.F. Peck, Analyst, 89, 662 (1964).
71. S. Przeszlakowski, Chem. Anal., 12, 57 (1967).
72. S. Przeszlakowski, Chem. Anal., 12, 321 (1967).
73. S. Przeszlakowski, Chem. Anal 12, 1071 (1967).
74. S. Przeszlakowski, Roczniki Chemii, 41, 1681 (1967).
75. S. Przeszlakowski, Roczniki Chemii, 42, 975 (1968).
76. S. Przeszlakowski, Roczniki Chemii, 43, 1337 (1969).
77. S. Przeszlakowski and E. Soczewinski, Chem. Anal., 11, 895 (1966).
78. S. Przeszlakowski, E. Soczewinski and A. Flieger, Chem. Anal., 13, 841 (1968).
79. L. Ramaley and W.A. Holcombe, Analytical Letters, 1, 143 (1967).
80. M.N. Sastri, A.P. Rao and A.R.K. Sarma, J. Chromatog., 19, 630 (1965).
81. M.N. Sastri, A.P. Rao and A.R.K. Sarma, Indian J. Chemistry, 4, 287 (1966).
82. S. Siekierski, Croatice Chem. Acta, 35, A11 (1963).
83. W.Smułek, Croatica Chem. Acta, 35, A12 (1963).
84. W.Smułek, Nukleonika, 11, 635 (1966).
85. W.Smułek and K. Zelenay, Proc. Anal. Chem. Conf. Application Physico-Chemical Methods in Chem. Anal., Vol. II, p. 47, Budapest (1966); Polish Inst. Nucl. Phys. Rept., INR-677/V/XII/C, Cracow, 1965.
86. V. Spěváčková, Chem. Listy, 62, 1194 (1968).
87. I. Stroński, Oesterr. Chemiker-Zeitung, 68, 5 (1967).
88. I. Stroński, Chromatographia, 2, 285 (1969); Polish Inst. Nucl. Phys. Rept., 675/C, Cracow, 1969.
89. I. Stronski, Radiochem. Radioanal. Letters, 1, 191, (1969).
90. W.S. Sijperda and G. de Vries, Geol. en Mijnbouw, 45, 315(1966).
91. W.S. Sijperda and G. de Vries, Geol. en Mijnbouw, 47, 197(1968).
92. C. Testa, J. Chromatog., 5, 236 (1961).
93. C. Testa, Anal. Chem., 34, 1556 (1962).

94. C. Testa, Com. Naz. En. Nucl. Rept., RT/PROT (65) 33 (1965).
95. C. Testa, Minerva Fisiconucleare Torino, 10, 202 (1966).
96. C. Testa, Health Physics, 12, 1768 (1966).
97. G. de Vries and U.A.Th. Brinkman, in preparation.
98. G. de Vries and U.A.Th. Brinkman, unpublished observations.
99. G. de Vries en W.S. Sijperda, Geol. en Mijnbouw, 45, 275 (1966).
100. A. Waksmundzki and S. Przeszlakowski, in Stationary Phase in Paper and Thin Layer Chromatography (K. Macek and I.M. Hais, eds.), Elsevier, Amsterdam, 1965, p. 199.
101. A. Waksmundzki and S. Przeszlakowski, Chem. Anal., 11, 159 (1966).
102. K. Watanabe, XXth Conf. Chem. Soc. Japan, 1967.
103. G.A. Yagodin and A.M. Chekmarev, Radiokhimiya, 11, 234 (1969).
104. G. Bagliano, G. Grassini, M. Lederer and L. Ossicini, J. Chromatog., 14, 238 (1964).
105. L.S. Bark, G. Duncan and R.J.T. Graham, Analyst, 92, 31 (1967).
106. S.E. Bryan and M.L. Good, J. Inorg. Nucl. Chem., 21, 339(1961).
107. J.I. Bullock, S.S. Choi, D.A. Goodrick, D.G. Tuck and E.J. Woodhouse, J. Phys. Chem., 68, 2687 (1964).
108. C.C. Chang, Hua Hsueh Tung Pao, 1963, 624.
109. C.F. Coleman, Nucl. Sci. Eng., 17, 274 (1963).
110. C.F. Coleman, C.F. Blake, Jr. and K.B. Brown, Talanta, 9, 297 (1962).
111. C.F. Coleman, K.B. Brown, J.G. Moore and D.J. Crouse, Ind. Eng. Chem., 50, 1756 (1958).
112. R.M. Diamond, in Solvent Extraction Chemistry (D. Dyrssen, J.O. Liljenzin and J. Rydberg, eds.), North-Holland Publ. Co., Amsterdam, 1967, p. 358.
113. R.M. Diamond and D.C. Whitney, in Ion Exchange (J.A. Marinsky, ed.), M. Dekker, New York, 1966, Vol. I, p. 316 ff.
114. M.L. Good, S.E. Bryan, F.F. Holland, Jr. and G.J. Maus, J. Inorg. Nucl. Chem., 25, 1167 (1963).
115. T. Goto, J. Inorg. Nucl. Chem., 31, 1111 (1969).
116. E. Högfeldt, in Ion Exchange (J.A. Marinsky, ed.), M. Dekker, New York, 1966, Vol. I, p. 156 ff.
117. T. Ishimori, E. Akatsu, W. Cheng, K. Tsukuechi and T. Osakabe, JAERI Rept., 1062 (1964).
118. T. Ishimori and E. Nakamura, JAERI Rept., 1047 (1963).
119. K.A. Kraus and F. Nelson, Proc. Intern. Conf. Peaceful Uses At. Energy, Geneva, 1955, 7, 113 (1956).
120. D. Kuiper, Thesis, Amsterdam, 1969.
121. M. Lederer and L. Ossicini, J. Chromatog., 13, 188 (1964).
122. Y. Marcus, Coord. Chem. Rev., 2, 195, 257 (1967).
123. Y. Marcus, J. Appl. Chem., in the press.
124. Y. Marcus and A.S. Kertes, Ion Exchange and Solvent Extraction of Metal Complexes, Wiley, New York, 1968.
125. F.L. Moore, U.S. At. Energy Comm. Rept., NAS-NS 3101, (1960).
126. N.L. Olenovich, E.A. Mazurenko, V.N. Ermilov and M.M. Rogachko, Zavod. Lab., 30, 489 (1964).
127. S. Przeszlakowski, Roczniki Chemii, 43, 151 (1969).
128. S. Przeszlakowski, Roczniki Chemii, 43, 1113 (1969).
129. E.L. Smith and J.E. Page, J. Soc. Chem. Ind., 67, 48 (1948).
130. J.M. White, P. Kelly and N.C. Li, J. Inorg. Nucl. Chem., 16, 337 (1961).

CONTINUOUS SAMPLE FLOW DENSITY GRADIENT CENTRIFUGATION

George B. Cline
Department of Biology
College of General Studies
University of Alabama at Birmingham
Birmingham, Alabama 35233

I. INTRODUCTION

Zonal centrifugation procedures are useful for the isolation and purification of a wide variety of particulates ranging in size from whole cells (1,2) down through smaller components such as subcellular particulates (3, 4, 5) and viruses (6, 7, 8) to macromolecules such as the gamma globulin proteins of human and animal sera (9, 10). While zonal separations in density gradients can be performed in a variety of centrifuge rotors from the angle head (11) to the swinging bucket types (12), the recently developed series of zonal centrifuge rotors have filled a need by providing high resolution separation and purification of gram quantities of materials from contaminants in a single operation.

Zonal centrifuge rotors have a variety of shapes and descriptions and can be classified in several ways. The simplest distinction is to call those rotors which hold a fixed volume of sample as "batch-type" and those rotors in which the sample can either be a fixed volume or continuously varied as "continuous-sample flow-with-isopycnic-banding" (flo-band) type (13). While most batch-type zonal rotors can be operated in existing preparative-type centrifuges, the newer flo-band rotors have a different shape and configuration and require special centrifuge drive systems. It is this last class of zonal centrifuge rotors and systems which will be discussed here. This presentation will cover how the flo-band rotors work, most of the important facets to consider in using these rotors for separations and some selected separations of a wide variety of particulates.

II. BASIS FOR FLO-BAND SEPARATIONS

The rate at which a particle sediments through a density gradient depends primarily on the size and secondarily on the density of the particle. Zonal separation methods which rely strongly on distinction of size generally are quite satisfactory as high resolution techniques but the mass of sample recovered is small. These so-called rate separations require the zones of particulates to be removed from the gradient before they sediment through and hit the rotor wall.

If the proper gradient material is used, particulates can be banded isopycnically in the density gradient. But as soon as the particulates have sedimented from the starting sample zone, the sample can be replaced and the procedure repeated. The particulates continue to sediment into the gradient and band isopycnically. By replacing the sample at intervals we have a semi-batch type of separations method. It is generally easier to continuously move a thin layer of sample through the rotor at a rate which permits the particulate of interest to be sedimented from the flowing sample. This is the method of operation of the flo-band rotors.

By varying the sample flow rate through a flo-band rotor, it is possible to select particulates on the basis of size. Faster flow rates permit only the largest of particulates to be trapped while slower flow rates permit both large and smaller materials to be collected. Both types are then banded isopycnically. This two-dimensional separation (size vs. density) is made continuously and is a highly useful feature of the flo-band rotors.

Figures 1 and 2 show theoretical removal (cleanout) for several speeds of the K-II and J-I rotors respectively. These data are calculated from the physical characteristics of a particle, the geometry and gravitational force produced by the rotor (14). Measured cleanout is generally about 10% less than predicted by the theoretical curves.

III. CENTRIFUGE DESCRIPTIONS

The Model K and RK zonal centrifuges are air turbine driven systems which can spin rotors in a refrigerated vacuum up to 35,000 rpm in the Model K and up to 60,000 rpm in the Model RK. The rotors have access lines to both ends and are suspended from the drive system. The top access line goes through the center of the air turbine while the bottom access line goes through a journal bearing. Direct access is made to the contents of the rotor through enclosed seals. The seals are bathed in a closed loop coolant system which offers a high degree of safety when processing biologically hazardous materials.

For the reograd process the rotors must be accelerated slowly from rest and decelerated slowly to rest to permit the gradient to reorient without mixing. Electronic ramping circuits automatically control the air supply to the turbine for controlled changes in rotor speed. The ramp will accelerate the rotor in the range of from 1 rpm/sec to 10 rpm/sec up to 10,000 rpm. A second ramp will permit deceleration starting anywhere below 10,000 rpm and reduce the speed at any selected rate between 1 rpm/sec and 10 rpm/sec.

The acceleration rate of the rotor, once it is beyond the control range of the ramping circuit, varies with the type of rotor unit. The J-I rotor accelerates at about 13,000 rpm/min. while the larger K-III accelerates at about 500 rpm/min.

IV. TYPES OF FLO-BAND ROTORS

Table 1 shows a partial list of flo-band rotors which have been built and used for a variety of separations. Except for length, the rotors for the Model RK centrifuge have exactly the same dimensions as the Model K rotors and thus carry the same numeral. The J-I rotor is the latest in the series to be designed by the Molecular Anatomy Program of the Oak Ridge National Laboratory. This rotor is only made of titanium and has the highest speed and gravitational force of any of the flo-band rotors. It also holds the smallest volume of gradient.

Table II shows the numbering sequence and a brief description of the K- and RK- type rotors.

TABLE I

PHYSICAL CHARACTERISTICS OF REPRESENTATIVE K-, RK- and J-type ZONAL ROTORS

Parameters	K-II	K-X	RK-II	RK-III	RK-V	J-I
Maximum safe speed	35,000rpm	35,000rpm	35,000rpm	35,000rpm	35.000rpm	60,000rpm
Force (g-max) at safe speed	83,500	83,000	83,000	95,000	83,000	162,000
Capacity (liters) with core	3.6	6.8	1.8	1.7	3.3	0.95
Core radius average inches	3.8	2.354	3.8	4.3	1.65	1.352
Bowl outside diameter (Inches)	6.0	6.0	6.0	6.4	6.0	4.465
Bowl inside diameter (Inches)	4.8	4.8	4.8	5.2	4.8	3.5
Inside length (Inches)	30.0	30.0	14.1	14.1	14.1	16.0

TABLE II

K AND RK ROTORS (CORES AND SHELLS*)

Rotor No.	Titanium or Aluminum	Characteristics
I.	Ti	Reorienting gradient. Continuous-flow-with-banding. Small Volume. (Prototype only).
II.	Al	Reorienting gradient. Continuous-flow-with-banding.
III.	Ti	Reorienting gradient. Continuous-flow-with-banding.
IV.	Al	Dynamic loading and unloading in II shell. For rate separations.
V.	Al	Static and dynamic loading and unloading in II shell. For rate separations.
VI.	Al	Similar to II but with center debris trap.
VII.**	Ti	Dynamic loading and unloading in III shell. For rate separations.
VIII.	Ti	Static and dynamic loading and unloading in III shell. For rate separations.
IX.**	Ti	Similar to III but with center debris trap.
X.	Al	Reorienting gradient. Continuous-flow-with-banding (6.8 liters of gradient) in II shell.
XI.	Ti	Continuous-flow-with-pelleting core in III shell.

** Under construction

* Rotor shells are designated as K-II, RK-II or J-I.

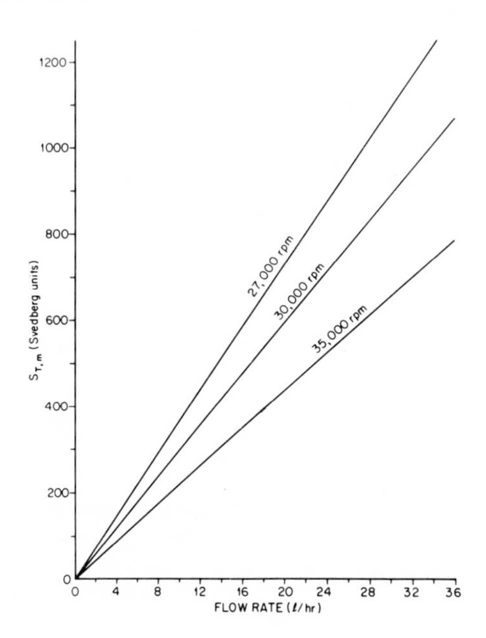

Figure 1

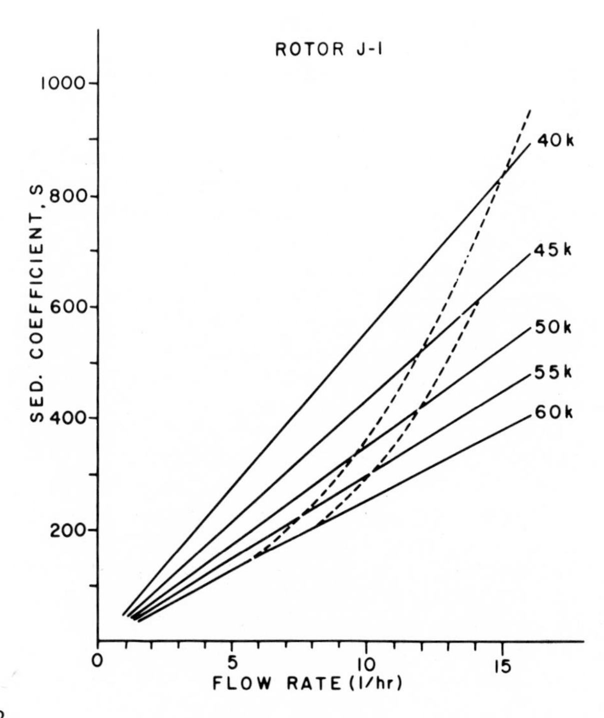

Figure 2

Figure 1. Calculated values for removal of particles from a moving sample by the K-II rotor as a function of sedimentation coefficient, rotor speed and sample flow-through rate. (Courtesy of Academic Press).

Figure 2. Calculated values for removal of particles from a moving sample by the J-I rotor as a function of sedimentation coefficient, rotor speed and sample flow-through rate. The dotted lines show the maximal flow rates when flow goes from bottom to top (top line) and flow from top to bottom (bottom line).

V. GRADIENT MATERIALS

Many particulates of biological origin are sensitive to many factors such as extremes in pH, changes in osmotic pressure, shear and hydrostatic pressure. Such factors must be kept in mind when choosing a gradient material for the separation of particulates of interest. Some important questions to consider are:

1. Does the gradient material affect viability of the product to be isolated?
2. Can the gradient material be readily removed from the product?
3. Does the gradient material affect assay procedures for the product?
4. Can the gradient material be sterilized?
5. Can it be buffered across the range of concentrations used?
6. Is it expensive and can it be recovered for reuse?
7. Does it have the density range to band the particulates isopycnically?
8. Does it have a high diffusion rate?
9. Is it corrosive to the metals in the rotor?

There does not appear to be one universally acceptable gradient material for all particulates. Although sucrose is cheap and perhaps the most widely used material, it has a high osmotic pressure and cannot be made dense enough to band many particulates.

VI. FORMING DENSITY GRADIENTS IN ZONAL ROTORS

Most of the gradients now being used in the flo-band rotors are of the discontinuous or step-type. These gradients are useful for either batch or continuous flow separations and are made simply by pumping into the rotor a series of solutions which differ in density. (This technique is commonly used to make gradients in swinging bucket tubes.) For zonal rotor separations we do not want the material to diffuse extensively and smooth out the gradient but rather remain in "steps." Gradient solutes always diffuse to some degree, and the diffusion interfaces between any two adjacent solutions give us ideal places in the gradient to stop and band individual species of particles. By using several solutions we have "gradients" within a gradient, and each diffusion interface can be used to stop and band separate types of particulates. Discontinuous gradients made of some materials have some interesting characteristics. When solutions of potassium citrate or cesium chloride or other ionizable material are layered on top of each other, it is possible to get liquid junction potentials across the interfaces. This potential can interrupt the normal sedimentation and banding of some particulates by causing aggregation. For gradient materials like these, linear density gradients are used where the concentration of the solute is varied continuously across the volume of the gradient in the rotor. To make linear gradients one must either use a gradient forming device or make discontinuous gradients and wait until the gradient diffuses. Each manufacturer of zonal rotors produces its own unique type gradient former which is geared for the volumes required by its respective zonal rotors. The choice of gradient shapes is large but basically the formers offer gradients which are either linear with rotor volume or gradients which are linear to rotor radius. Each type of gradient has its own separations characteristics and has been compared for highest resolution and capacity (15).

VII. THE REORIENTING GRADIENT TECHNIQUE

Most of the flo-band rotors for the K-, RK- series centrifuge utilize the reorienting gradient principle for simplicity of operation. This principle (16) utilizes the inherent stability of density gradients to keep the gradient oriented in line with the gravitational field. Gradients are formed in the rotors by pumping in a series of solutions which differ in density with the lightest density on top and the highest density in the bottom of the rotor. As the rotor is accelerated from rest, the gradient begins to slide up the rotor wall to stay aligned with the direction of gravitational force. At about 2,000 rpm the gradient is within 1 degree of being vertical and the gradient becomes more nearly verticle as the rotor speed is increased.

Separations are made in the gradient while it is spinning at high speed. The gradients are recovered by reversing the acceleration process. As the rotor is slowed, the gravitational force is decreased radially below that necessary to keep the gradient oriented on the wall, and the gradient slides down and reorients to the original horizontal position. The reorienting gradient technique is called the reograd process.

Low viscosity gradients by theory (17) reorient with little or no mixing. Higher viscosity gradients due to the higher shear forces involved therefore should not reorient as well. In practice there seems to be little difference between the two types of solutions; both reograd with little or no mixing if speed changes are smooth and slow below about 2,000 rpm. A recent study has shown that zone width with the reograd process is equal to or narrower than zones recovered by the dynamic loading and unloading process. (18)

VIII. SAMPLE SIZE

Although the Model K and RK zonal rotors were designed for large scale operations, it is difficult to say what the upper limits of sample size might be because of the many factors which affect it. One factor is the accumulation of material in the zones; it is possible to pack virus so tightly that severe aggregation problems occur and the product is difficult to recover from the rotor. The sample size would thus be limited by the concentration of the particulate in the starting sample. These factors can be listed:

1. Concentration of particulates in the starting sample.
2. Zone capacity of the gradient used.
3. Maximum density required to band the particulates.
4. Sample flow-through speed and gradient washout.
5. Type of gradient material (rate of back diffusion).
6. Temperature of sample and rotor.

Approximate sample volumes for the K-11 rotor using sucrose gradients and influenza virus is over 200 liters. Sample volume for the K-X rotor and the same materials is over 300 liters.

While most investigators will be concerned about the maximum amounts of sample which can be processed, some will be concerned with how little can be processed and not lost in the gradient. While it is difficult to state definite lower minimums, it would be probably not be warranted to flow a liter of sample through the 6.8 liter K-X rotor. These smaller volumes are easily covered by the smaller volume rotors such as the RK-II (1.8 liter gradient) or the J-I (950 ml gradient).

IX. SEPARATIONS PROCEDURE IN THE K-TYPE FLO-BAND ROTORS

Figure 3 shows a series of diagrams to outline the various stages in the operation of a K-type zonal rotor for a flo-band separation. To outline these steps briefly (omitting many of the small operational steps such as clamping lines or bubble removal), we can divide a separations into six stages.

Stage 1. Rotor set-up. This involves the selection of the right type of rotor for the separation, connecting it to the centrifuge drive and checking for possible seal leaks.

Stage 2. Gradient loading. The choice of gradient material and the gradient shape are dictated by the type(s) of particulates to be isolated. Generally, if one has a single class of particulate to be isolated, the gradient can be made of two solutions; and if two types of particulates are to be isolated, then a three step gradient is used. The gradient is loaded into the rotor by the bottom access line with the lightest solution first and followed by the more dense solution.

Stage 3. Rotor acceleration. This step is executed by the automatic ramping function of the control panel. The rotor is slowly accelerated to about 3,000 rpm to orient the gradient on the rotor wall. The flow-through process is established at this time using buffer or water. The rotor is then accelerated to the top selected speed for the separation.

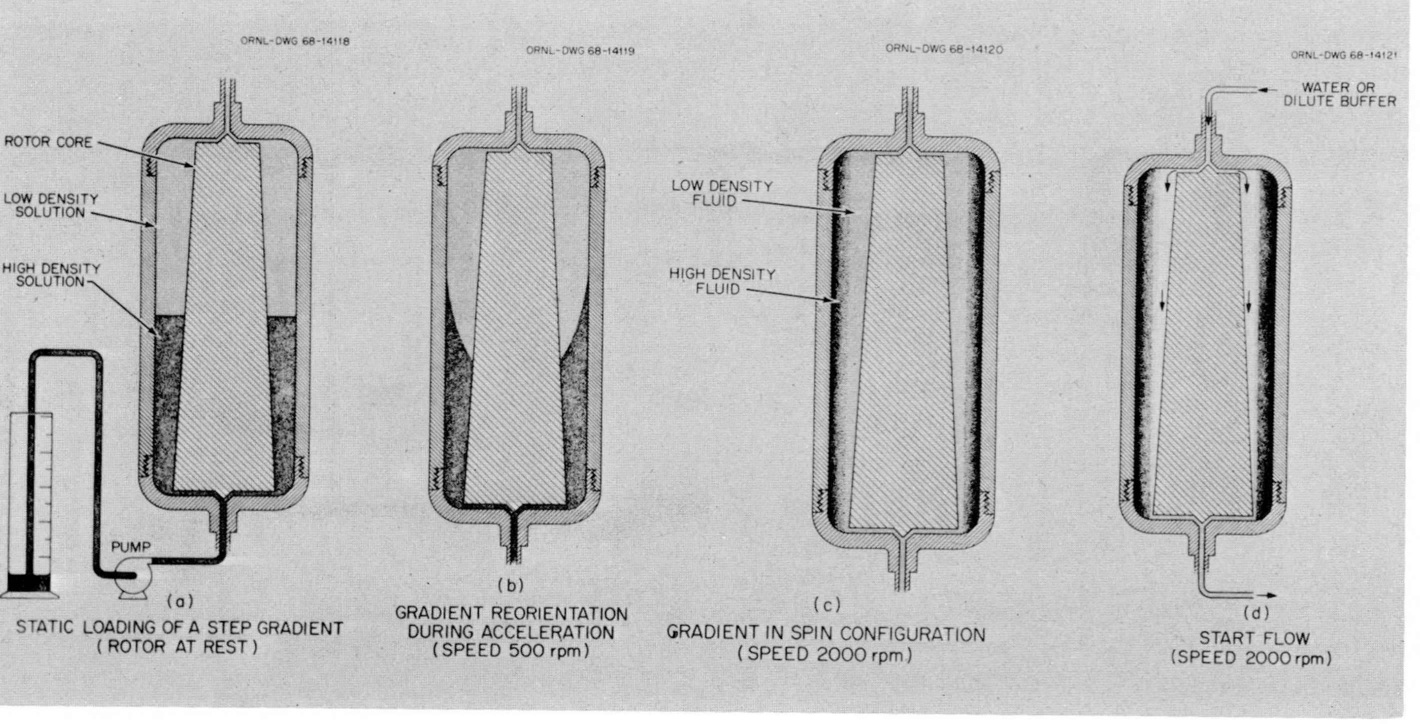
ORNL-DWG 68-14118
ROTOR CORE
LOW DENSITY SOLUTION
HIGH DENSITY SOLUTION
PUMP
(a)
STATIC LOADING OF A STEP GRADIENT
(ROTOR AT REST)
ORNL-DWG 68-14119
(b)
GRADIENT REORIENTATION
DURING ACCELERATION
(SPEED 500 rpm)
ORNL-DWG 68-14120
LOW DENSITY FLUID
HIGH DENSITY FLUID
(c)
GRADIENT IN SPIN CONFIGURATION
(SPEED 2000 rpm)
ORNL-DWG 68-14121
WATER OR DILUTE BUFFER
(d)
START FLOW
(SPEED 2000 rpm)

Figure 3

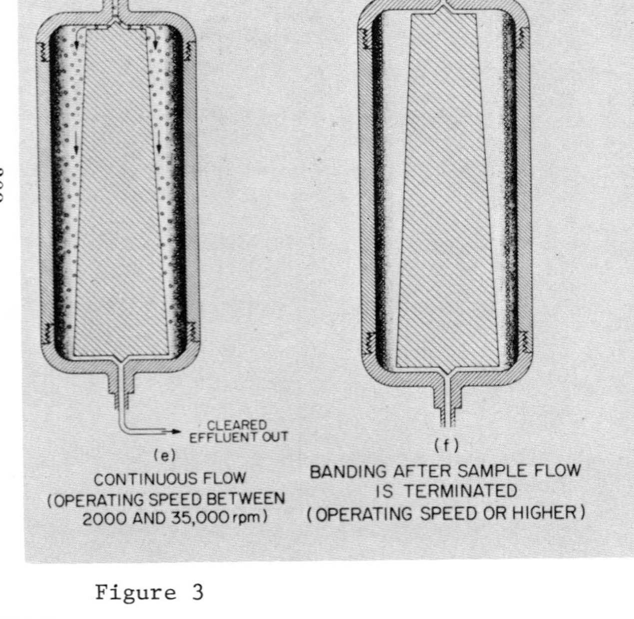
ORNL-DWG 68-14122
PARTICLE SUSPENSION IN
CLEARED EFFLUENT OUT
(e)
CONTINUOUS FLOW
(OPERATING SPEED BETWEEN 2000 AND 35,000 rpm)
ORNL-DWG 68-14123
(f)
BANDING AFTER SAMPLE FLOW IS TERMINATED
(OPERATING SPEED OR HIGHER)
ORNL-DWG 68-14124
(g)
GRADIENT REORIENTATION DURING DECELERATION
(SPEED 500 rpm)
ORNL-DWG 68-14125
(h)
ROTOR AT REST BEFORE UNLOADING

Figure 3

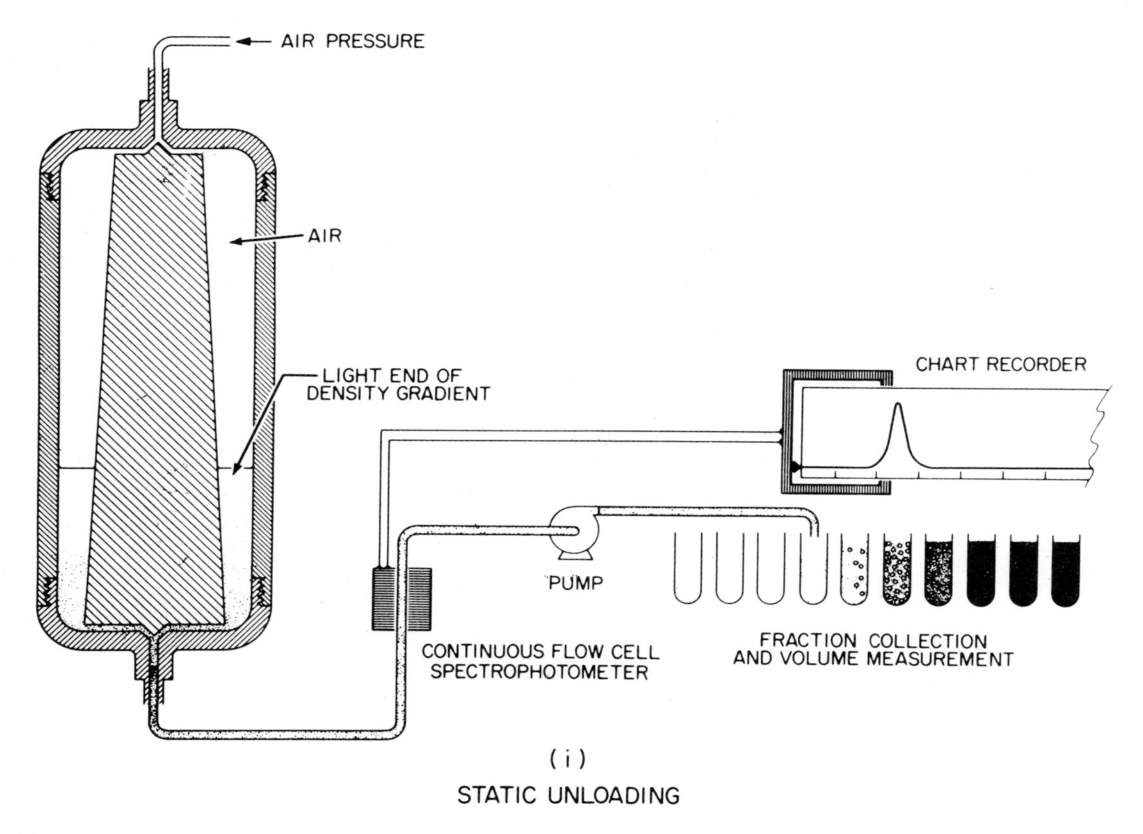

(i)
STATIC UNLOADING

Figure 3i

Figure 3. Schematic drawings of the operation of the K-II centrifuge using a static gradient loading and unloading procedure.

(*a*)- Static loading of a step gradient. The rotor, at rest, is filled with water or dilute buffer in the upper half and a high-density sucrose solution in the lower half.

(*b*)- Gradient reorientation during acceleration. The density gradient changes from a horizontal to a vertical orientation during acceleration from rest up to about 1600 rpm. Surfaces of isodensity are described by parabolas of revolution. The diagram shows gradient configuration for a speed of about 500 rpm.

(*c*)- Gradient in spin configuration. At speeds above 2000 rpm the density gradient is essentially in a vertical orientation.

(*d*)- Initiation of flow. At about 2500 rpm water or dilute buffer solution is pumped past the narrow end of the core downward, displacing slightly denser solution out at the large lower end of the core. When the taper volume is completely filled with the water or buffer, the fluid continues out through the lower shaft and seal. Flow is maintained during acceleration to the chosen operating speed.

(e)- Continuous-sample flow. When the chosen operating speed is reached, the flow is switched from dilute buffer or water to the particle suspension.

(*f*)- Banding after sample flow is terminated. After all of the particle suspension has moved through the rotor, centrifugation is continued for an interval sufficient to allow all particles, including those which entered the rotor last, to band.

(*g*)- Gradient reorientation during deceleration. Reverse air flow through the turbine is used to brake the rotor to about 2000 rpm. A controlled deceleration program is then used to bring the rotor gradually to a stop, with very gradual reorientation of the gradient to the rest configuration.

(*h*)- Rotor at rest before unloading. The concentrated particles are in a narrow band at their isopycnic level.

(*i*)- Static unloading of the gradient. Using air or water pressure and a small peristaltic pump to control flow, the gradient and suspended particles are recovered as a series of discreet fractions. Unloading may also be done through the upper line using dense fluid to displace the rotor volume. (Courtesy of Academic Press).

Stage 4. Sample flow through. When the rotor reaches the top operating speed the buffer flow rate is adjusted to the calculated rate for the sample. The buffer is then stopped and the sample quickly substituted. Small adjustments in flow rate are made and the sample is pumped through the rotor. The rotor is allowed to continue to spin for a time after completion of flow-through.

Stage 5. Rotor deceleration and gradient recovery. The rotor is braked to approximately 5,000 rpm and allowed to coast to about 3,000 rpm. It is then ramped down to rest and the gradient recovered by either of two methods. The easier method is to pump water to the top access line of the rotor to displace the gradient out the bottom. The second method is to recover the gradient from the top of the rotor by displacement with a dense solution to the bottom. The gradient is monitored continuously at one or more wavelengths by a spectrophotometer and collected into conveniently sized aliquots.

Stage 6. Sample assay and gradient plotting. The density is determined for each aliquot recovered and the data plotted on the chart tracing of the absorbance curves. The gradient shape will indicate if serious mixing or other problems are present. Also, density will indicate where particulates should be banded even though the spectrophotometer may not detect them. The aliquots containing the zone of particulates are then either centrifuged in an angle head rotor to pellet the particulates (if they will withstand pelleting) or dialyzed to remove the gradient material. Analysis of the fractions may be practical either before or after solute removal.

X. ZONE ASSAY PROCEDURES

The purpose of separating particulates in a flo-band rotor is both to handle large volumes of starting sample and to recover a purified zone of particulates. Assessment of the amount of particles trapped and banded as well as their purity is required. Flow through rates can be calculated from the gravitational force produced by the rotor and the size and density of the particle to be isolated. In practice, theoretical calculations have been close to empirical data, and have suggested the approximate flow through rate for optimal removal (cleanout) of the particles from the sample stream. It is still important to determine a cleanout curve for the particulate in question and this is done by comparing the amount coming through the rotor in the effluent against what was going in. When such determinations are made for a variety of flow through rates, the data can be plotted to show what percentage of the particulates are being trapped by the rotor. In practice, for small particulates like animal viruses 95% removal is acceptable, and this flow rate should be determined.

Although 95% of the initial concentration of particulates may be trapped by the rotor, the same amount of particulates will not be recovered in the particulate zone for a variety of reasons. The most prominent reasons are that the particulates do have a range in density and some will be outside of the major zone and in low concentration. Other particulates are adsorbed or trapped by debris and may end up elsewhere in the gradient. It is thus important to assay each recovered gradient aliquot to determine the actual distribution of material. When very large amounts of material are present, as in industrial separations of some viruses, the virus zone is so prominent that assay of other than the zone fractions is unwarranted.

The Model K-X rotor holds 6.8 liters of gradient and sample. Generally the gradient in research applications is collected in 100 ml aliquots, and most of these 68 fractions must be analyzed for a variety of components. Analyses by hand are almost prohibitive. Rapid analysis by computer-coupled analyzer is one of the solutions and the developer of the zonal centrifuge (19) has designed an analyzer to handle the zonal rotor fractions. The analyzer is called the GeMSAEC in credit for the development program supported by the National Institute of General Medical Sciences and the Atomic Energy Commission (20). Although other analyzer systems are useful for handling the fractions, the GeMSAEC analyzer and associated computer programs were designed for the zonal rotors and rapidly speed up the assay of each aliquot of the gradient.

XI. APPLICATIONS

The Model K, RK- and J-type flo-band rotors can be used for the isolation of any particulate which falls into the size range from about 50 microns down to macromolecules such as globular serum proteins. While discussion thus far about these rotors has been about their use for continuous sample flow, it is also important to show that the rotors are not limited only to continuous flow. Not flowing the sample through but holding it on top of the gradient will permit rate separations in these rotors, and indeed, the large gradient volume makes these rotors good for large batch-type samples (e.g., 600 ml.). A variety of replacable cores is available to change the function of the rotors. The rotors can also be operated efficiently with no density gradient to pellet material on the wall.

The applications presented below are divided into two groups; the first group contains selected flo-band applications and the second smaller group contains rate separations.

A. Viruses

Influenza virus. The Model K centrifuge and the type II rotor were originally designed for efficient isolation of influenza virus. Most of the major vaccine producers in the United States, Japan and several Eastern European countries are using the Model K and RK zonal centrifuges for the isolation of influenza virus and encephalitis virus (21). Since pharmaceutical companies are using the machines to make proprietary products, little information is published.

One early study on influenza virus was done in a K-II rotor operating at 27,000 rpm rather than the top speed of 35,000 rpm (8). At the reduced speed and using a barium precipitated virus preparation (partially purified and concentrated), a flow through rate of about 5 liters per hour resulted in approximately 70% to 80% of the chicken cell agglutinating (CCA) activity being collected by the rotor. The starting sample was the allantoic fluid from 15,000 eggs. Depending on the strain of virus processed, the purity of the isopycnically banded virus ranged from 8,500 CCA/mg protein to about 22,000 CCA/mg protein. Up to two-thirds of the virus collected was contained in a 200 ml aliquot of the gradient.

The degree of purity and efficiency of trapping (cleanout) of the virions from the sample depends in large part on the nature of the sample. Most influenza virus for human vaccines is grown in embryonated chicken eggs and the virus is obtained from the egg in the chorioallantoic fluid. Some by-products of breaking the eggs open, such as yolk platelets and fibrous proteins (denatured), can cause flow restrictions in the rotors. For this reason, several commercial processors are doing a prior clarification step to rid the virus suspension of these unwanted particulates and other compounds which are highly corrosive to the aluminum rotors.

B. Insect viruses

Most of the known types of insect viruses can be isolated and purified in sucrose density gradients in the Model K-, RK-, and J-type flo-band rotors. Most of these viruses fall into four classes called: nuclear polyhedral inclusion bodies, cytoplasmic polyhedral inclusion bodies, granulosis and pox. Some specific types within these classes have been found to band at different densities: DNA-containing nuclear type at 56% or 58% (22); the DNA-containing granulosis type at 58% (23); the RNA-containing cytoplasmic type at 60% to 64% (23); and the DNA-containing Pox type at 66% sucrose (23). Discontinuous density gradients are recommended for the isolation and purification of these viruses since density differences are small between the viruses and contaminating microbes and cell components, and adequate separations are not easily made by other types of gradients (linear with volume or linear with rotor radius).

Figure 4 shows a K-X rotor separation of the nuclear and cytoplasmic viruses of the Tussock moth. The volume of the rotor was 6700 ml and the discontinuous gradient was designed to band the viruses and bacteria equidistantly apart. While there is incomplete separation between the nuclear zone and the bacterial zone in this experiment, we now know that a longer centrifugation after flow-through gives much improved separation. The gradient is so shallow and the concentration of sucrose so close to the density of the nuclear particles that the observed sedimentation rate of the nuclear type particles is tremendously reduced (from about 30,000 to less than 25). The cytoplasmic particles are banded at 60%, the nuclear at 56% and the bacteria at 54%.

Figures 5 and 6 show two other separations on viruses from the cotton boll worm *Heliothis zea* and the Gypsy moth. These separations were made for high capacity rather than high resolution and show more overlap between the zones. K-X zonal rotor separation of the nuclear polyhedral inclusion bodies of *Neodiprion sertifer*, the European pine saw fly, has recently been reported (24).

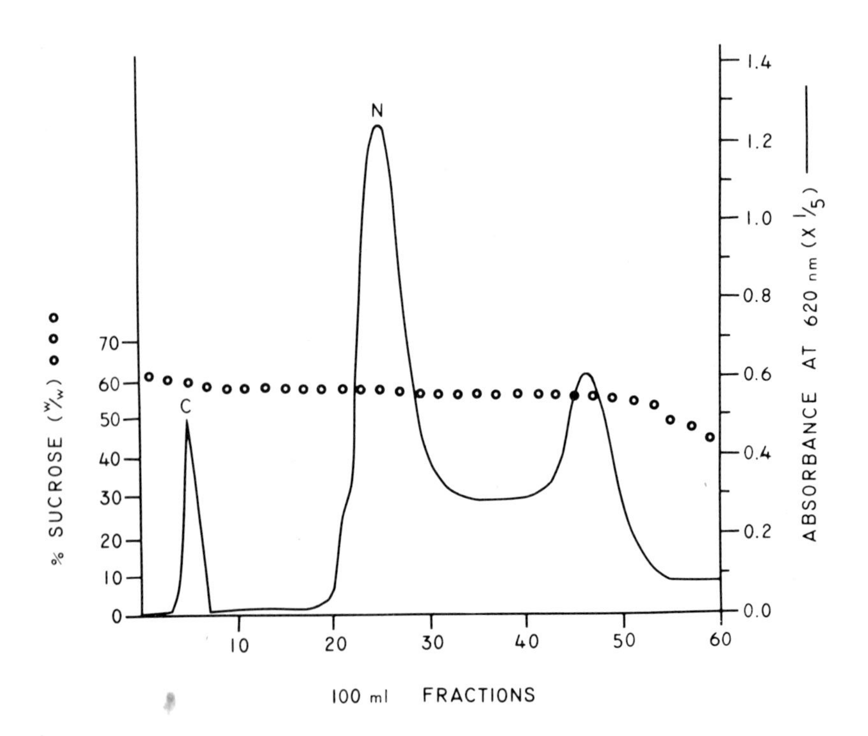

Figure 4. K-X separation of Cytoplasmic Polyhedral Inclusion Bodies (CPIB) from the Nuclear Polyhedral Inclusion Bodies (NPIB) from bacteria and cell components of the Tussock Moth, *Hemerocampa pseudotsugata*. Small differences in banding density require a shallow gradient for high resolution separations.

C. Cell components

Many components of plant and animal cells can be separated by continuous sample flow centrifugation. Some components, such as nuclei, are too dense to band isopycnically in sucrose and penetrate the gradient to the rotor wall. Albright (25) has used this property to collect about 500 grams of rat liver nuclei in one experiment in the K-X rotor. The rotor was used in the continuous flow mode with a gradient to hold all of the other components off the wall except the nuclei. The nuclei were subsequently recovered by washing the rotor wall.

Other components such as mitochondria, chloroplasts, smooth and rough endoplasmic reticulum, plasma membrane, lysosomes, and peroxisomes can be isolated in the flo-band rotors. Only preliminary data is presented here since work is in progress and most separations were made in non-refrigerated gradients. Refrigeration will improve the separations by holding the gradients in shape and by reducing band spread of the particulate fractions.

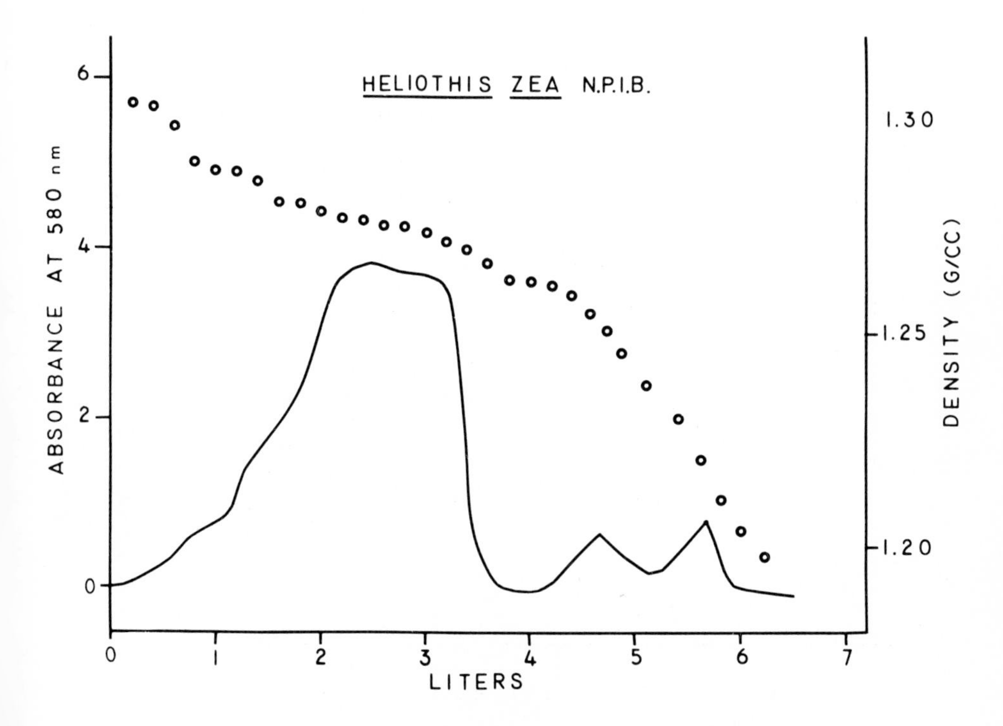

Figure 5. K-X separation of NPIB from a lypholyzed preparation of the virus infected cotton boll worm Heleothis zea. The NPIB's are banded in the gradient between fractions 2 and 3. (Courtesy of Dr. Julian Breillatt, Oak Ridge National Laboratory).

D. Rat liver

Figure 7 shows a K-II rotor separation of components from a homogenate of rat liver prepared in 8.5% sucrose 0.01M phosphate buffer containing 0.001M $MgSO_4$. All gradient solutions were made up in the same buffer solution. The figure legend shows the volumes and concentrations of the solutions used to make the gradient. The sample flow rate was 21 liters per hour at a rotor speed of 20,000 rpm, and the rotor was spun at 35,000 rpm for 30 minutes after completion of the flow through. The mitochondria were recovered in 45.7% sucrose while the adjacent zones at 46% and 47.3% contained rough endoplasmic reticulum. The lighter two zones were identified by light microscopy as membranous.

Best separations of mitochondria are made when no ions are added either to the starting homogenate or to the gradient. It is also suggested that an initial step of differential centrifugation be used to separate many of the contaminants from the mitochondrial preparation.

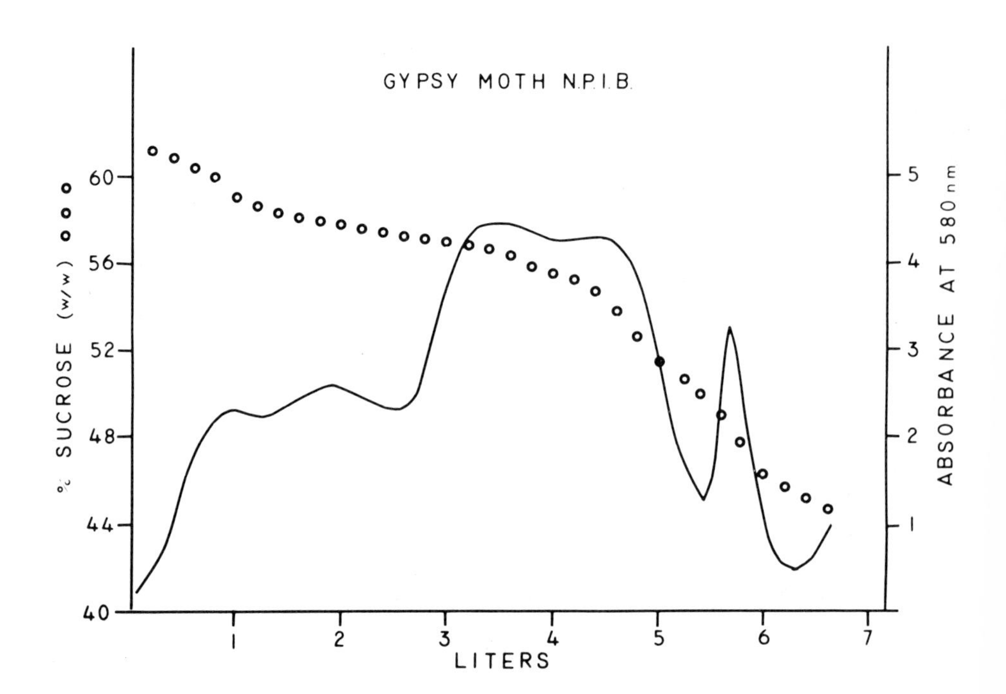

Figure 6. K-X separation of NPIB's from an homogenate of virus infected Gypsy moth. The polyhedra were banded isopycnically in two adjacent zones centered in approximately 55% and 57% sucrose. (Courtesy of Dr. Julian Breillatt, Oak Ridge National Laboratory).

E. Plant cells

Figure 8 shows a study of the trapping efficiency (cleanout) of the K-II rotor for chloroplasts from collard plant leaves. The chloroplast concentration was determined by hemocytometer count both on the starting sample and on the effluent after varying the flow-through rate. The rotor speed was 4,000 rpm.

The data from the cleanout experiment above was used to select the flow rate of 15 liters/hour for the separation shown in Figures 9 and 10. Figure 9 shows a separation of chloroplasts made in phosphate buffer while Figure 10 shows a similar separation of material prepared in Tris buffer. Two chloroplast zones were recovered in each case centered at about 39% and 46% sucrose. The gradients were slightly different in the two separations.

F. Microbe collection

The flo-band rotors are used without density gradients for the collection of bacteria from fermenters. In rotors such as the K-X where the internal volume is 6.8 liters, almost the entire capacity

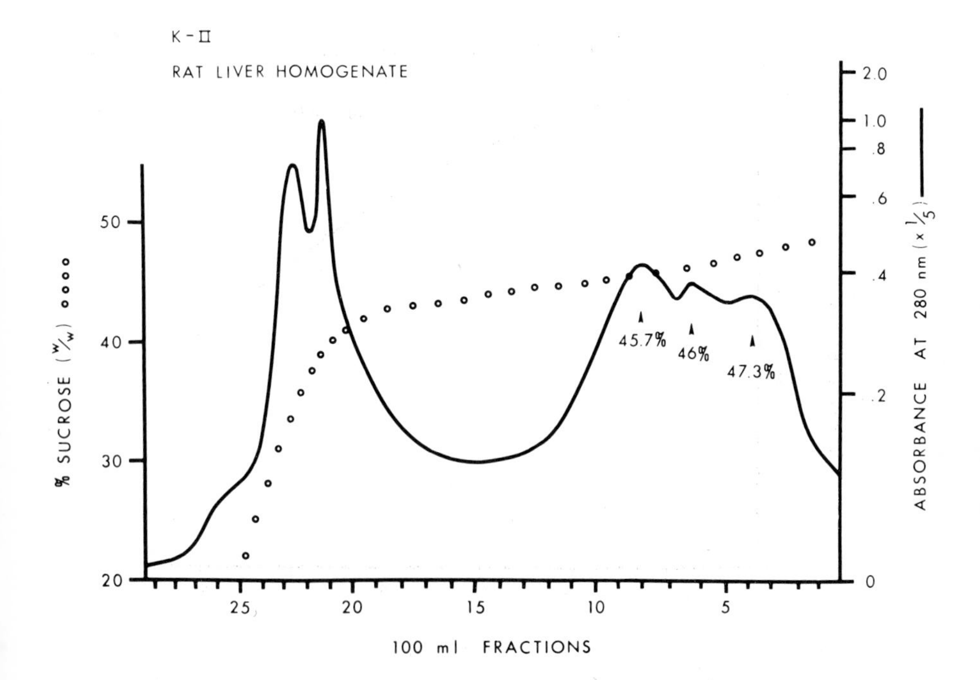

Figure 7. K-II separation of mitochondria and other large components from an homogenate of rat liver. The liver was homogenized in phosphate buffered saline (PBS), pH 7.25, and diluted with 2 liters of cold PBS. See text for details.

can be filled with cells. When the rotor is full, it must be dismantled and the contents recovered by scraping the "pellet" out. E. coli is collected at 30 liters per hour at 30,000 rpm in the K-X rotor.

G. Solution sterilization

Occasionally it is not only important to recover particulates from solution but to "purify" the solution of particulates. This is done by using the flo-band rotor to remove the particulates from the moving sample. One application is the removal of ribosomes from soluble phase of bacterial systems.

Since the various hepatitis "agents" are in the size range of Australia antigen or bigger (paramyxovirus candidates), it should be practical to "sterilize" plasma of these agents.

H. Water fractionation

The flo-band rotors have been used for the analysis of water borne particulates of both inorganic and organic origin (26).

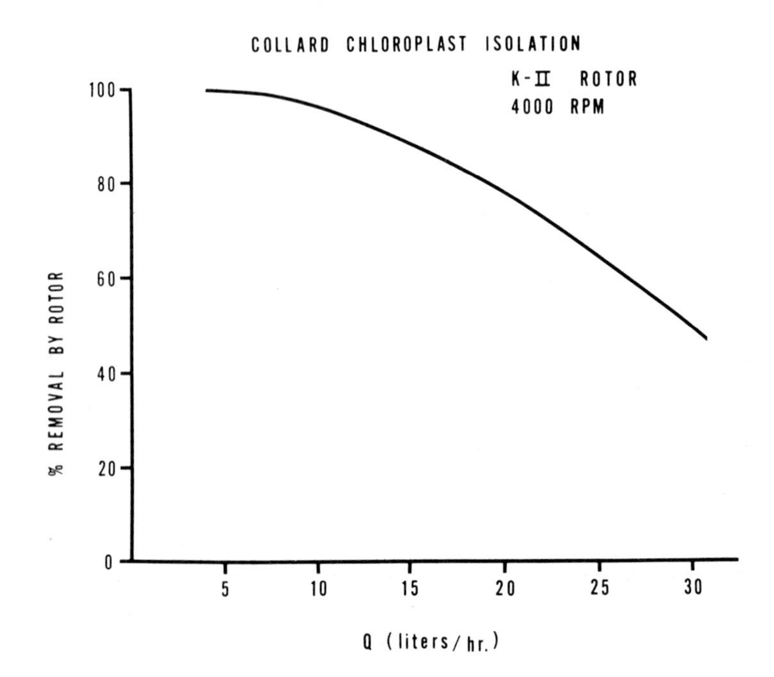

Figure 8. A cleanout curve for various types of chloroplasts from the collard plant in the K-II rotor. Cleanout represents the percent of the starting titer of particles which is removed by the rotor from the starting sample. Particles were counted by hemocytometer. Data points fell within ± 5% of the curve.

Figure 11 shows a fractionation scheme used for the analysis of sea water for the determination of the distribution of biomass, its viability and levels of pesticide contamination. While fresh water samples can be processed in aluminum rotors, the same rotors cannot be used for salt water because of the corrosion problems. Only titanium should be used for sea water.

The mineral prospecting capability of these rotors has not been exploited. During water fractionation studies mineral colloids are isolated. A systematic analysis of these permits not only a quantitative determination of the minerals present but also an analysis of radionucleides present in the water.

I. Rate separations in flo-band rotors

1. Serum proteins

Flo-band rotors were made by converting batch type rotors for continuous sample flow capabilities (13). The flow band rotors

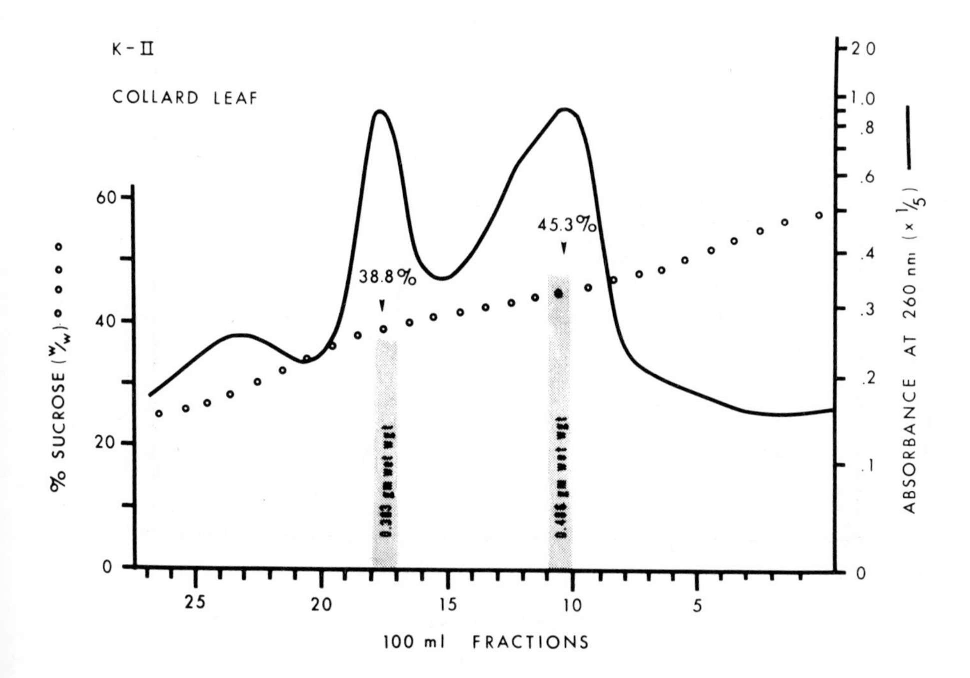

Figure 9. Gradient profile of chloroplasts collected from the experiment in Figure 8. The chloroplast peaks are shown along with the wet weight of chloroplasts in one fraction from the center of each peak. Phosphate buffer used throughout sample and gradient. See text for details.

can thus still be used for rate separations. One of the important applications is the separation of classes of proteins from both human and animal serum. The large capacity of the rotors permit the use of large samples, and separations are rapidly made because the sedimenting distance is relatively short. When warm gradients are used (10) separations of the 19 S macroglobulin fraction can be accomplished in about 4 hours. The sucrose gradient must contain enough salt to keep the macroglobulin in solution.

2. Fibrin sheets

Fibrin from whole, frozen or citrated blood can be isolated as a sheet in the Model K- and RK-type rotors. When calcium (0.03M) is added to citrated plasma and stirred for several hours in the cold, and then allowed to warm up, the fibrin molecules polymerize and form a tangled mass. When calcium treated plasma is centrifuged in a zonal rotor, the polymerized fibrin molecules sediment to the rotor wall and are pressed into a thin sheet (27). If a layer of 20% sucrose is used on the rotor wall, the fibrin

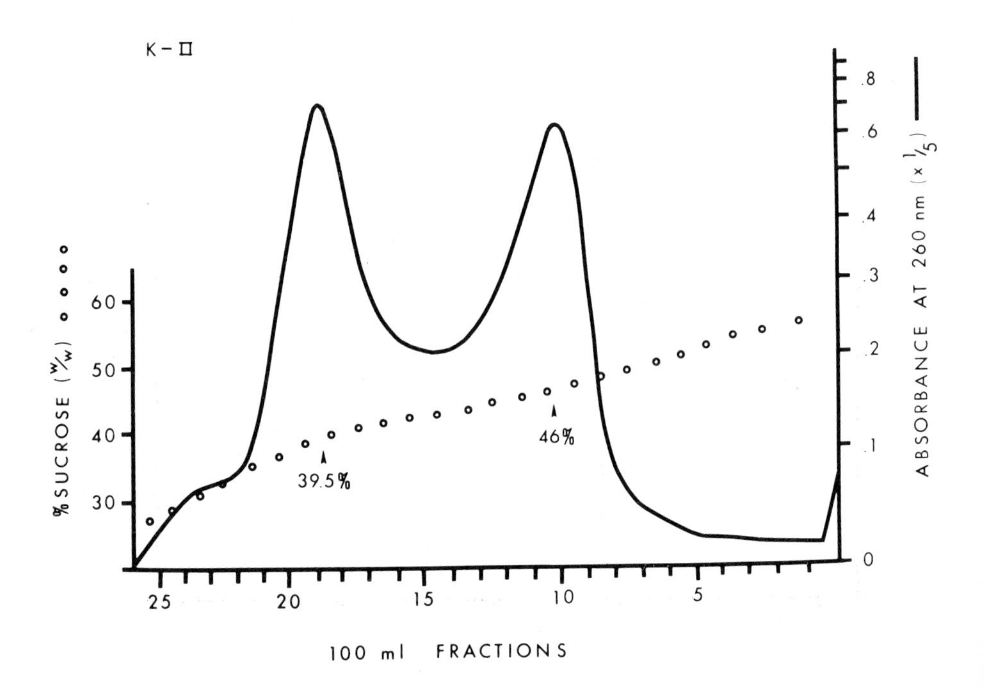

Figure 10. Gradient profile of a collard leaf chloroplast separation in the K-II rotor. Sample flow through rate was 15 liters/hour; rotor speed, 4000 rpm. Tris buffer used throughout sample and gradient. See text for details.

will pass through the sucrose and leave the serum proteins behind. The sheet is thus free of serum proteins and is non-antigenic. There are a number of suggested uses for the sheets and these include wrappings for burns, substrate for cell growth and others. The large size of the Model K rotor gives large sheets measuring 76 cm by 42 cm.

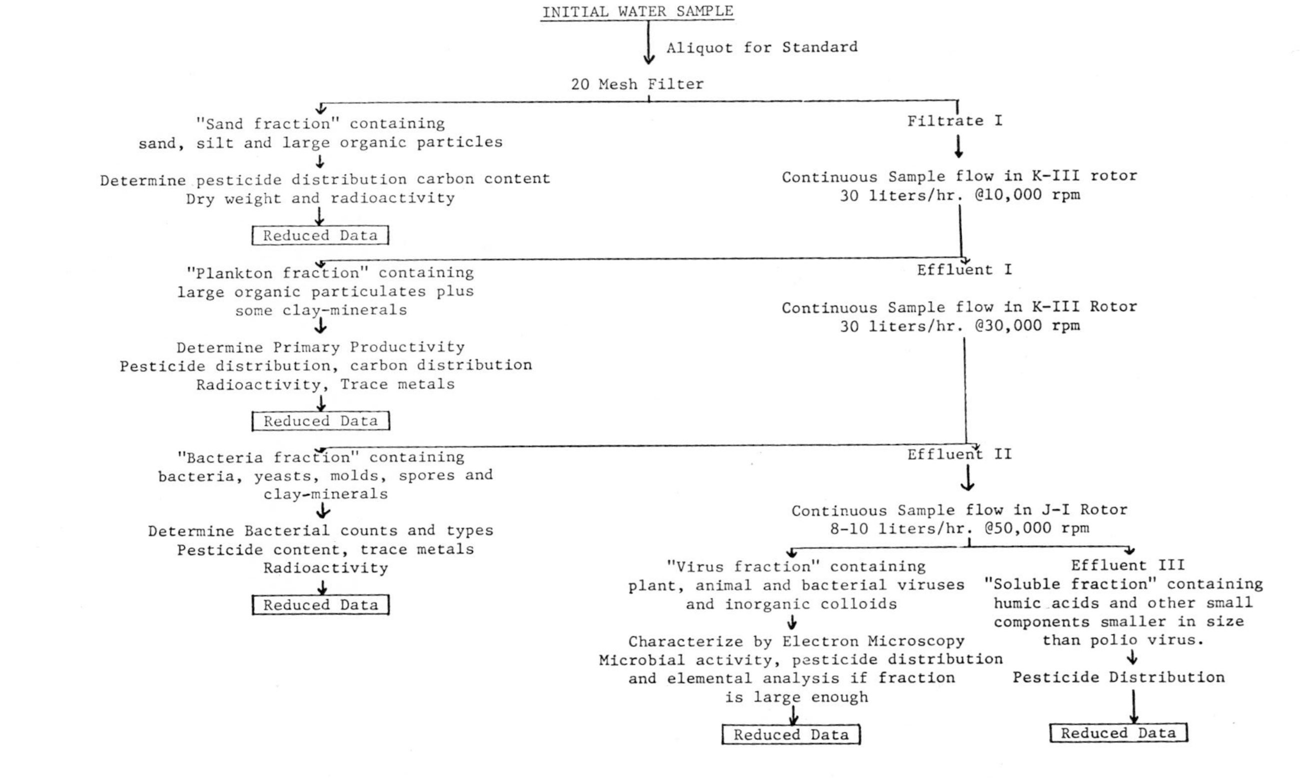

Figure 11. A fractionation scheme for the isolation of several classes of water-borne particulates. Material trapped at each step is separated on the basis of isopycnic banding density.

ACKNOWLEDGEMENTS

The work reported from the author's laboratory was kindly supported by a grant from Electro-Nucleonics, Inc., Fairfield, New Jersey and the Electro-Nucleonics Laboratory, Inc., Bethesda, Maryland. The author thanks Mr. R. Ryel, Mrs. E. Ryel and Mrs. M. Dagg for their assistance and thanks Dr. Julian Breillatt of the Oak Ridge National Laboratory for data of unpublished experiments.

REFERENCES

1. Shortman, K., in Progress in Separation and Purification, Vol. 2, (Theo Gerritsen, ed.) Wiley Interscience, p. 167 (1969).

2. Halvorsen, H. O., unpublished studies.

3. Anderson, N. G., and Cline, G. B., in Methods in Virology, (K. Marmarosch and H. Koprowski, eds.), Vol. II, pp. 137-178. Academic Press, New York, 1967.

4. Cline, G. B., Ryel, R. B. and Dagg, M. D., Microsymposium in Particle Separation from Plant Material, Oak Ridge National Laboratory, Oak Ridge, Tenn. (1970).

5. Cline, G. B., and Ryel, R. B., in Methods in Virology, (Colowick and Kaplan, eds.) Academic Press, 1971.

6. Cline, G. B., Coates, H., Anderson, N. G., Chanock, R. M. and Harris, W. W., J. of Virology 1, 659 (1967).

7. Cline, G. B., Nunley, C. E., and Anderson, N. G., Nature, 212, 487 (1966).

8. Reimer, C. B., Baker, R. S., Van Frank, R. M., Newlin, T. E., Cline, G. B. and Anderson, N. G., J. of Virology 1, 1207 (1967).

9. Fisher, W. D. and Canning, R. E., Natl. Cancer Inst. Monograph, 21, 403 (1966).

10. Cline, G. B., ORNL- 4171 Special, pp. 84-86 (1967).

11. Fisher, W. D., Cline, G. B. and Anderson, N. G., Anal Biochem. 9, 477 (1964).

12. Brakke, M. K., Adv. Virus Res. 7, 193-224 (1960).

13. Anderson, N.G., Waters, D. A., Nunley, C. E., Gibson, R. F., Schilling, R., Denny, E. C., Cline, G. B., Babelay, E. F., and Perardi, T. E., Anal Biochem. 32, 460 (1969).

14. Berman, A. S., Natl. Cancer Inst. Monogr. 21, 51 (1966).

15. Griffith, O. M., Cline, G. B., and Ryel, R. B., Unpublished studies.

16. Anderson, N. G., in Natl. Cancer Inst. Monogr., 21, 1 (1966).

17. Hsu, H. W. and Anderson, N. G., Biophy. J., 9, 173-188 (1969).

18. Cline, G. B., Brantly, J. N. and Gerin, J., Unpublished studies.

19. Anderson, N. G., Science, 166, 317-324 (1969).

20. Anderson, N. G., Anal. Biochem., 23, 207 (1968).

21. Fennell, R., Personal Communication.

22. Cline, G. B., Ryel, E., Ignoffo, C., and Shapiro, M., Proceedings of Internatl. Colloquim on Insect Pathology, University of Maryland, 1970.

23. Cline, G. B., Ibid.

24. Mazzone, H. M., Breillatt, J. P., and Anderson, N. G., Ibid.

25. Albright, C., Microsymposium on Separation of Plant Cell Particulates, Oak Ridge National Laboratory, Oak Ridge, Tenn., 1970.

26. Lammers, W. T., in Chemical and Microbiological Analysis in Water and Water Pollution, (L.L. Ciaccio ed.) Marcel Dekker, N.Y., 590-634 (1971).

27. Breillatt, J., Harrell, B. W., and Boling, K. W., Submitted to Science.

CONTINUOUS CHROMATOGRAPHIC REFINING

P. E. Barker
University of Aston
Birmingham, England

I. INTRODUCTION

Continuous chromatographic refining (C.C.R.) is a mass transfer separation technique which originated soon after the inception of analytical gas-chromatography by James and Martin in 1952. Whereas analytical gas-chromatography has emphasis on the separation of minute quantities of material, C.C.R. is an attempt to optimize the use of the main components of gas chromatography namely an inert gas, a solid carrier and a relatively non-volatile liquid for the separation of much larger quantities of feed mixtures ranging from light hydrocarbons to essential oils, metal chelates, etc. More recently the C.C.R. technique has been extended to include the continuous separation of non-volatile mixtures such as enzymes, proteins and carbohydrate polymers like Dextran, etc., where now the system consists of for example, an inert mobile liquid phase and a chromatographic medium such as controlled pore size silica beads.

In optimizing the chromatographic type system for larger scale separations, as the scale increases from gms/hr to kgms/hr operation, not only is it of importance to design the most efficient separation system, but the most efficient system at minimum overall operating and capital equipment cost. Different viewpoints of the way the chromatographic system can best be scaled up have been taken over the last fifteen years and these fall into two main categories (a) batch (b) continuous.

In (a) a direct scale-up of the analytical process is attempted by using larger diameter packed beds, incorporating perhaps a baffling system which gives strong radial unit mixing but minimum logitudinal mixing. In (b) the chromatographic bed is moved counter-current to the inert mobile phase so that by judicious choice of flowrates of bed and inert phase, a component or group of components move preferentially with the inert mobile phase while the other component or remainder of the components move in the direction of the chromatographic bed.

The batch process was the first to be tried in 1953/54 and the technique has been actively developed over the intervening years so that commercial units of 12 in. diameter or greater are now available.

The continuous mode of processing was introduced soon after (1955/56) and has taken several forms. In particular may be mentioned the vertical moving bed technique where attrition of particles occurs and later the rotating circular column technique which eliminated the mechanical handling and attrition of the chromatographic particles. While scientific publications have only given details of small diameter continuous units, the author has reason to believe that continuous units comparable to the batch units are in commerical operation.

Those investigators who choose the continuous type processing do so primarily because (1) the sheer volume occupied by large scale injections of feed into a batch column reduces the efficiency of the column although to some extent this can be offset by periodic injections of smaller quantities of feed. (2) With the continuous process loss of mass transfer efficiency with increasing column size is not as critical, since column length can be increased and purity of products readily achieved. Increasing column length in the batch process past a certain point will cause more back mixing and even poorer product purities. (3) To get band shift between two components in the batch process requires a greater number of plates than in the continuous, although some or all of the advantage may be lost if a mixture containing several components has to be separated, since N-1 columns are required to separate an N component mixture completely as in other continuous mass transfer processes. (4) Continuous processing is in general favored in the Chemical Industry in the majority of mass transfer separation processes, since experience has shown that greater throughputs, higher purities and lower costs are normally possible than an equivalent batch process.

II. THE PRINCIPLE OF CONTINUOUS CHROMATOGRAPHIC REFINING

This method, like other processes for separating different compounds is dependent upon the components distributing themselves differently between two phases, one mobile the other stationary. The mobile phase may be gas or liquid. The stationary phase may be a non-volatile solvent absorbed on the surface of a suitable solid such as keiselguhr, the solid acting as a carrier and taking no part in the separation, or alternatively a solid phase such as an adsorbent, molecular sieve, ion exchange resin, etc.

Each phase has a preferential selectivity for one or other of the solute components. Movement of two phases countercurrently will produce in each phase a relatively high concentration of one solute and a relatively low concentration of the other, so that given a sufficient height of column complete separation is possible.

The presence of the stationary non-volatile solvent phase in continuous chromatography acts in a similar manner to the solvent phase in extractive distillation, whereby gas-liquid equilibria are modified by the presence of the solvent, making a physical separation of the components of the mixture easier. Moreover by distributing these solvents as very thin films on specially prepared solids with a high surface to volume ratio, high mass transfer rates are obtained. When a solid phase alone is used a variety of surface phenomenon which may be adsorbtive, diffusive or electronic come into play thereby producing a selective retention of one component over another.

Most of the work todate in continuous chromatography has involved gas-liquid systems and for this reason the following theoretical treatments refer to such systems. However, in the subsequent treatment it is hoped the reader will not forget the generality of the technique and its application to other two phase systems.

The suitability of a non-volatile solvent for a given separation together with conditions of operation may be judged from a chromatogram (see Figure 1) of the binary mixture (AB) obtained on a gas chromatography apparatus.

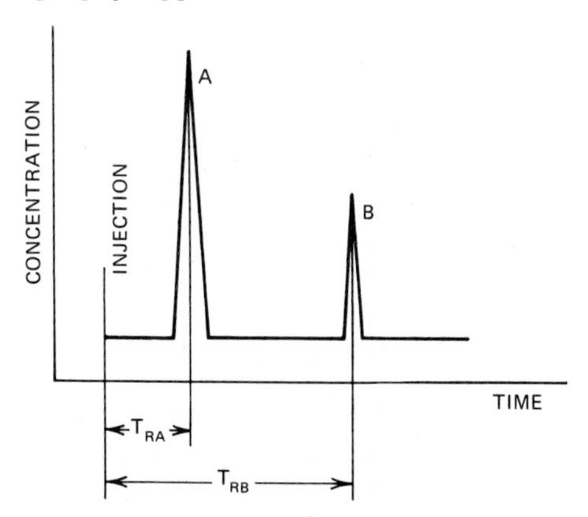

Figure 1. Chromatogram of the binary mixture AB.

The linear velocity of component A through the column is L/t_{RA} where L is the length of the column and t_{RA} the retention time of component A. Similarly the velocity of B is L/t_{RB}. Thus if the column packing is moved countercurrently to the carrier gas stream at a velocity greater than L/t_{RB} but less than L/t_{RA} component B will be carried in the direction of the packing absorbed in the liquid phase. Component A will be eluted in the carrier gas.

It is more convenient to express the conditions of separation in terms of the volume flow of solvent and carrier gas rather than linear velocities.

The partition coefficient K, for a particular component may be defined as:

Weight of component per unit volume of solvent phase to weight of component per unit volume of carrier gas phase. Let the partition coefficient of components A and B at the temperature of the rectification section be K_A^R and K_B^R respectively (see Figure 2). Let the volume flow rates of solvent and carrier gas in the rectification section be S and F^R respectively. These may be considered constant when the temperature of the column is kept constant, and the pressure drop across the column is small.

A material balance on component A around the feed point gives

$$M_A = F^R Y_A + SX_A \qquad [1]$$

where M_A is the feed rate of component A to the column, $F^R Y_A$ denotes the amount of component traveling upwards in the gas stream and SX_A the amount traveling downwards in the solvent stream. Y_A, X_A are the weight of component A per unit volume of carrier gas phase and solvent phase respectively.

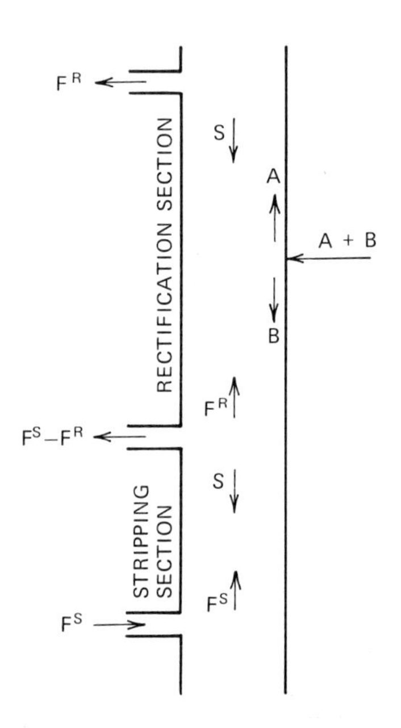

Figure 2. Gas and solvent flows.

The condition for resultant movement of component A up the column is that

$$F^R Y_A > SX_A \qquad [2]$$

i.e.

$$\frac{F^R}{S} > \frac{X_A}{Y_A} \quad . \qquad [3]$$

Now assuming the amount of feed entering the column is very small, so that infinite dilution conditions obtain in the gas and liquid phases

$$\frac{F^R}{S} > K_A{}^R \quad . \qquad [4]$$

Similarly for component B to travel preferentially down the column

$$\frac{F^R}{S} < K_B{}^R \quad . \qquad [5]$$

Hence, separation will occur only if the ratio of gas to liquid flows is between the partition coefficients under investigation

$$K^R{}_A < \frac{F^R}{S} < K^R{}_B \quad . \qquad [6]$$

Component B will be stripped from the solvent if

$$K^{S}_{B} < \frac{F^{S}}{S} \quad [7]$$

where K^{S}_{B} is the partition coefficient of component B at the temperature of operation of the stripping section, and F^{S} is the carrier gas flow rate in that section.

A. Effect of Finite Concentration of Solutes in the Mobile Phase

The preceding relations are true only at the infinite dilution and for an infinite number of plates in the packed section. As neither of these conditions is realized in practice, the real operating ranges will be slightly different.

It was shown by Barker & Lloyd (1) for hydrocarbon systems that the effect of finite concentration can either increase or decrease the partition coefficient depending on the system and the concentration, but in general for chromatographic systems it is found either to have negligible effect or to increase the value of the partition coefficient.

Assuming K to increase with component concentrations, the actual operating range from equation [6] will be according to Fitch et al. (2)

$$(K_{A}^{R} + \gamma_{A}) < \frac{F^{R}}{S} < (K_{B}^{R} + \gamma_{B}) \quad [8]$$

where γ_A, γ_B are factors accounting for the effect of finite concentrations on partition coefficient. The effect of a finite column length i.e., a finite number of theoretical plates is to narrow the operating range.

$$(K_{A}^{R} + \gamma_{A} + \delta_{A}) < \frac{F^{R}}{S} < (K_{B}^{R} + \gamma_{B} - \delta_{B}) \quad [9]$$

where δ_A, δ_B are factors accounting for effect of finite column length on partition coefficient.

B. Probabilistic Approach

Sciance & Crosser (3) have developed a relationship between the required column length, the feed location and the degree of separation of a binary mixture based on the probability theory. Assuming no longitudinal diffusion, and that the rates of adsorption and desorption are in equilibrium (i.e. their ratio is approximated by the partition coefficient) the relations derived for a column having the feed at its center are

$$\mathrm{Ln} \quad (\mu_Z)_A = \frac{L.R_A}{2\ F^{R}} \quad (K_{A}^{R} - \emptyset) \quad [10]$$

$$\mathrm{Ln} \quad [1 - (\mu_Z)_B] = \frac{-L\, R_B}{2\, F^R} (K_B^R - \emptyset) \qquad [11]$$

where $(\mu_Z)_A$ = mass ratio component A in bottoms/feed,
$1 - (\mu_Z)_B$ = mass ratio component B in tops/feed,
$\emptyset$ = operating F^R/S,
L = column length, and
R_A, R_B = rate constants of desorption.
If the system is designed to give equal product purity, equations [10] and [11] can be combined and simplied.

$$\frac{R_B}{R_A} = \frac{\emptyset/K^R_A - 1}{SF - \emptyset/K^R_A} \qquad [12]$$

where SF = separation factor (K^R_B / K^R_A).

To find the required column length for a given separation, knowledge of the values R_A and R_B is necessary. The only way these values can be determined is by curve fitting on the elution curve obtained from an analytical column (39). The same theoretical operating conditions are obtained by this probabilistic approach for the separation of binary mixtures as by the method used to derive equation [6].

III. GENERAL DESCRIPTION OF A MOVING BED APPARATUS FOR BINARY SEPARATIONS

A typical apparatus for moving bed chromatography is shown in Figure 3. The vertical column which may be of metal or glass and usually about 1 in.diameter is fed by chromatographic type packings from a hopper A, the flow of solids being controlled by a variable orifice B at the column base and removed by a feed table C. The solids pass through a rectification section D and into a stripping section E, which is surrounded by an electrically heated air jacket. The section between the base of the stripper and the orifice enables the solid to cool down prior to passing to the atmosphere or alternatively to be gas-lifted back to hopper A. The column is vibrated to ensure a continuous steady stream of solids. The lengths of D and E are shown in Figure 3 to be 5ft and 2ft 6 in respectively, although D could be any length depending on the difficulty of separation. A dry gas enters a heater H, passing to the column at a pressure slightly above atmospheric to prevent leakage of moist air into the column.

The liquid or gas mixture to be separated is injected into the column by a micropump J. The more strongly absorbed component passes down in the liquid phase and is removed in the heated stripper by the gas. The less strongly absorbed component passes up in the gas stream. Both products pass through a recovery system which usually consists of cold traps containing solid

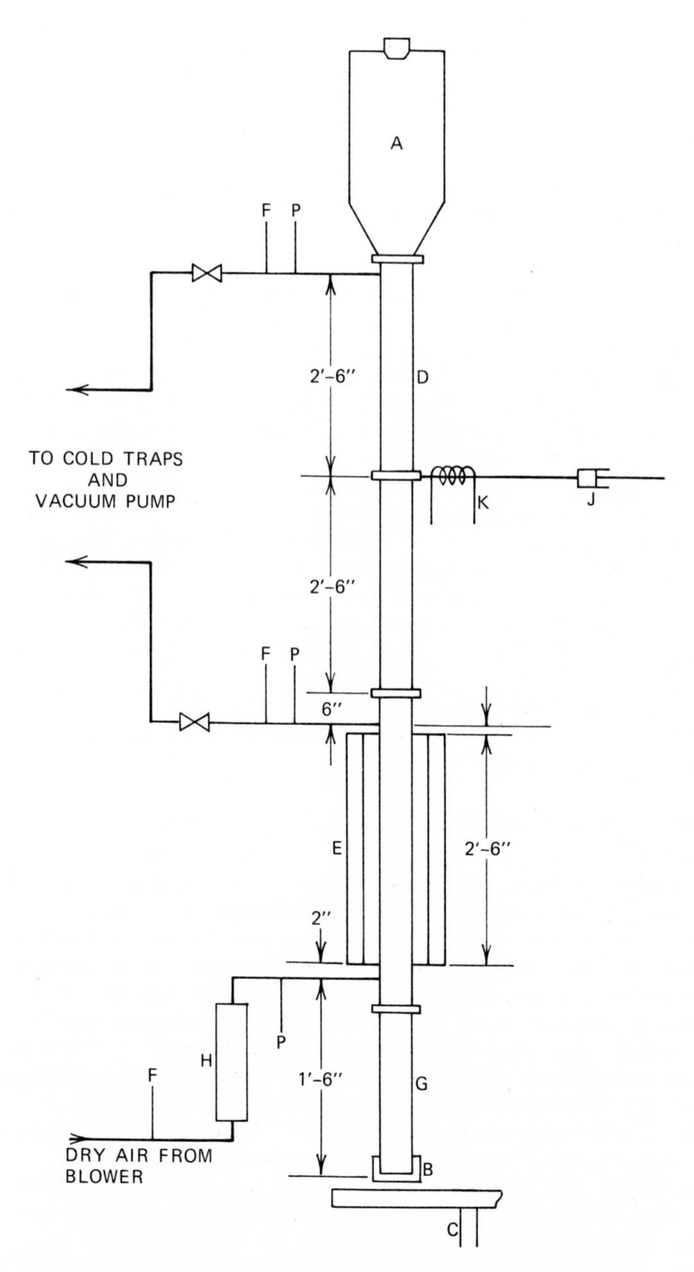

Figure 3. Apparatus for moving bed chromatography

carbon dioxide and acetone, although with less volatile systems water or even air cooled condensers are adequate. Rarely is it possible to achieve greater than 90% recovery of the components by heat exchange methods for volatile solutes, although this is usually regarded as adequate for laboratory purposes. To recover the remaining amount of component from the gas stream, adsorbents or molecular sieves are necessary. Such methods are only used when recirculating inert carrier gases on production scale chromatographic equipment.

The gas used is normally nitrogen, although Barker et al (8,11,20) have successfully used air for non-decomposing separations. Air is used for cheapness but should only be used on small scale equipment, taking particular care that all the equipment is properly bonded and earthed, and no source of ignition can occur on the equipment. Heated surface temperatures should be kept below the ignition temperature of the components being separated. If in doubt use nitrogen.

The jacket temperature of the stripper is usually kept about 50°C above the temperature of D which in the diagram is shown unjacketed and therefore at ambient conditions. For light hydrocarbon and volatile solvent mixtures, such an arrangement is satisfactory, but for comparatively non-volatile mixtures D is jacketed and operated at a temperature approximating that at which successful separation in an analytical G.C. apparatus could be achieved.

Although liquid feeds may be injected into the column, it is preferable to vaporize the feed by a heater K to achieve more efficient column operation.

A. The Separation of Binary Mixtures Using Moving Bed Chromatography

One of the advantages of this process over simple distillation is that separations can be achieved by differences in polarity as well as boiling point. Polar molecules are those in which the charge density is not uniform over the surface of the molecule but is concentrated at one or two points. The method can therefore be appropriately demonstrated by separating two liquids with approximately the same boiling point but of different polarity. Such a system is benzene (B.Pt. 80.1°C) and cyclohexane (B.Pt. 80.7°C) using a polyglycol derivative (polyoxyethylene 400 dirinoleate) as stationary phase absorbed on 10/20 mesh kieselguhr. Typical operating conditions on a 1in diameter column when separating a 50/50 v/v 30 ml/hr liquid feed mixtures are shown in Table I, while concentration profiles within the column corresponding to the three runs are given in Figure 4.

The ideal separation range for cyclohexane and benzene is for values of $\frac{F^R}{S}$ from 750 to 300 which are the approximate partition coefficients of benzene and cyclohexane respectively at 20°C. The actual range of separation is slightly narrower than the predicted value, see Figure 5.

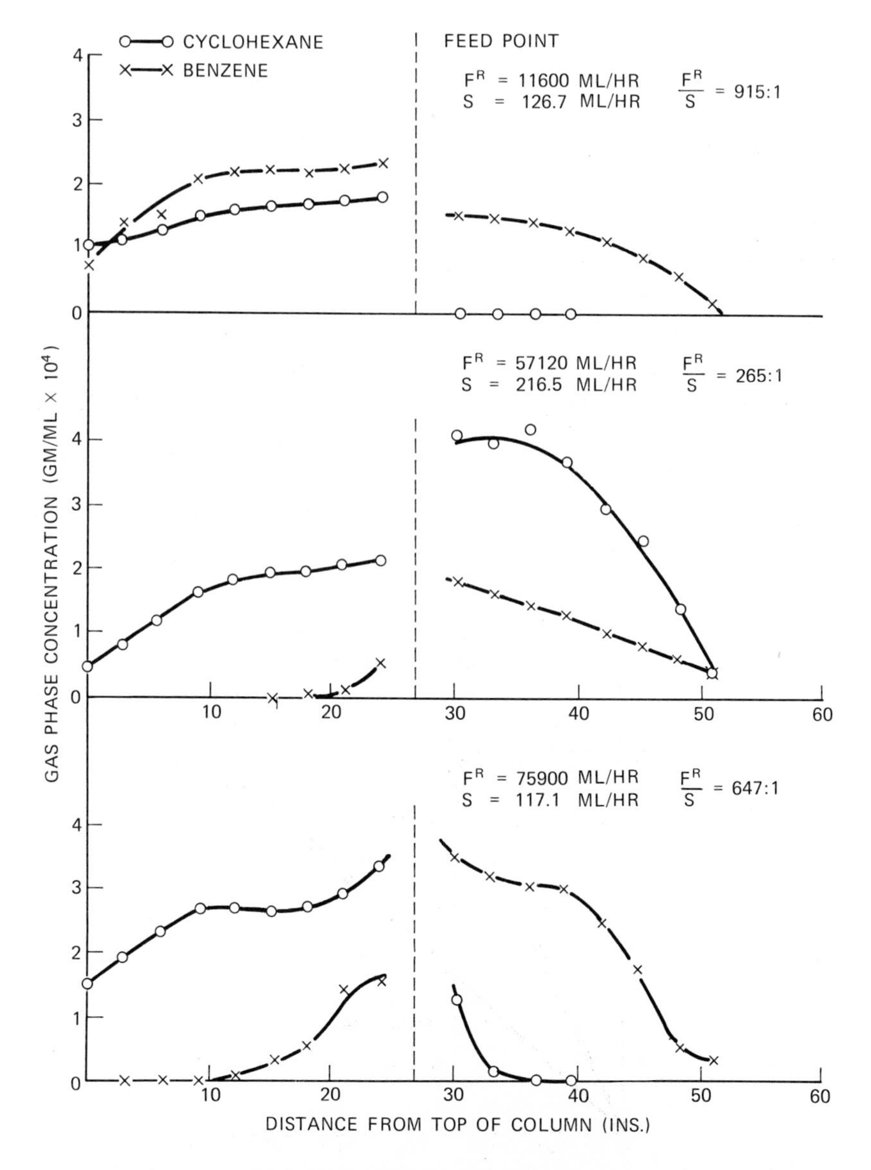

Figure 4. Concentration profiles of benzene and cyclohexane separated by moving bed chromatography.

TABLE I

Operating Data

System examined Run number	Cyclohexane/Benzene 1	 2	 3
Rectification outlet pressure (mm Hg)	730.5	749.3	725.5
Stripper outlet pressure (mm Hg)	749.7	758.1	737.9
Mean column pressure (mm Hg)	740.1	753.7	731.7
Flow-rate of pure air, rectification section			
(a) at N.T.P. (ml/h)	105,200	52,800	67,800
(b) at column conditions (ml/h)	116,000	57,120	75,900
Flow-rate pure air, stripping section at N.T.P. (ml/h)	92,800	93,500	91,900
Flow-rate of solids (gm/h)	432	739	400
Flow-rate of solvent (ml/h)	126.7	216.5	117.1

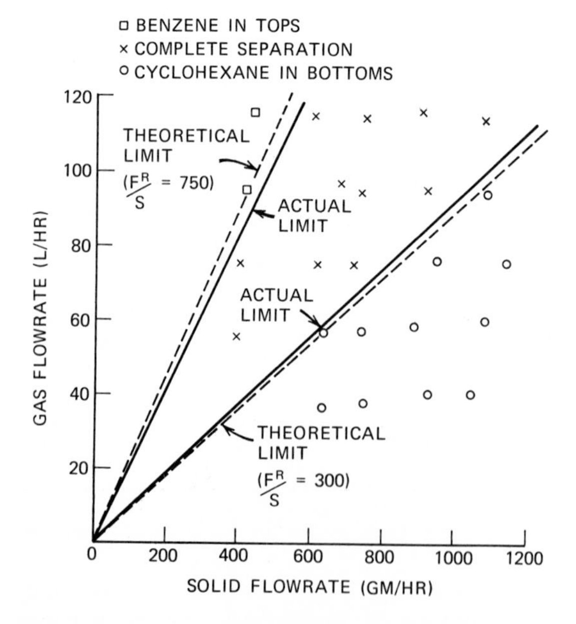

Figure 5: Actual and theoretical separation ranges for cyclohexane and benzene.

The theoretical range of flow ratios corresponds to

$$K^R_{\text{Cyclohexane}} < \frac{F^R}{S} < K^R_{\text{Benzene}} \qquad [13]$$

which is true at infinite dilution and for an infinite number of theoretical plates. It was found in the work on equilibrium determinations by Barker & Lloyd (1) that the partition coefficients of benzene were unaffected by the presence of cyclohexane or benzene. Cyclohexane behaved differently in that the partition coefficients were increased by the presence of benzene but not cyclohexane, while the effect of a finite number of mass transfer stages is to narrow the operating range to give an actual range according to equation [9]

$$K^R_{\text{Cyclohexane}} + \gamma_C + \delta_C < \frac{F^R}{S} < (K^R_{\text{Benzene}} - \delta_B). \qquad [14]$$

A further illustration of the technique when separating the system cyclohexane/methylcyclohexane is shown in Figures 6 and 7 with the corresponding operating data recorded in Table II. It will be observed that the operating range for this system is narrower than for the benzene-cyclohexane system. The reason for this is that none of the three solutes have permanent dipoles, but benzene is more polarizable per unit volume than either cyclohexane or methylcyclohexane due to the presence of the highly mobile electrons in the molecular structure. Thus in the case of a polar solvent of the type used, a dipole will be induced by the energy field of the solvent molecules. This effect will be greater for benzene, making it less volatile, and explains the large difference in partition coefficient values between cyclohexane and benzene, although they boil at virtually the same temperature.

Schultz (4) using a single column from which both products are collected has evaluated in terms of velocities the operating conditions of the column. He investigated the separation of cis and trans butene-2 on a column 100 cm long and 1 cm diameter, using 30% dibutyl phthalate on 0.3 - 0.4 mm Sterchamol particles. At a feed rate of 78 ml/hr consisting of 37.6% by volume of trans and 62.4% cis, the purity of trans 32 cm above the feed was 99.73% and the purity of cis at 32 cm below the feed was 99.4% by volume. The column was operated at 22°C which was well above the boiling point of both materials. Using a larger column, 138 cm long and 2.6 cm diameter, 21 g/hr of a 38.8 mol % 2.2 dimethyl butane and 61.2% mol % cyclopentane mixture was separated using 7.8 benzoquinoline as non-volatile liquid, with the column operating at the boiling point of the materials. Product purities in excess of 99.999 mol % were claimed to have been obtained.

Tiley and co-workers (2) using a 2 cm diameter column, 4 ft long and a packing of 20% $^w/_w$ dinonylphthalate on 44-60 BSS mesh celite, investigated the separation of 1.1 binary feed mixtures of diethyl ether, dimethoxy methane and dichloromethane.

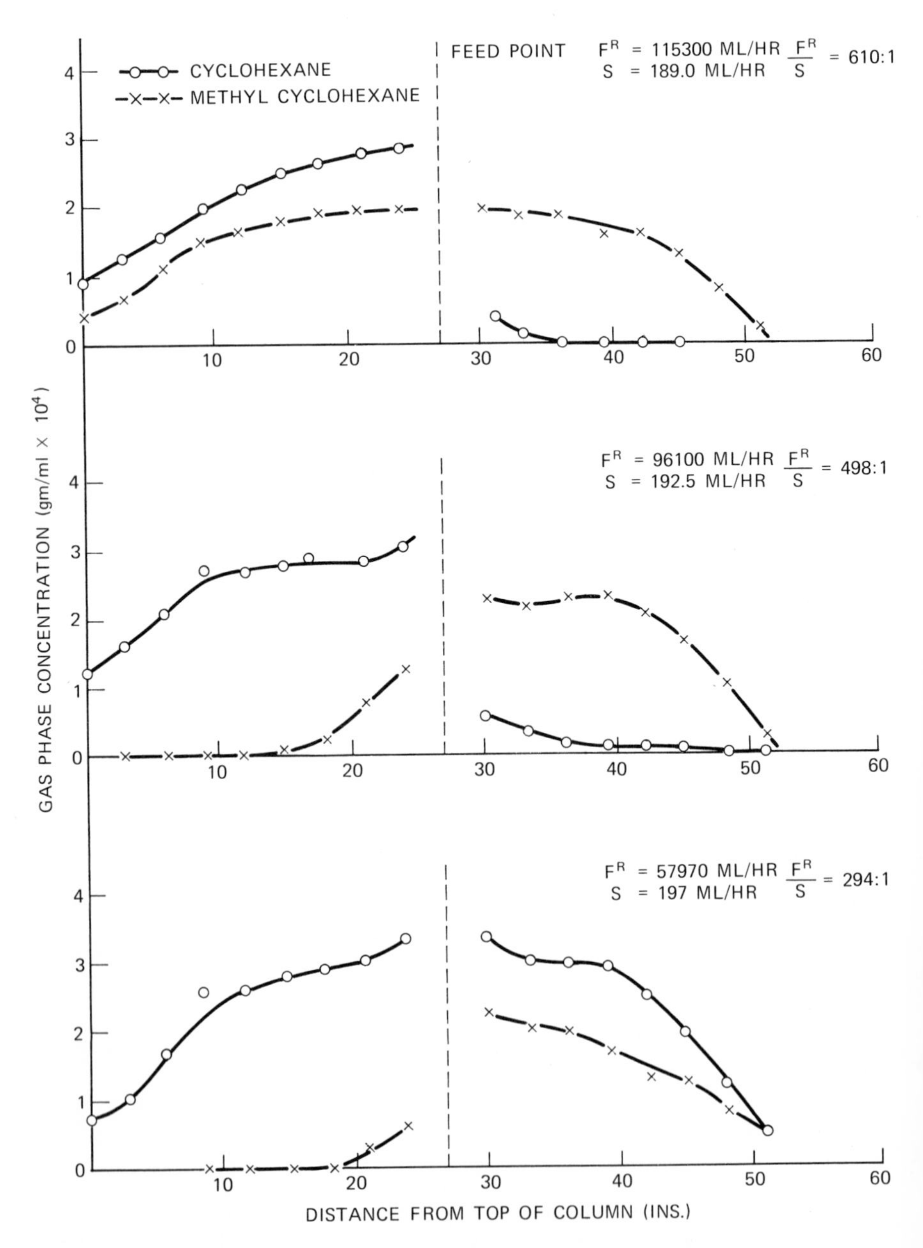

Figure 6. Concentration profiles of cyclohexane and methylcyclohexane separated by moving bed chromatography.

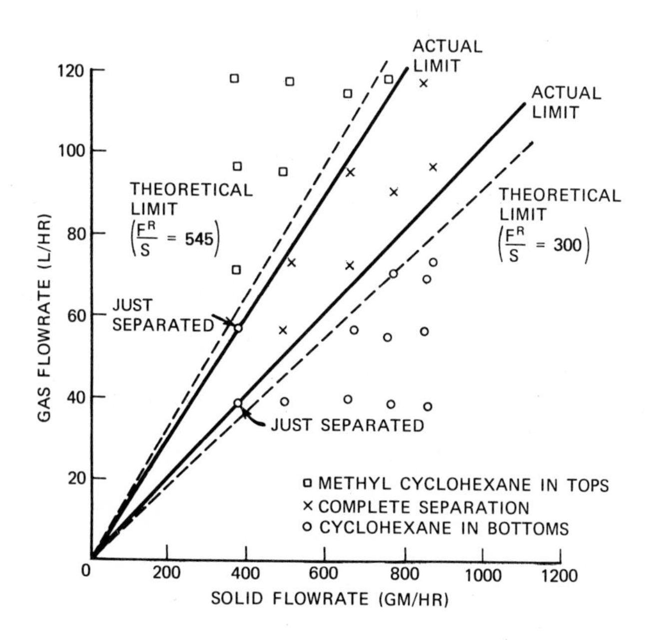

Figure 7. Actual and theoretical separation ranges for cyclohexane and methylcyclohexane.

TABLE II

Operating Data

System Examined Run number	Cyclohexane/Methyl Cyclohexane 1	 2	 3
Rectification outlet pressure (mm Hg)	720.8	721.8	726.6
Stripper outlet pressure (mm Hg)	739.6	739.2	735.4
Mean column pressure (mm Hg)	730.2	730.5	731.0
Flow-rate of pure air, rectification section			
(a) at N.T.P. (ml/h)	103,800	86,200	52,000
(b) at column condition (ml/h)	115,300	96,100	57,970
Flow-rate pure air, stripping section at N.T.P. (ml/h)	90,600	92,300	92,100
Flow-rate of solids (gm/h)	645	657	674
Flow-rate of solvent (ml/h)	189.0	192.5	197.0

The feed material was vaporized into a stream of nitrogen before entering the column, and although this resulted in low feed rates of 3-5 ml/hr, it gave rapid attainment of equilibrium around the feed point.

Scott (5) has also used the technique to separate aromatic hydrocarbons from coal gas.

B. The evaluation of the column efficiency in terms of H.T.U.'s

Since moving bed chromatography is a steady state counter-current differential process the efficiency may be expressed by the transfer unit concept proposed by Chilton and Colburn (6).

A material balance for a component over a column height dL gives

$$S\,dX = F^R\,dY \qquad [15]$$

$$= K_G a\,(Y - Y_\epsilon)\,dL \qquad [16]$$

where $K_G a$ = overall gas phase mass transfer coefficient

Y = solute concentration in gas phase (gm/ml)

Y_ϵ = equilibrium value of solute concentration in gas phase (gm/ml)

The height of the column may be calculated

$$L = \frac{F^R}{K_G a} \int_{Y_2}^{Y_1} \frac{dY}{Y-Y_\epsilon} \qquad [17]$$

This may be written

$$L = H_{OG} \times N_{OG} \qquad [18]$$

where H_{OG} is the height of a transfer unit ($= F^R/K_G a$)

and N_{OG} is the number of transfer units $\int_{Y_2}^{Y_1} \frac{dY}{(Y - Y_\epsilon)}$

Where it is possible to assume that the operating and equilibrium lines are straight it can be shown that

$$\int_{Y_2}^{Y_1} \frac{dY}{Y - Y_\epsilon} = \frac{Y_1 - Y_2}{(Y-Y_\epsilon)_1 - (Y-Y_\epsilon)_2} \times \ln \frac{(Y - Y\epsilon)_1}{(Y-Y\epsilon)_2} = \frac{LK_G a}{F^R} \qquad [19]$$

This may be written:

(a) for investigations above the feed point

$$N_{OG} = \frac{1}{F^R / (K^F S - 1)} \quad \ln \left\{ \frac{M_T / K^F S - Y_2 \; (F^R/K^F S - 1)}{M_T / K^F S - Y_1 \; (F^R/K^F S - 1)} \right\} \qquad [20]$$

K^F = partition coefficient at finite concentrations on solute free basis

(b) for investigations below the feed point

$$N_{OG} = \frac{1}{(1 - F^R/K^F S)} \quad \ln \left\{ \frac{M_B/K^F S - Y_2 \; (1 - F_R/K^F S)}{M_B/K^F S - Y_1 \; (1 - F_R/K^F S)} \right\} \qquad [20a]$$

where M_T = flow rate of solute leaving in the top product stream

M_B = flow rate of solute leaving in the bottom product stream.

Applying equations [20, 20a] to the runs shown in Figures 4 and 6 gave results as shown in Tables IIIa and IIIb. The values of H_{OG} obtained for cyclohexane above the feed point and for methyl cyclohexane and benzene below the feed point are shown in Figure 8. A

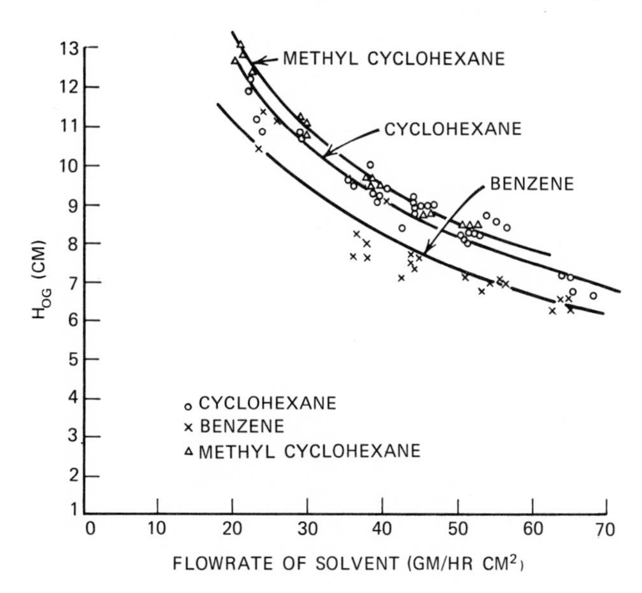

Figure 8: Height transfer units versus solvent flow rate for methylcyclohexane, cyclohexane and benzene.

TABLE IIIa

Investigation of top 9 in of column

Gas flow-rate (solute free) (ml/h) F^R	Liquid flow-rate (solute free) (ml/h) S	Column temp. (C°)	Mean solute concentration (g/mlX10^4)	Partition coefficient Solute basis	Partition coefficient Solute free K
		System:	Cyclohexane/benzene		
116000	126·7	20·3	1·21	297	327
116000	126·7	20·3	1·30	769	845
57120	216·5	20·8	1·12	291	295
75900	117·1	21·0	2·10	289	293
	System:	Cyclohexane/methylcyclohexane			
115300	189·0	19·0	1·52	328·8	329
115300	189·0	19·0	1·00	572	574
96100	192·5	19·8	1·90	294	300
57970	197·0	19·5	1·10	296·9	302

TABLE IIIa (cont)

Investigation of top 9 in. of column

Wt. of solute in tops (g) M_T	Gas Phase Concentration: At top of column (g/ml X 10^4) y_2	Gas Phase Concentration: 9 in. from top of column (g/ml X 10^4) y_1	N_{OG}	H_{OG} (cm)	H_{OG} (cm)	Solute investigated
		System: cyclohexane/benzene				
11·73	1·011	1·560	2·025	4·44	11·29	cyclohexane
8·41	0·80	2·14	2·043	4·40	11·19	benzene
2·36	0·41	1·611	2·558	3·52	8·94	cyclohexane
11·57	1·52	2·67	2·05	4·39	11·15	cyclohexane
		System: cyclohexane/methyl cyclohexane				
11·71	1·016	2·04	2·288	3·93	9·98	cyclohexane
5·50	0·413	1·47	2·366	3·80	9·66	methylcyclo-hexane
11·52	1·20	2·66	2·513	3·58	9·09	cyclohexane
4·37	0·755	1·59	2·491	3·61	9·16	cyclohexane

TABLE IIIb

Investigation of Section 42-1/4 in - 51-1/4 in from Top of Column

Gas flow-rate (solute free) (ml/h) F^R	Liquid flow-rate (solute free) (ml/h) S	Column temp. (C°)	Mean solute concentration (g/mlX10^4)	Partition coefficient Solute basis	Partition coefficient Solute free K
		System: Cyclohexane/Benzene			
116000	126.7	20.3	0.74	769	790
57120	216.5	20.8	1.51	295	312
57120	216.5	20.8	0.62	758	802
75900	117.1	21.0	1.31	764	835
	System:	Cyclohexane/methylcyclohexane			
115300	189.0	19.0	0.85	603	624
96100	192.5	19.8	1.06	585	620
57970	197.0	19.5	0.75	591	630
57970	197.0	19.5	1.40	319.8	340

TABLE IIIb (cont)

Investigation of Section 42-1/4 in - 51-1/4 in from Top of Column

Wt. of solute in bottoms (g) M_B	Gas Phase Concentrations 42-1/4 in from top of column (g/ml X 10^4) y_1	51-1/4 in from top of column (g/ml X 10^4) y_2	N_{OG}	H_{OG} (in)	H_{OG} (cm)	Solute investigated
System: cyclohexane/benzene						
5.00	1.406	0.15	2.04	4.40	11.19	Benzene
9.15	2.84	0.32	2.323	3.87	9.84	Cyclohexane
13.27	1.03	0.381	2.97	3.03	7.70	Benzene
12.81	2.360	0.284	2.09	4.31	10.93	Benzene
System: cyclohexane/methyl cyclohexane						
7.04	1.595	0.264	2.369	3.80	9.66	Methylcyclo-hexane
11.61	2.03	0.296	2.375	3.79	9.60	Methylcyclo-hexane
11.60	1.356	0.329	2.405	3.74	9.48	Methylcyclo-hexane
7.31	2.413	0.350	2.32	3.88	9.86	Cyclohexane

logarithmic plot of H_{OG} against liquid flow rate of solvent flow yielded a straight line relationship for each of the three components.

The corresponding empirically determined equations are

For cyclohexane $$H_{OG} = \frac{60.82}{S^{0.514}} \qquad [21a]$$

For methylcyclohexane $$H_{OG} = \frac{60.24}{S^{0.501}} \qquad [21b]$$

For benzene $$H_{OG} = \frac{50.47}{S^{0.495}} \qquad [21c]$$

C. Location of resistance to mass transfer.

The general equation for mass transfer is

$$\frac{1}{K_G a} = \frac{1}{k_g a} + \frac{1}{K^F k_L a} \qquad [22]$$

K^F = partition coefficient at finite concentration on solute free basis,

where $\frac{1}{K_G a}$ represents the total resistance to mass transfer,

$\frac{1}{k_g a}$ represents the gas phase resistance to mass transfer,

$\frac{1}{K^F k_L a}$ represents the liquid phase resistance to mass transfer.

Now $\frac{1}{k_g a} = f(F)$ namely function (f) of F, the gas flow rate transfer the gas film thickness and hence the gas film resistance will vary with gas flow-rate. We assume that $1/k_L a$ is not a function of gas flow rate i.e. that the gas flow will not tend to mix the liquid. This is reasonable if the solvent is situated mainly in the pores of the packing.

Thus the above equation can be arranged

$$\frac{1}{K_G a} = \frac{1}{f(F)} + \frac{1}{K^F k_L a} \qquad [23]$$

At infinite gas velocity there will be no resistance in the gas phase i.e. when $\frac{1}{F} = 0$. In a plot of $\frac{1}{K_G a}$ against $\frac{1}{F}$ at constant liquid flow S^1, Barker and Lloyd (40) found a straight line relation for S^1 which extrapolated back to the origin as shown in

Figure 9.

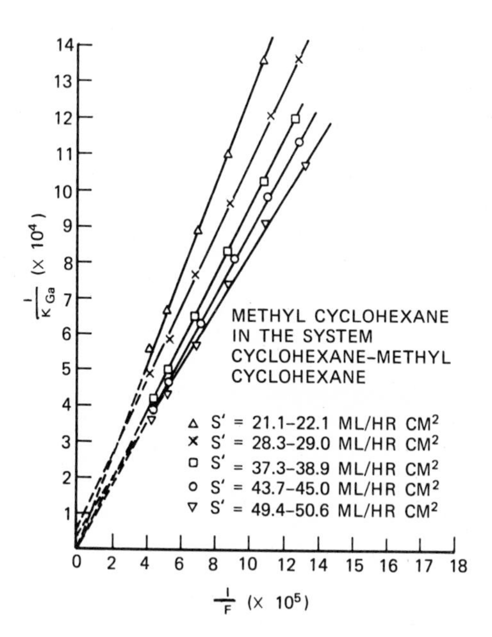

Figure 9. Resistance to mass transfer as a function of gas flow.

A statistical treatment of their results indicated a negligible interception on the ordinate axis. This is to be expected since the values of H_{OG} showed no measurable dependence on gas flow rate.

Thus as $\frac{1}{F} = 0$, $\frac{1}{K_G a} = \frac{1}{K^F k_L a} = 0$. This is assumed to be constant since $k_L \neq f(F)$ and S^1 is constant.

It would appear therefore that for moving bed columns under the conditions used by Barker and Lloyd that the resistance to mass transfer lies in the gas film.

D. <u>The evaluation of column efficiency in terms of overall H.E.T.P.'s</u>.

Separation of a binary mixture by continuous gas chromatography is a process similar to two solvent countercurrent extraction , in which one solvent is the inert carrier gas, and the other is the non-volatile stationary phase. The only difference is that in chromatographic separations the heavier component is stripped from the stationary phase before the latter leaves the column.

Following the approach of Alders (7) in the treatment of liquid-liquid extraction where the process is considered as a large number of discrete stages, a relationship for the number of equivalent theoretical plates necessary for the separation of a binary mixture into products of a given purity can be derived. Tiley and co-workers (2), assuming a column consisting of a large number of plates with a central feed point and equal product purities, have obtained an approximate relation for the number of plates in the column. A more general relationship which does not assume equal product purities nor restrict the application to a 1:1 feed mixture has been derived by Barker and Huntington (8,9). If difficulty is experienced in following the derivation below it is suggested the reader consult Chapter 5 of the book by Alders (9).

Considering a two solvent countercurrent extraction in which a binary feed mixture enters at a point between the two ends of the column; one component to one offtake in the first solvent (gas stream), while the other component travels to the other offtake in the second solvent (non-volatile liquid). From a binary feed mixture containing 1 part of component i, assume that x parts of i are removed in the gas stream and (1 - x) parts removed in the solvent stream, the fraction of i not extracted $\Psi = x$.

If in unit time y parts of i pass from the feed to the $(m+1)^{th}$ stage and z parts pass to the $(m-1)^{th}$ stage, where m is the feed stage, then $E_1 = z/y$.

From the Kremser equation (44)

$$\text{hence} \quad \frac{x}{y} = \frac{E_1 - 1}{E_1^n - 1} \qquad [24]$$

$$\frac{1 - x}{z} = \frac{1/E_2 - 1}{1/E_2^m - 1} \qquad [25]$$

where E_1 and E_2 are the extraction factors for each component above and below the feed, n and m are the number of stages above and below the feed respectively.

From equations [24] and [25]:

$$\frac{1 - x}{z} = \frac{\dfrac{1/E_2 - 1}{1/E_2^m - 1} \cdot \dfrac{x}{y}}{(E_1 - 1)/(E_1^n - 1)} \qquad [26]$$

$$\text{Thus} \quad \frac{1}{x} - 1 = \frac{(z/y) \times (E_2^{m-1} - E_2^m)(E_1^n - 1)}{(E_1 - 1)\ (1 - E_2^m)}$$

from which by substitution of $\Psi = x$ and $E_1 = z/y$ gives

$$x = \Psi = \frac{(E_1 - 1)\ (E_2^{\ m} - 1)}{E_1E_2^{\ m-1}(E_2^{\ m} - 1)(E_1^{\ n} - 1) + (E_1 - 1)(E_2^{\ m} - 1)} \qquad [27]$$

If $E_1 = E_2 = E$ i.e., at infinite dilution when quantities of solvent are large compared with quantity of feed

$$\Psi = \frac{E^m - 1}{E^{m+n} - 1} \qquad [28]$$

Considering components A and B under conditions such that

$E_A < 1 < E_B$ then if $n \gg 1$, using equation [28]

$$\Psi_A = \frac{E_A^{\ n} - 1}{E_A^{\ 2n} - 1} \rightarrow 1 - E_A^{\ n} \qquad [29]$$

$$\Psi_B = \frac{1 - E_B^{\ n}}{1 - E_B^{\ 2n}} \rightarrow E_B^{\ -n} \qquad [30]$$

For a 1:1 by weight feed mixture $\Psi_A = 1 - \Psi_B$

hence $E_A \cdot E_B = 1.$

As $E = K/\emptyset$, the required flow ratio $\emptyset$ is $(K_A \cdot K_B)^{1/2}$

If in unit time the feed mixture contains W_A parts of component A and W_B parts of component B

$$\text{then}\quad \Psi_A = 1 - \frac{\delta_{W_A}}{W_A} \qquad [31]$$

$$\Psi_B = \frac{\delta_{W_A}}{W_B} \qquad [32]$$

where Ψ_A and Ψ_B are the fractions not extracted at the top of the column. δ_{W_A} is the amount of component A which is extracted (i.e., appears in the bottom product) and δ_{W_B} is the amount of component B which is not extracted (i.e., appears in the top product).

Combining equations [29] and [31]

$$1 - E_A{}^n = 1 - \frac{\delta_{W_A}}{W_A} \qquad [33]$$

$$1 - (K_A / \emptyset)^n = {}^{M_{AT}}/_{W_A} \qquad [34]$$

Where M_{AT} is the collection rate of component A at the top of the column.

$$n \log K_A - n \log \emptyset = \log (1 - {}^{M_{AT}}/_{W_A}) \qquad [35]$$

Similarly combining equations [30] and [32]

$$E_B{}^{-n} = \delta_{W_A} / W_B \qquad [36]$$

$$(K_B / \emptyset)^{-n} = M_{BT} / W_B \qquad [37]$$

Where M_{BT} is the collection rate of component B at the top of the column.

$$- n \log K_B + n \log \emptyset = \log M_{BT} / W_B \qquad [38]$$

For a specified purity of products there will be a range of flow ratio, depending on the number of theoretical plates. The maximum will be represented by equation [38] and the minimum by equation [35].

As the ratio of partition coefficients is likely to vary to a lesser extent than the actual values, a combined equation will be more accurate.

$$\log \frac{\emptyset_{max}}{\emptyset_{min}} = \log \frac{K_B}{K_A} + \frac{1}{n} \left[\log \left(1 - \frac{M_{AT}}{W_A}\right) + \log \frac{M_{BT}}{W_B}\right] \qquad [39]$$

In the operation of the vertical moving bed column a single operating value $\emptyset$ is used, i.e., $\emptyset_{max} = \emptyset_{min}$ and the equation reduces to

$$\log SF = \frac{-2}{N_c} \left[\log \left(1 - \frac{M_{AT}}{W_A}\right) + \log \frac{M_{BT}}{W_B}\right] \qquad [40]$$

Where N_c is the total number of theoretical plates ($\simeq 2n$) in the continuous column. SF = ratio K_B/K_A.

For a 1:1 feed mixture and equal product purities

$$N_c = \frac{-4}{\log SF} \log I, \qquad [41]$$

where I = product impurity fraction.

In the case of the circular chromatographic column described later in the chapter where the operating values of $\emptyset$ does vary slightly above and below the feed owing to the bleed of carrier gas injected at the feed point, a more accurate estimate of N_c is obtained by taking the value of $\emptyset$ above the feed as $\emptyset_{max}$ and the value below as $\emptyset_{min}$.

IV. SEPARATION OF THREE COMPONENT MIXTURES

A more common problem of separation is where one component or group of components has to be removed from components appearing on a chromatogram on either side of the single component or group of components.

A patented apparatus (see Figure 10) has been devised by Barker & Lloyd (10) for carrying out such separations, while Barker & Huntington (11) report experimental findings using such equipment.

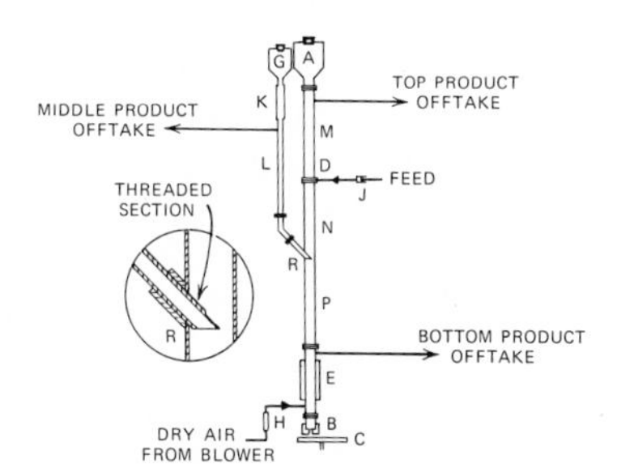

Figure 10: Moving bed column for ternary separations.

The column used in their experimental work was constructed from copper tubing with brass flanges and neoprene gaskets.

The packing was placed in the hoppers A and G from which it passed by gravity through the columns and the controlling orifice B on to the rotating table C, finally being removed by a scraper arm onto the pan of a weighing machine. In some initial experiments 13% $^w/_w$ polyoxyethylene 400 diricinoleate on 14-18 B.S mesh Johns Manville C22 firebrick was used.

The control of the solids in the side arm was effected by the intrusion of a 1.27 cm diameter side arm into the 2.54 cm diameter main column. The position of the side arm R altered the effective cross sectional area allotted to each part of the column and hence the ratio of the solid flows in the main column to the side-arm. It was found necessary to have the intrustion otherwise no solids flow was obtained in the side-arm.

The solids leaving the main hopper (A) passed through the rectification section (MNP) and then into the stripping section (E). The distances M,N,P,E were 68.6, 63.5, 68.6 and 52.1 cm respectively. The solids leaving the hopper (G) served to reflux back the heaviest component entering the side-arm (R) thus permitting a pure side-product to be obtained. The effective length of L for mass transfer was 82.5 cm. The stripper (E) was surrounded by an electrically heated air-jacket.

Air was supplied to the column below the stripping section by a blower and was drawn off under a slight vacuum from three product offtake points. The air was dried and preheated to 50°C in a heater (H). It entered the column at a pressure slightly above atmospheric to prevent leakage of moist air into the system through the orifice (B).

A 1:1:1 $^{v}/_{v}$ liquid mixture of benzene, cyclohexane and methylcyclohexane was injected unheated into the column at the feed point D by a micropump (J). All the product streams were passed through cold traps containing carbon dioxide and acetone to condense out most of the separated products.

Gas sampling points were situated at 7.62 cm intervals in the rectification sections of the columns.

These side tubes were fitted with serum caps so that gas samples could be taken and analysed in a conventional G.L.C. column.

A. General operating conditions for moving bed ternary separations

When using the equipment as a single column and without the side-arm in use, product purities of 99.5% were obtained at the top and bottom of the column and maximum purities of 78.6% by volume of methylcyclohexane contaminated by benzene from the middle section.

Table IV shows complete operating data for six of the runs when the side-arm was brought into use. A typical concentration profile is shown in Figure 11. These data show how benzene entering the side-arm can be refluxed back thus enabling a pure side-stream to be produced. Due to insufficient column length below the junction R a pure bottoms product of benzene could not be obtained simultaneously with high purity overhead and side stream products.

Extending the principles determined for binary separations (see equations [6,7]) to ternary systems, in the sections M and N of the column, the ideal separating range for cyclohexane to be carried up the column and the methylcyclohexane and benzene downwards corresponds to:-

TABLE IV

Operating Data for Ternary Separation with Side-Arm in Operation

	1	2	3	4	5	6
Upper Column						
Pure air flow at column conditions l/hr	40.6	43.8	52.3	49.8	52.2	60.5
F^R/S	455	475	557	532	559	641
Side Arm						
Pure air flow at column conditions l/hr	49.3	45.8	44.9	39.5	38.6	36.7
F^R/S	787	764	703	658	644	605
Middle Column						
Pure air flow at column conditions l/hr	89.9	89.6	97.2	89.3	90.8	97.2
F^R/S	592	587	630	581	592	628
Mean Column						
Pressure mm.Hg	731.4	721.9	735.4	713.0	726.1	722.9
Temperature °K	295.2	295.0	295.2	295.3	295.2	295.2
Upper Column						
Solids flowrate gms/hr	705	736	712	740	738	745
Solvent flow mls/hr	89.2	92.8	90.3	93.7	93.3	94.5
gms/hr.cm^2	18.75	19.52	18.98	19.69	19.65	19.86
Side Arm						
Solids flowrate gms/hr	495	474	503	474	474	478
Solvent flow mls/hr	62.7	60.0	63.8	60.0	60.0	60.0
gms/hr.cm^2	51.1	48.8	52.0	48.8	48.8	49.4
Mole Percentages						
1) Top product						
Cyclohexane	99.5	99.5	97.3	97.5	98.7	91.6
Methylcyclohexane	-	-	2.7	2.5	1.3	8.4
2) Side Arm product						
Cyclohexane	43.2	-	-	-	-	-
Methylcyclohexane	53.3	93.7	99.5	99.5	99.5	99.5
Benzene	12.5	6.3	-	-	-	-
3) In gas phase at point above bottoms product offtake						
Methylcyclohexane	33.1	31.3	28.2	30.0	26.4	20.7
Benzene	66.9	68.7	71.8	70.0	73.6	79.3

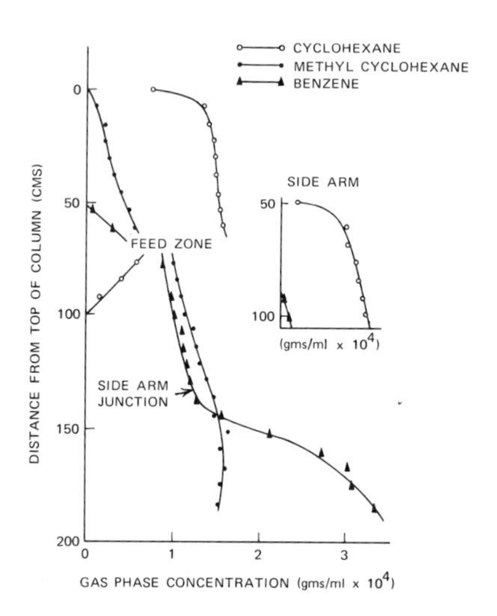

Figure 11: Concentration profile in moving bed column.

$$K^R \text{ cyclohexane} < \frac{F_1^R}{S_1^R} < K^R \text{ methylcyclohexane} < K^R \text{ benzene} \qquad [42]$$

where subscript 1, refers to flows in section M, N of the column. In the side-arm for benzene to be refluxed back down the column the ideal separating range corresponds to

$$K^R \text{ methycyclohexane} < \frac{F_2^R}{S_2^R} < K^R \text{ benzene} \qquad [43]$$

where subscript 2, refers to flows in section L of the column.
Similarly in section P of the column

$$K^R \text{ methylcyclohexane} < \frac{F_1^R + F_2^R}{S_1^R + S_2^R} < K^R \text{ benzene} \qquad [44]$$

One of the limiting factors in the operation of the column with the side-arm is that the ratio F_2^R / S_2^R must be of the same order as the ratio

$$\frac{F_1^R + F_2^R}{S_1^R + S_2^R}$$

i.e., both must be between the partition coefficients of methylcyclohexane and benzene. These ideal operating ranges are modified by the presence of the other components and by using columns of finite length as previously described for binary separations.

The principle could be extended to multicomponent separations or alternatively a series of single columns in series operating at different F^R/S ratios could be used to separate multicomponent systems.

The efficiency data in terms of H.T.U's are shown in Table V for the series of runs given in Table 4. This table shows that the mass transfer efficiency of the column is reduced when the % of liquid phase is reduced presumably due to the poorer coverage of brick giving a reduced surface area for mass transfer.

TABLE V

HTU Data for Operating Conditions in Table IV

Rectification Section gas phase concentrations of Cyclohexane (Gm/ml$\times10^4$) 22.9 cms from top	1.478	1.151	1.174	1.123	.916
Partition coefficient cyclohexane	277	279	275	276	285
HTU cyclohexane 0-22.9 (cms)	13.21	14.40	15.60	14.81	14.02
From equation [21a] $H_{OG} = 60.82/S^{0.514}$ for 1 in. diameter columns	13.31	13.40	13.20	13.17	13.08
Side Arm 1/2 in. diameter columns gas phase concentrations of methylcyclohexane (gms/ml$\times10^4$)					
At top of column	.282	.220	.271	.235	.280
22.9 cm from top	.803	.725	.945	.855	.963
Partition coefficient methylcyclohexane	534	535	537	542	538
HTU methylcyclohexane 0-22.9 (cms)	6.48	5.69	6.40	6.37	7.30
From equation [21b] $H_{OG} = 60.24/S^{0.501}$ for 1 in. diameter columns	8.58	8.32	8.58	8.58	8.55

The H_{OG} values obtained for methylcyclohexane in the side-arm given in Table V, illustrate the considerable beneficial effect of using smaller diameter tubes. The effect of tube size reduction from 2.54 to 1.27 cm more than offsets the increase in H_{OG} due to decreasing the liquid phase percentage from 29.6% to 13%. It will be recalled that the emperical equation (namely [21b] $H_{OG} = 60.24/S^{0.501}$) was determined for 1 in.diameter columns separating methylcyclohexane from benzene and using 29.6% polyglycol liquid phase on C22 Johns Manville fire-brick.

V. CIRCULAR COLUMNS FOR CONTINUOUS CHROMATOGRAPHIC REFINING

It is inevitable that with the moving bed column previously described, some breakdown of the solid through attrition will occur leading to loss of brick and solvent on elutriation. To eliminate this loss and to give greater ease of operation without the problem of physically moving solids, circular chromatography columns have been proposed by a variety of inventors, which will enable separations to be carried out without the disadvantages mentioned.

The operation of the circular column is based on the moving bed principle. The column which is in the form of a circle, rotates passed fixed inlet and outlet ports. This means that the main disadvantages of the moving bed process viz attrition and solids flow control, are obviated by having no relative movement between the packing and the column wall. Only sufficient packing to fill the column is required as no recirculation is necessary. The use of this type of column was first suggested by the schematic diagrams of Pichler (12) and in the patent by the Gulf Research and Development Co., (13). The first actual equipment constructed on these lines was by Luft (14).

In all these arrangements the flow of carrier gas within the column has to be controlled by pressure drop, and hence much of the column which could be used for separation, has to be employed in just creating a pressure drop. In a recent article Glasser (15) has described a similar piece of equipment where again flow is controlled by pressure drop. In the scheme proposed by Barker (16) a column was designed which gave a unidirectional flow of carrier fluid by using cam operated valves and thereby giving efficient use of column length and carrier fluid.

In the first three arrangements shown in Figure 12 the effective separation length cannot exceed half of the total circumference, and there will be an excess of carrier gas travelling in the same direction as the column rotation thereby reducing the offtake concentration. Glasser (15) in arrangement (C) uses only about 20% of the total carrier gas input for stripping the absorbed component or passing through the column used. Littlewood (17) points out that excess gas in the column will reduce the concentration and may permit higher throughputs, as the throughput is limited by gas phase saturation. In fact

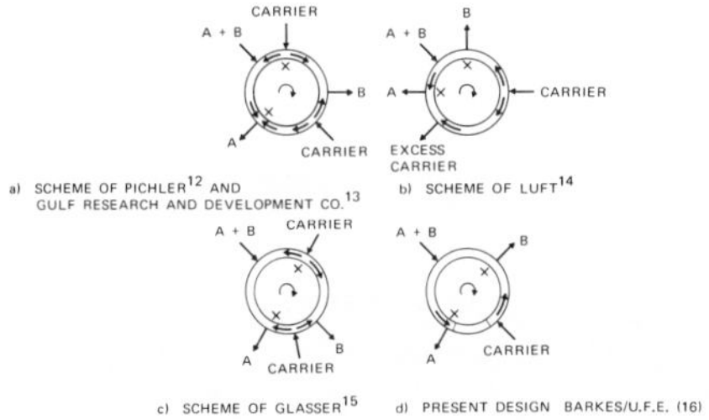

Figure 12: Circular columns for continuous separations.

in this scheme the excess air will only reduce the concentration in the co-current section of the column and not in the separation section itself. In scheme (d) the design proposed by Barker (16), as the carrier gas travels only in one direction, the length of the column which can be used for separation purposes is limited only by the amount required for stripping and to provide a gas seal.

A. General operating principle of the Barker - Universal Fisher Group five feet diameter circular chromatographic machine

The prototype illustrated in Figure 13 and shown schematically in Figure 14 consists essentially of a 1 in. square cross-section chamber in the form of a circle of 5 ft diameter.

The chamber is divided into eight equal sections connected by means of external valves, each section containing a copper helix through which can be passed a heating or cooling medium. By suitable cam arrangements, as the chamber (or column) rotates, different temperatures may be achieved in the packed sections, so that parts may act as a stripping section. A hot, dry inert carrier gas enters the column at D and passes counter-current to the direction of rotation of the column, leaving through the offtake ports B and C. To prevent the carrier gas traveling from D to B, the valve chamber segments, which is in the arc DB, is always closed.

The feed mixture to be separated enters continuously at A, the lighter (less strongly absorbed) component traveling in the gas stream towards product offtake port B, while the heavier liquid with which the support is coated travel in the direction of rotation of the column and are stripped by heat in the section CD and leave through the product offtake port C.

To permit the gas flow into and out of the column through the inlet and outlet ports there are 180 gas passages equally spaced over the chamber face, each gas passage being closed by

Figure 13: Circular chromatographic machine.

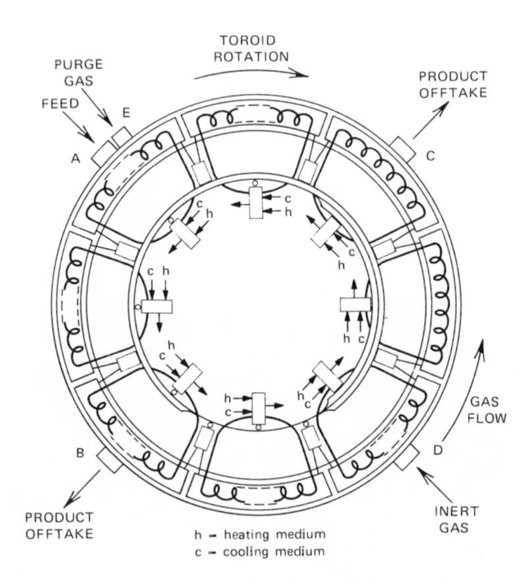

Figure 14: Cross section of circular chromatographic machine.

a self sealing valve. These valves are automatically opened when they pass under one of the inlet or offtake ports. Sealing is achieved on the face of the chamber by means of 0 rings that cover at least 3 of the 3/64 in diameter gas passages at any time under each port. The 0 rings are sealed against the face by means of spring-loaded plates.

The feed mixture entering the column at A is vaporized before entry and to ensure that any which condenses on entry does not remain in the valve chambers, a "bleed" of carrier gas is fed in just after the feed port at E. Sampling points are incorporated in the lines between the chamber sections so that the concentration profile within the packed bed could be obtained. The circular packed bed is rotated by a toothed wheel driving a chain attached to the periphery of the column, at speeds between 1 and 10 revolutions per hour.

The commercial machine (see Figure 15) carried many improvements over the prototype including separate entry and exit ports for feed and product offtake ports respectively. Also electrically heated elements in lieu of the copper helices giving higher temperature capability and easier temperature control and a refined method of sealing between the stationary ports and the moving wheel.

B. Separations using the five feet diameter circular chromatograph

1. Hydrocarbon separations.

a. 97% cyclopentane purification

Figure 15: Commercial circular chromatographic machine.

The purification of a 97% pure cyclopentane containing five detectable impurities as shown in Figure 16 illustrates the high throughput capability of this type of machine. The machine was packed with a Johns Manville C22 firebrick coated with 30 per cent polyoxyethylene 400 diricinoleate. Although more favorable stationary phases might have been chosen Barker & Huntington (18) chose this phase because it had been used in earlier work on 1 inch diameter moving bed columns and they were interested in seeing the relationship between the two types of equipment. This phase however did have the advantage of bringing all the detectable impurities to one side of the main cyclopentane peak, so that the impurities could be removed in one pass rather than two through the machine.

The operating behavior of this type of equipment is similar to the moving bed column, so the gas and liquid rates to achieve separation can again be approximately determined from the partition coefficients of the two adjacent components one is interested in separating. For the stationary phase used in these experiments the values of the partition coefficients at 20°C were as follows:

$K_{cyclopentane} = 119.1$; $K_{impurity} = 78.5$, hence the separation factor given by the ratio of these K values is 1.518.

Table VI shows the evaluation of the operating conditions in terms of gas flow that will just allow a pure bottom product (cyclopentane) to be realized. In all these runs it is observed

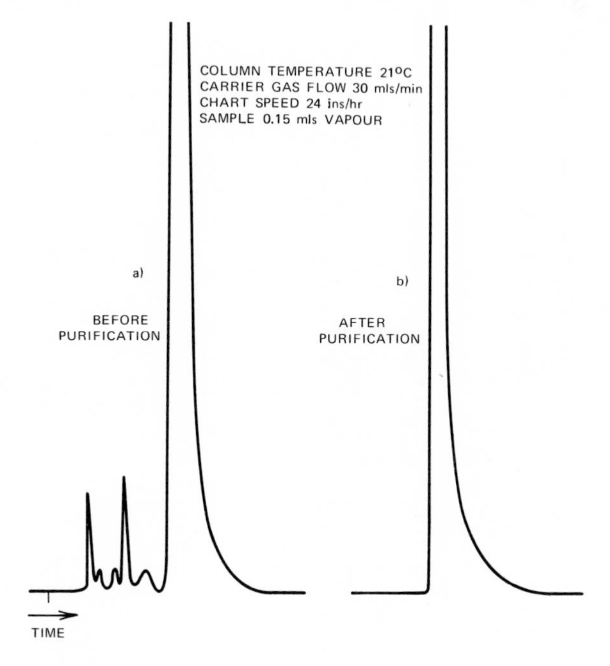

Figure 16: Chromatogram of cyclopentane before and after purification.

TABLE VI

Evaluation of operating conditions for cyclopentane purification

Column operating temperature was 20°C, and feed rate was 72 ml per hour

Flow at ambient conditions, liters per hour			Mean column pressure p.s.i.g.				Top product, per cent of cyclopentane (on inert-free gas basis)	Bottom product cyclopentane
upper offtake flow rate	lower offtake flow rate	bleed carrier flow rate	upper column	lower column	upper column F^R/S	lower column F^R/S		
291	504	45	9.2	11.3	160	124	96.4	Pure; limit 34 inches from offtake
235	504	47	9.8	12.0	126	92.7	93.1	Pure; limit 8 inches from offtake
199	504	47	10.3	12.1	104	76.2	90.8	Pure
164	504	51	11.1	12.7	84	56.7	88.1	Trace impurity
131	504	56	11.6	13.1	67	37.6	68.2	Trace impurity 99 per cent pure

that the top product, which ideally should be the impurities, always contains a large percentage of cyclopentane. This is because of the small amount of impurity initially present in the mixture, and the limited effective length of 107 inches for mass transfer between the two offtake ports.

As the feed rate is increased, the gas flow also has to be slightly increased to maintain pure cyclopentane as bottom product. Table VII shows the proportion of the feed leaving at each offtake port for varying feed rates. The saturation limit of the carrier gas which was air had been exceeded at all the feed rates used in Table VII and accounts for the high proportion of cyclopentane in the bottom product. This indicates that the column can be operated successfully above the saturation limit but to prevent liquid pockets occurring within the column it is preferable to increase the gas flow rates so that saturation does not occur at such low feed rates.

b. 99% cyclohexane purification

The 99% cyclohexane fraction chosen had one detectable impurity see Figure 17 when using a 1.8 m long x 4 mm.i.d. column packed with 52 to 60 mesh Johns Manville C22 firebrick coated with 6% $w/_w$ of polyoxyethylene 400 diricinoleate, and flame ionization type detector. The partition coefficients for the cyclohexane and impurity at 20°C are $K_{cyclohexane} = 306$; $K_{impurity} =$ 193.6, and hence the separation factor is 1.582. Although the separation factor is of a similar order to that of the cyclopentane system, it can be seen that the actual gas flow requirements will be much greater, by about a factor of three.

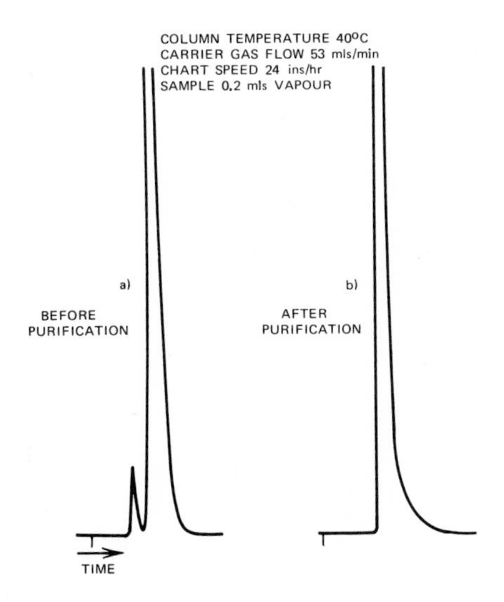

Figure 17: Chromatogram of cyclohexane before and after purification.

TABLE VII

Variation of cyclopentane-feed distribution with feed rate

Column operating temperature was 20°C, and at all feed rates pure bottom product (cyclopentane) was obtained

Feed rate ml per hour	Flows at ambient conditions, liters per hour			Mean column pressure, p.s.i.g.				Per cent. of feed mixture leaving	
	upper offtake	lower offtake flow rate	"bleed" carrier flow rate	upper column	lower column	upper column F^R/S	lower column F^R/S	top product offtake	bottom product offtake
120	208	504	47	10.3	12.1	109	80	17.4	82.6
202	208	504	47	10.3	12.1	109	80	16.2	83.8
300	214	504	47	10.2	12.1	112	83	17.5	82.5
410	232	504	47	9.9	12.0	122	92	15.2	84.8

TABLE VIII

Variation of cyclohexane-feed distribution with feed rate

Column operating temperature was 21°C

	Flows at ambient conditions, liters per hour			Mean column, pressure, p.s.i.g.				% feed mixture leaving		
Feed rate, ml per hour	upper offtake flow rate	lower offtake flow rate	"bleed" carrier flow rate	upper column	lower column	upper column F^R/S	lower column F^R/S	top offtake	bottom offtake	Limit of impurity from bottom port
108	648	600	216	3.5	8.0	467	249	40.5	59.5	6 inches
160	780	492	254	3.0	7.6	583	310	55.3	44.7	22 inches
208	840	492	254	2.8	7.5	623	345	55.5	44.5	14 inches

Table VIII shows the proportion of the feed leaving the column at both offtake ports, maintaining a pure bottom product, but in this instance the column was not saturated at any of the feed throughputs. Table VIII also shows a high loss of the cyclohexane product into the impurity fraction, necessitating the rerunning of this fraction through the machine to reduce excessive wastage.

c. The separation of the azeotropic system cyclohexane/benzene

The cyclohexane/benzene system has been extensively studied by Barker & Huntington (8) and their results for this system show the effect of column temperature, feed rate and gas/liquid ratio on the behavior of a circular continuous chromatography machine.

The four column temperatures employed in their experiments were 20° - 21°, 31° - 32°, 35° - 36° and 43° - 44°. Feed rates were varied from 66 to 100 ml/hour while F^R/S ratios varied from 168 to 892. In all cases the feed was heated to about 87°C before entering the column, so that the feed entered as a vapor.

The operating conditions of each of the runs are presented in Table IX. Upper column refers to the section between the top offtake and the feed, and the lower column the section between the bottom offtake and the feed zone.

All the runs were operating in the laminar region for gas flow except for the column operation at 20° - 21°C when the maximum value of Reynolds number based on the empty column was 13.6 which is in the transition region (Re10 - 30).

Plots of concentration against distance show the same characteristic features as those obtained for vertical moving bed columns, and are illustrated for the runs at 20° - 21°C in Figure 18. There is a sharp drop in concentration at the top of the column, due to the incoming fresh packing, and the concentration of benzene below the feed increases as the column is des-

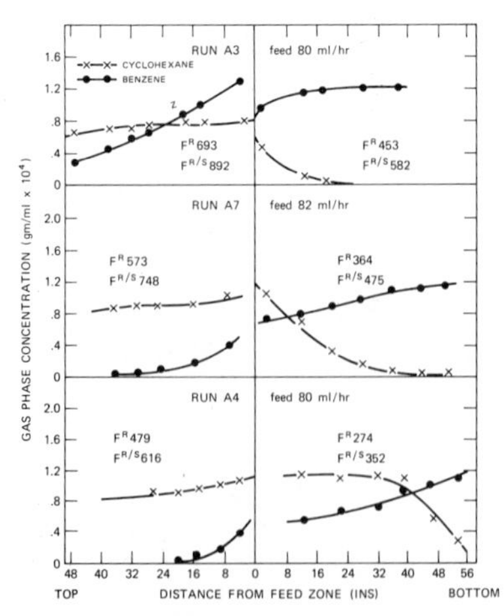

Figure 18: Concentration profile for the separation of cyclohexane/benzene at 20° - 21°C.

TABLE IX

Operating conditions for the separation of the mixtures Cyclohexane/Benzene at increasing temperatures

Run Number	Feed rate ml/hr	Solvent Flow ml/hr	Offtake carrier flow at ambient conditions Upper	Bleed	Lower	Mean Column pressure p.s.i.g. Upper	Lower	Mean Column Operating F^R/S Upper	Lower
20-21°C									
A1	80	1111	690	188	560	2.7	5.6	524	326
A2	80	1111	778	186	546	2.5	5.3	596	390
A3	80	1778	828	192	546	2.7	5.8	892	582
A4	80	778	634	211	560	4.8	8.2	616	352
A5	82	767	740	200	557	2.4	5.5	836	485
A6	82	767	664	200	557	4.1	6.4	674	420
A7	82	767	717	198	613	3.6	6.1	748	475
A8	110	767	731	196	602	3.2	5.8	780	497
A9	110	767	697	202	597	4.0	6.3	712	449
31-32°C									
B1	66	758	512	67	762	3.9	7.5	553	403
B2	66	758	446	71	762	4.5	7.9	464	334
B3	66	758	358	73	762	5.3	8.5	362	248
B4	66	758	388	73	688	5.1	8.4	397	274
B5	81	758	398	73	732	4.9	8.5	407	282
B6	89	758	388	72	732	5.0	8.2	398	277
B7	89	758	378	70	732	4.7	7.8	391	273
B8	100	758	388	72	732	5.0	8.4	398	274

TABLE IX (cont.)

Run Number	Feed rate ml/hr	Solvent flow ml/hr	Offtake carrier flow at ambient conditions			Mean Column pressure p.s.i.g.		Mean Column Operating F^R/S	
			Upper	Bleed	Lower	Upper	Lower	Upper	Lower
35-36°C									
C1	66	767	316	102	567	5.1	8.4	321	189
C2	65	767	395	99	602	4.9	8.3	405	260
C3	65	767	363	104	602	5.4	8.7	362	226
C4	86	767	274	94	573	5.5	8.7	274	158
C5	83	758	318	96	877	4.7	8.0	378	244
C6	74	758	294	98	874	4.8	8.1	340	212
C7	74	758	330	107	877	4.3	7.6	398	245
43-44°C									
D1	94	758	502	98	563	4.7	7.0	539	388
D2	94	758	343	107	626	5.5	8.6	350	211
D3	94	758	291	105	612	6.0	9.0	289	163
D4	64	758	322	99	585	5.4	7.9	332	204
D5	64	758	297	103	573	5.7	8.4	308	174

cended. The partition coefficients of cyclohexane and benzene at this temperature are 302 and 775 respectively giving a separation factor of 2.57.

At 20° - 21° the concentration level within the column is much less than in the vertical column. When plotted in terms of mole fractions, again similar profiles are observed see Figure 19.

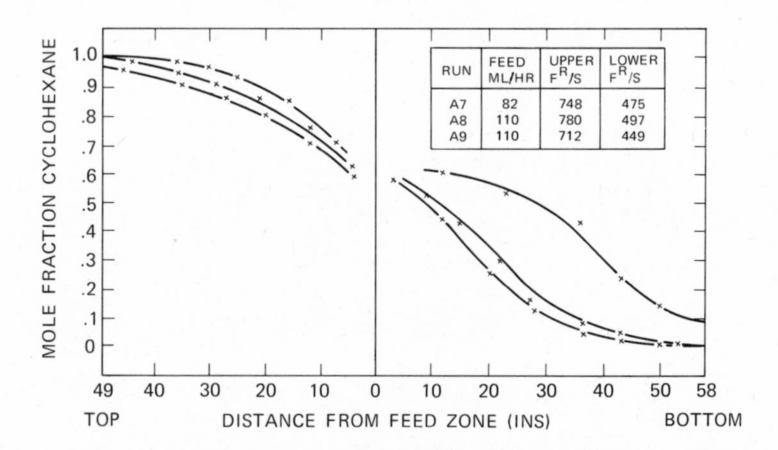

Figure 19: Mole fraction of cyclohexane in column at 20-21°C.

As the ratio increases the length of column necessary for the separation increases, particularly below the feed zone. In Figure 19 it would be expected that curve A7 should lie between A8 and A9 from F^R/S considerations, but it can be seen that because of the higher feed rate (110 ml/hr compared with 82 ml/hr) in fact an impure bottom product was obtained. The maximum feed rate which can be used to give pure products is about 90 ml/hr. The maximum and minimum operating values of F^R/S in the experiments which permitted realization of pure products were about 712 and 485 respectively, but it must be remembered that these are the mean values over the sections.

As the column temperature increases the partition coefficients of the components decrease, so that the gas flow must also decrease as the column is being operated at a constant liquid rate. However, the separation factor also decreases and makes the separation a little more difficult.

Decreasing the gas flow rate will increase the concentration level within the column for a constant feed rate. As the temperature increases, the gas phase saturation also increased quite rapidly, so that it is not expected that reducing the gas flow rate because of increased temperature will cause column saturation.

At column temperatures of 31° - 32° and 35° - 36° the concentration profiles were of the same form as at the lower temperature.

The highest column temperature used was 43° - 44° C, at which temperature the partition coefficients of cyclohexane and benzene are 145 and 345 respectively giving a separation factor of

2.38. The gas flow rate to perform the separation at this temperature was very low, giving much higher concentrations within the column (see Figure 20). Run D5 was the only run to give two pure

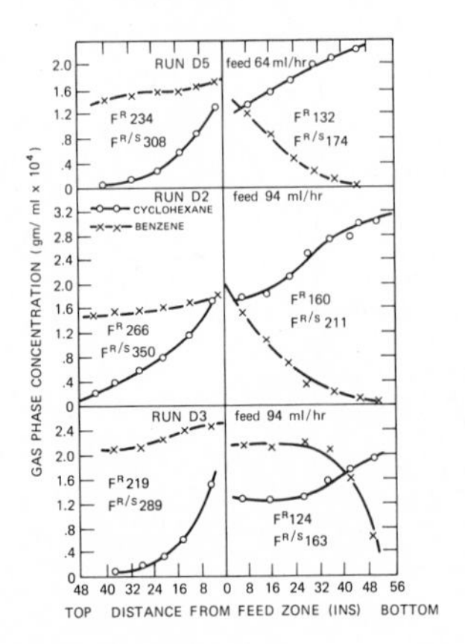

Figure 20: Separation at 43 - 44°C (see text).

products and indicates that the maximum feed rate is about 70 ml/hr. The maximum value of $F_{R/S}$ to give pure products is about 308 and the minimum 174, which shows that the theoretical range is narrowed slightly in practice. Also it would appear that the range is lessened as the feed rate is increased.

d. Column efficiency

The column efficiency for the benzene/cyclohexane system has been expressed in terms of both H.T.U. and the overall H.E.T.P. for the runs where operating conditions satisfy the relationships. For the evaluation of the H.T.U. a mean value of F^R/S was employed in each of the sections and the value of H.T.U. determined over 5 inch sections along the column (see Table X).

At 20° - 21° the mean value H.T.U. is about 1 inch the same as that found in the vertical moving bed columns. At a temperature of 31° - 32° the H.T.U. value is in general between 2 and 3.5 inches. The same range is observed at a column temperature of 35° - 36°C but at 43° - 44°C the value drops to a mean value of about 1.5 inches. The reason for the low values at 20° - 21° C may be because turbulence was beginning to occur in the bed, which promotes mass transfer.

The calculated H.T.U. values below the feed are shown in Table XI. The H.T.U. values are in general much higher than those observed for benzene above the feed, although it can be seen that as the distance from the feed zone decreases, the value of H.T.U. gradually decreases and approaches the value obtained for benzene. The value of benzene below the feed zone is considered to be due to the equilibrium being established more slowly below the feed than above, because as the column operating temperature is below the boiling point of the mixture some will condense on entry and be carried down the column as liquid.

TABLE X

Efficiency based on Benzene above the feed zone

H.T.U. (inches) Benzene

Run Number	Section above the Feed Zone (inches)					Mean
	5-10	10-15	15-20	20-25	25-30	
A4	1.11	1.02	-	-	-	1.06
6	0.93	0.85	-	-	-	0.89
9	0.82	0.90	0.85	-	-	0.86
B3	2.92	2.36	2.24	-	-	2.51
4	3.62	2.82	2.99	2.86	2.29	2.91
5	3.08	3.20	3.34	3.28	2.78	3.14
6	2.70	3.14	3.00	2.51	2.16	2.70
7	4.72	4.35	3.50	3.15	2.06	3.55
C1	3.97	3.27	2.83	-	-	3.36
4	2.86	2.49	-	-	-	2.67
6	3.55	3.07	2.92	2.86	2.16	2.91
D3	1.79	1.63	1.70	1.87	1.83	1.76
5	1.88	1.53	1.38	1.48	1.14	1.48

TABLE XI

Efficiency based on Cyclohexane below the feed zone

H.T.U. (inches) Cyclohexane

Run Number	Section below the Feed Zone (inches) 5-10	10-15	15-20	20-25	25-30	30-35
A3	6.71	4.92	-	-	-	-
5	8.15	7.02	7.12	5.72	4.37	-
7	11.2	8.48	6.83	7.04	5.44	4.59
8	15.6	14.0	10.8	8.06	7.48	5.25
B1	13.3	12.6	9.45	5.98	-	-
2	7.63	6.25	5.75	4.13	-	-
3	4.77	4.0	3.33	2.21	1.98	1.44
4	4.07	3.57	3.44	2.93	2.45	-
5	4.90	4.29	3.55	2.72	2.96	-
6	5.25	4.66	3.45	3.18	3.24	2.76
C2	8.24	6.73	6.08	6.33	5.23	5.12
3	6.42	5.74	4.72	4.21	3.44	3.73
5	6.17	5.40	5.01	4.27	4.06	4.13
7	5.31	4.50	4.61	5.62	4.45	4.07
D1	9.75	9.31	7.82	-	-	-
2	18.3	8.55	8.55	6.80	5.62	4.95
4	6.97	5.87	5.15	4.75	3.82	3.09
5	4.21	4.03	3.18	2.56	2.38	2.02

The overall mass transfer efficiency of the circular column in terms of H.E.T.P. for those operating conditions satisfying equation [40], are presented in Table XII, where it is seen that the mean value for a feed rate of 65 ml/hr is about 2 inches.

TABLE XII

Overall efficiency of the separation of the mixture Cyclohexane/Benzene

Run Number	Feed rate ml/hr	Distance from feed zone (inches) above	below	Total (inches)	$\frac{\emptyset \text{ Upper}}{\emptyset \text{ Lower}}$	N_C	HETP (inches)
A6	82	30	58	88	1.604	44.2	1.99
A7	80	40	47	87	1.575	42.6	2.40
B3	66	28	52	80	1.46	37.2	2.15
B4	66	40	36	76	1.45	36.8	2.06
B5	81	49	39	88	1.44	36.4	2.42
B6	89	46	46	92	1.44	36.4	2.53
B7	89	44	51	95	1.434	36.1	2.63
C1	66	42	58	100	1.698	51.7	1.93
C3	65	48	56	104	1.602	43.2	2.41
D5	64	47	55	102	1.770	59.9	1.70

From investigations on analytical columns it would be expected that the efficiency should rise with temperature to reach a maximum about 10° below the boiling points of the components being separated. In the operation of the circular column this has not been observed, probably due to the use of a bleed stream of gas which must be kept above a certain minimum value. This gas stream becomes increasingly important as the temperature rises because the gas flow in the upper section is reduced. The flow in the lower column is different from that in the upper section by the amount equal to this bleed stream.

2. The separation of the azeotropic system Diethylether - dichloromethane

This system is a further example of the ease with which azeotropic systems separate into their constituents using the continuous chromatographic refining technique. The partition coefficients at 20°C using the polyglycol phase previously mentioned are $K_{diethylether} = 101$; $K_{dichloromethane} = 482$, giving a separation factor of 4.77. The boiling points of the two materials are 34.8° and 41°C respectively.

The acutal operating conditions are shown in Table XIII which indicates the effective use of the column at different rates, with the upper and lower sections of the column being used to an equal extent. It can be seen that the maximum throughput consistent with pure products is about 220 ml of feed stock per hour under these conditions. Products were defined as pure at > 99.5 per cent. Table XIV shows the values of H.E.T.P. for this separation, indicating an increase in H.E.T.P. with feed rate due presumably to the higher concentration levels within the column.

3. The separation of the close-boiling dimethoxymethane-dichloromethane mixture

The boiling points of the two constituents are 42° and 41° respectively and using the polyglycol phase previously mentioned the partition coefficients at 20°C are $K_{D.M.M.} = 172$, $K_{D.C.M.} = 482$ giving a separation factor of 2.80. This system illustrates another advantage of the continuous chromatographic refining technique, how by suitable choice of stationary phase, the vapor liquid equilibria of a system can be modified, making the system much easier to separate than if no third phase were used (Table XV). Used in this way the technique is analogous to extractive distillation, however on a small scale, extractive distillation columns need to be carefully and expensively instrumented and are normally more difficult to operate successfully than the equipment described here.

The efficiency of the circular column in terms of H.E.T.P. is shown in Table XVI and again shows the reduction in the efficiency of the column with increasing feed rate.

C. General operating principle of the Barker-Universal Fisher Group Ltd., Compact circular chromatographic machine.

One of the main deficiencies of the 5 feet diameter circular chromatographic machine was the limited separating power due to the insufficient length of mass transfer zone available for separation purposes (9 feet). To maintain the basic concept of a circular machine and yet obtain long mass transfer zones makes the machine very cumbersome. The difficulty of compactness has been overcome by the Barker-Universal Eng. Group Ltd. (19) by forming a closed loop cylindrical nest of straight tubes held between a top and bottom ring, as shown in the prototype model Figure 21 and schematically in Figure 22.

The circular chromatography machine consists essentially of 44, 1 inch bore stainless steel tubes, 9 inches in length forming a closed loop cylindrical nest of tubes disposed between top and bottom rings. This tube bundle may be rotated at speeds between 0.2 and 2.0 r.p.h. by a variable speed drive. Four stationary ports are located around the circumference of the upper ring available for service with clockwise rotation of the tube bundle as carrier gas inlet (I), slower fraction outlet (II), feed mixture inlet (III), and faster fraction outlet (IV).

TABLE XIII

Evaluation of the operating conditions for separating the diethyl ether-dichloromethane mixture

Column operating temperature was 20°C

Feed rate ml per hour	Flows at ambient conditions, liters per hour			Mean column pressure, p.s.i.g.				Product purity and limit of the impurity	
	upper offtake flow rate	lower offtake flow rate	"bleed" carrier flow rate	upper column	lower column	upper column F^R/S	lower column F^R/S	top offtake DEE	bottom offtake DCM
64	437	348	85	8.2	11.0	252	179	Pure limit 31 inches from offtake	Pure limit 38 inches from offtake
117	464	348	85	7.6	11.1	268	193	Pure limit 25 inches from offtake	Pure limit 22 inches from offtake
176	488	336	80	7.2	11.2	282	202	Pure limit 9 inches from offtake	Pure limit 10 inches from offtake
202	437	348	86	8.0	11.0	252	178	Pure	Pure
220	451	348	86	8.0	11.0	260	186	Pure	Pure
267	437	348	86	7.7	11.0	274	200	99 per cent pure	98.7 per cent pure
267	476	348	86	7.3	11.2	274	200	92.6 per cent pure	Pure

TABLE XIV

Overall efficiency for the separation of the mixture

Diethylether/Dichloromethane

Feed rate ml/hr	Distance from Feed zone (inches) above	below	Total inches	Ø Upper / Ø Lower	N_C	HETP (inches)
64	18	20	38	1.41	17.45	2.18
117	24	36	60	1.39	17.20	3.59
176	40	48	88	1.395	17.27	5.10
202	49	58	107	1.42	17.51	6.11
220	49	58	107	1.398	17.29	6.19

TABLE XV

Evaluation of the operating conditions for separating the dimethoxymethane-dichloromethane mixture

Column operating temperature was 20°C

Feed rate ml per hour	Flows at ambient conditions, liters per hour			Mean column pressure, p.s.i.g.				Product purity and limit of the impurity	
	upper offtake flow rate	lower offtake flow rate	"bleed" carrier flow rate	upper column	lower column	upper column F^R/S	lower column F^R/S	top offtake DMM	bottom offtake DCM
93	600	504	89	5.0	8.9	396	282	92 per cent pure	Pure limit 28 inches from off-take
48	443	432	76	5.6	9.2	290	202	Pure limit 33 inches from off-take	Pure limit 10 inches from off-take
48	525	432	72	5.2	9.0	347	250	Pure limit 18 inches from off-take	Pure limit 16 inches from off-take
70	525	432	72	5.2	9.0	347	250	Pure	Pure limit 8 inches from off-take
88	525	432	72	5.2	9.0	347	250	Pure	Pure
130	525	432	72	5.2	9.0	347	250	98 per cent	Pure
172	488	432	72	5.2	9.0	323	230	Pure	97.5 per cent pure

TABLE XVI

Overall efficiency for the separation of the mixture Diemethoxymethane/Dichloromethane

Feed rate ml/hr	Distance from Feed zone (inches) Above	Below	total (inches)	Ø Upper / Ø Lower	N_C	HETP (inches)
48	31	42	73	1.39	30.2	2.42
70	49	50	99	1.39	30.2	3.28
88	49	58	107	1.39	30.2	3.54

Figure 21: Compact circular chromatographic machine.

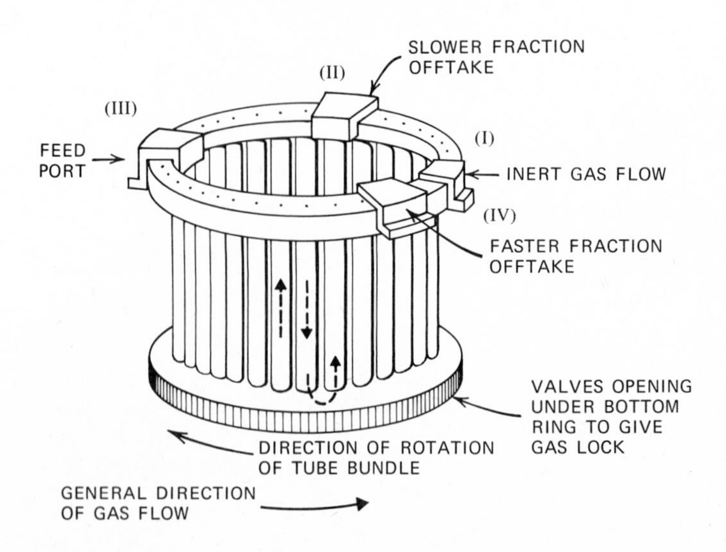

Figure 22: Representation of the cylindrical nest of closed tubes, also visible on the photograph in Figure 21.

To permit a unidirectional flow of inert gas countercurrent to the rotation of the bundle, poppet valves are closed in turn by a cam as they pass between the carrier gas inlet (I) and faster product outlet (IV).

The tube bundle and ports are housed in an electrically heated thermostatically controlled oven operating to 200°C. The oven is vented and continuously purged with preheated air as a precaution against the accumulation of explosive vapors in the event of accidental leakage. An explosion disc is also fitted in the top of the oven as an extra precaution. A photograph of the commerical model is shown in Figure 22a.

The principle of operation is similar to that described for the 5 feet diameter circular machine. A hot dry inert carrier gas enters as Port I passing up and down tubes in a general counter-current direction to the rotation of the tube bundle and leaving offtake ports (II) and IV). The feed mixture to be separated enters continuously at (III), the lighter (less strongly absorbed) components traveling preferentially in the gas stream towards product port (IV). The heavier (more strongly absorbed) components travel in the direction of rotation of the tube bundle, being stripped off by the hot inert gas in the zone between ports (I) and (II) and leaving through the offtake port (II).

1. The production of high purity hydrocarbons

a. Cyclopentane purification

The improved separation capability of the compact circular chromatography machine compared with the 5 feet circular chroma-

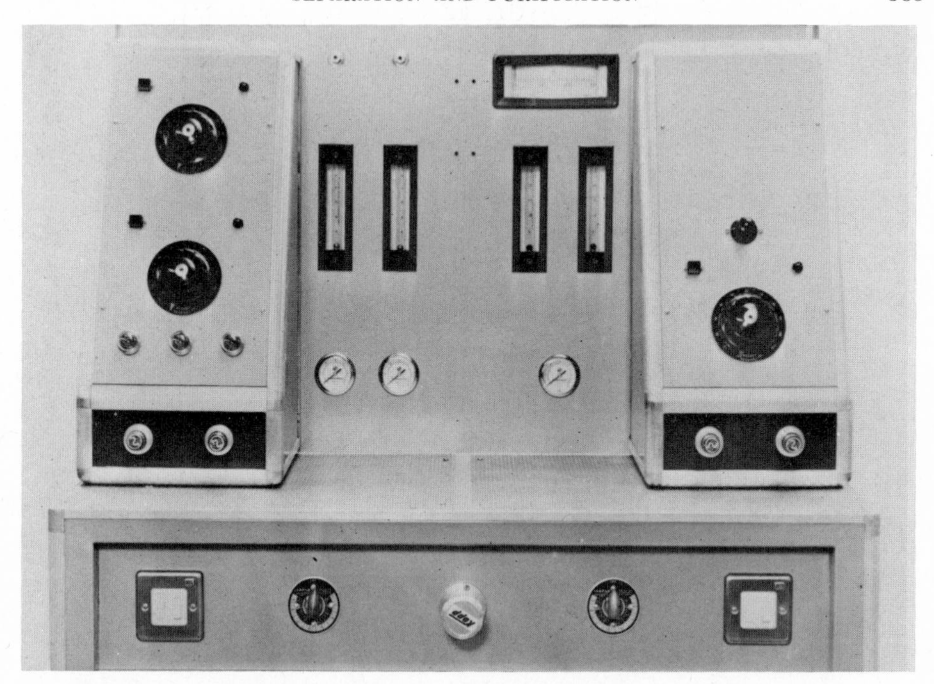

Figure 22a: Commercial compact circular chromatographic machine.

tograph is demonstrated by the results published by Barker & Al-Madfai (20) on cyclopentane and cyclohexane purifications. A 97% cyclopentane with four detectable impurities as shown in Figure 23 was used. The nearest impurity to the cyclopentane was identified as 2.2 dimethylbutane having a partition coefficient of 78.5 compared with 119.1 for cyclopentane. Hence the separation factor was 1.518. The tubes of the machine were packed with 14/18 mesh Johns Manville C22 firebrick coated with 30% $^{w}/_{w}$ polyoxyethylene 400 diricinoleate. The total weight of support was 2956.7 gm. of which 887.0 gm. were liquid phase.

Table XVII shows the operating data to achieve the column performance given in Table XVIII, the experiments being designed to obtain pure cyclopentane, with minimum loss of this material in the impurity fraction. Very high purity (99.999%) with virtually no loss of cyclopentane in the impurity fraction was achieved (see Figure 23). It should be noted that the purities recorded throughout this chapter refer to detectable purity based on gas chromatography with flame ionization detectors, rather than absolute purity.

The reduced throughput of the compact circular machine is to be expected because the cross-sectional area of the packed section is only one third that of the packed section used in the 5 feet diameter circular machine. It is seen that throughput is approximately proportional to cross-sectional area, while the length of mass transfer section (14 feet) compared to 9 feet for the 5 feet diameter circular machine considerably improves the purity and yield of the desired products.

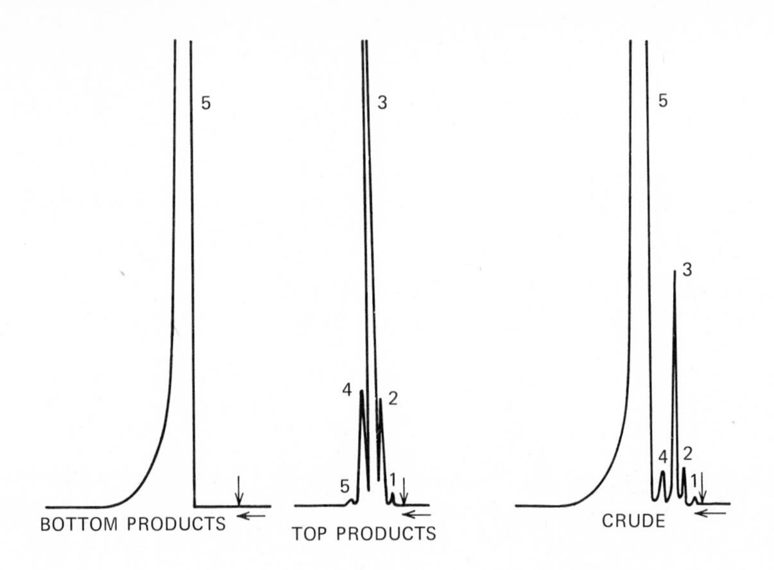

Figure 23: Purification of cyclopentane.

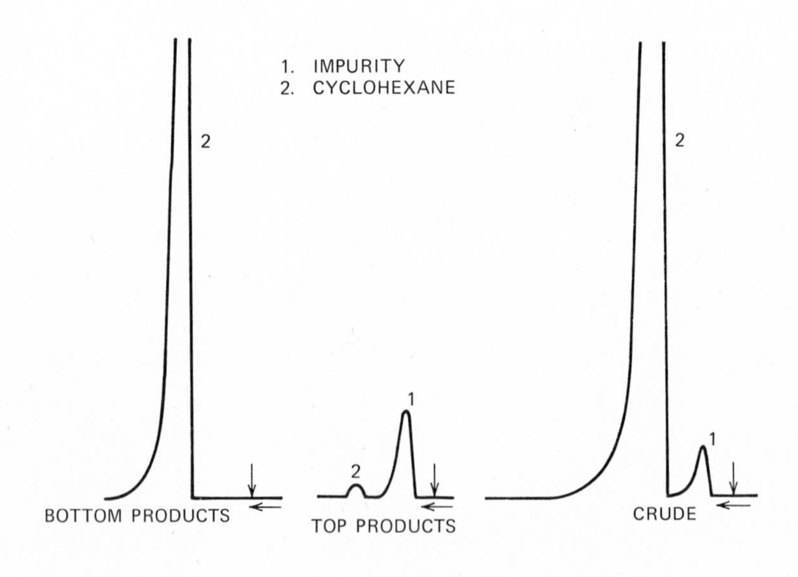

Figure 24: Purification of cyclohexane.

TABLE XVII

Operating Conditions for Cyclopentane Purification

Run	Feed rate ml/hr	Carrier Gas flow Rates at ambient conditions liters/hour			Mean Column Pressure p.s.i.g.		F^R/S	
		upper offtake	lower offtake	feed purge	upper column	lower column	upper column	lower column
1	26.9	94.9	167	19.9	5.5	6.0	80.3	61.9
2	41.6	94.9	167	19.9	6.1	6.3	77.8	60.9
3	60.3	94.9	167	19.9	6.1	6.3	77.8	60.9
4	154.1	94.9	360	19.9	3.5	4.5	89.9	67.1
5	154.4	112.9	360	19.9	3.5	4.5	105.85	83.2

All runs made at 1.0 R.P.H. = 0.8625 liter per hour stationary phase temperature = 22°C

TABLE XVIII

Column Performance for Cyclopentane Purification Runs

Run	Feed rate ml/hr	% Recovery of cyclopentane in lower offtake	% Purity of lower offtake product	Limit of impurities from lower offtake (ft)	% effective mass transfer length used in resolving mixture (14 ft available)
1	26.9	100	99.999%	12	30%
2	41.6	99.84	99.999%	9	40%
3	60.3	99.84	99.999%	3	50%
4	154.1	99.9	99.18%	-	100%
5	154.4	99.9	99.999%	1	90%

b. Cyclohexane purification

Further evidence of the superior separating power of the compact circular machine is shown in Figure 24 and Table XIX when purifying 99% cyclohexane. The one detectable impurity was identified mainly as methylcyclopentane having a partition coefficient of 193.6 at 20°C. Hence compared to 306 for cyclohexane, the separation factor is 1.582.

c. Separation of n-hexane from crude hexane

The separation of n-hexane from a crude hexane containing four main impurities, illustrates the ability of the compact circular machine to separate isomers and also the separation of a component from impurities distributed on either side of the desired product as shown in Figure 25.

A gas chromatographic analysis of the crude n-hexane showed four main peaks on the chromatogram, these peaks were identified and taken to be in order of elution from the column as 2-methylpentane (2 mp), 3-methylpentane (3 mp), n-hexane (n-hex) and methylcyclopentane (3 mcp). The partition coefficients of these compounds at 20°C were found to be $K_{2mp} = 101.8$, $K_{3mp} = 119.1$, $K_{n\text{-}hex} = 142.0$ and $K_{mcp} = 200.8$ giving separation factors of 1.170 for 3mp/2mp, 1.192 for n-hexane / 3 mp and 1.414 for mcp/n-hexane.

In the first series of runs (1-4) Table XX, methylcyclopentane was removed from the crude n-hexane with comparative ease (see Figure 25b) because of the separation factor of 1.414 for mcp/n-hexane and the long length of column available for mass transfer. The material collected from the top product port, i.e., n-hexane, 2mp, 3mp was now re-run to separate the n-hexane from the 2mp and 3 mp (see Figure 25c). Run 5 shows typical conditions for a feed rate of 12.8 ml/hr. The separation now is between two isomers n-hexane/3mp where the separation factor is 1.192, hence the lower feed rate compared with runs 1 to 4.

Runs 6 and 9 show the versatility of the machine in being able to cut at any chosen point on a chromatogram.

Typical component distribution within the column at steady state conditions are shown in Figures 26a and b.

d. The separation of Linalol from Rosewood oil

A chromatogram of the Rosewood oil see Figure 27 showed the following compounds or groups of compounds in order of elution from an analytical chromatogram terpene hydrocarbons, cineole, benzaldehyde, cis and trans linalol oxides, the main peak of linalol followed by terpene alcohols. The aim of these experiments was to try to separate compounds such as linalol without degradation of the mixture. Table XXI shows operating conditions and performance data for the experiments made with this oil.

In Run 3 complete removal of linalol oxides and light hydrocarbons was achieved (see Figure 27). In Run 4, Table XXI shows conditions when cutting on the other side of the linalol peak.

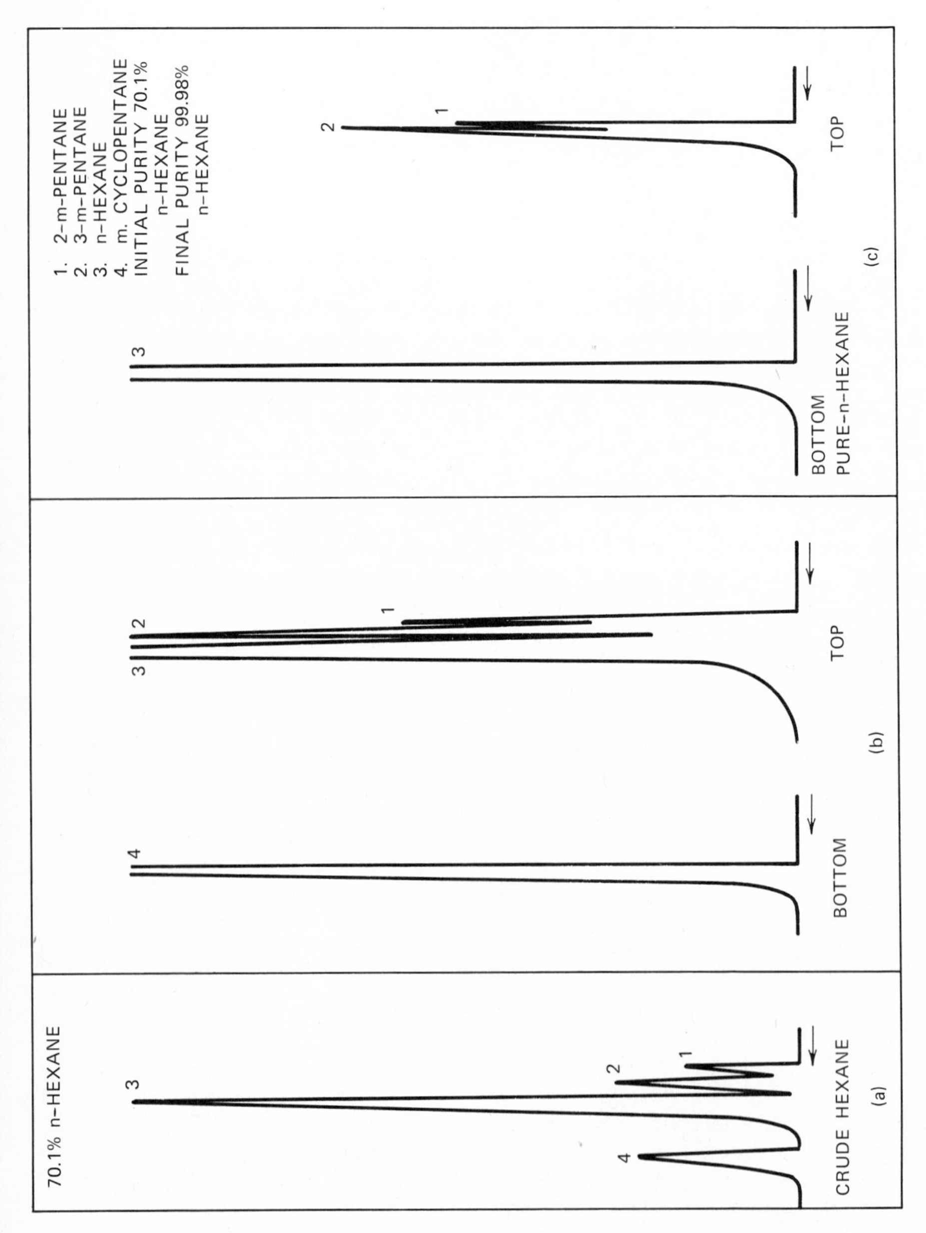

Figure 25: Separation of n-hexane from crude hexane.

TABLE XIX

Operating Conditions for the Cyclohexane Purification Runs

Run	Feed rate ml/hr	Carrier gas flow rates at ambient conditions liters/hour			Mean column pressure p.s.i.g.		$F^R/\underline{S}$	
		upper offtake	lower offtake	feed purge	upper column	lower column	upper column	lower column
1	36.8	242	648	29	2.4	3.5	241.7	200
2	64	242	648	29	2.4	3.5	241.7	200
3	76.4	242	648	29	3.5	4.5	226.7	197.7
4	152	242	648	29	2.4	3.5	241.7	200
5	126.7	343.2	648	31.2	3.3	5.1	324.6	269

All runs made at 1.0 R.P.H. = 0.8625 liter per hour
Temperature = 21°C

TABLE XX

The Separation of Crude Hexane into its Main Constituents

Run No.	Feed rate ml/hr	Flow at ambient conditions liters per hr upper offtake	upper offtake	feed purge	Mean Column pressure p.s.i.g. upper column	lower column
Separation between n-hexane and methylcyclopentane						
1	12.7	160.8	251	0	2.5	3.5
2	24.7	213	252	57	2.5	3.5
3	52.8	179.9	228	20.9	3.5	4.5
4	78.3	307	285	37.1	3.6	5.2
Separation of pure n-hexane						
5	12.8	241.4	186	28.4	5.7	7.3
Separation between 2 m pentane and 3 m pentane						
6	11.6	99	156	15	3.5	4.5
7	21.2	78	186	0	4.3	4.8
Separation between 3 m pentane and n-hexane						
8	14.8	115	246	15	5.25	5.75
9	26.1	109.8	186	0	4	4.6

TABLE XX (Cont)

F^R/S					
upper column	lower column	column temp. °C	RPH	upper offtake products	lower offtake products
159.5	150.8	20	1.00	2 mp.,3 mp.,n-hex	mcp
				complete separation	
211.1	147.3	18	1.00	2 mp.,3 mp.,n-hex	mcp
				mcp < 1%	
168.6	132.4	19.5	1.00	2 mp.,3 mp.,n-hex	mcp
				complete separation	
167.1	135.3	21	1.71	2 mp.,3 mp.,n-hex	n-hex.,mcp
				Loss 7% n-hex.from lower offtake	
117.9	92.27	17	1.71	2 mp.,3 mp.,n-hex	pure n-hexane
				Loss 3% n-hex.from upper offtake	
92.9	74.8	17.5	1.00	2 mp.	3 mp.n-hex.,mcp
				complete separation	
70.2	68.5	22.5	1.00	2 mp	3 mp.n-hex.,mcp
				complete separation	
98.5	83.46	17	1.00	2 mp.,3 mp	n-hex.,mcp
				complete separation	
109.2	105.8	21.5	1.00	2 mp.,3 mp	n-hex.,mcp
				complete separation	

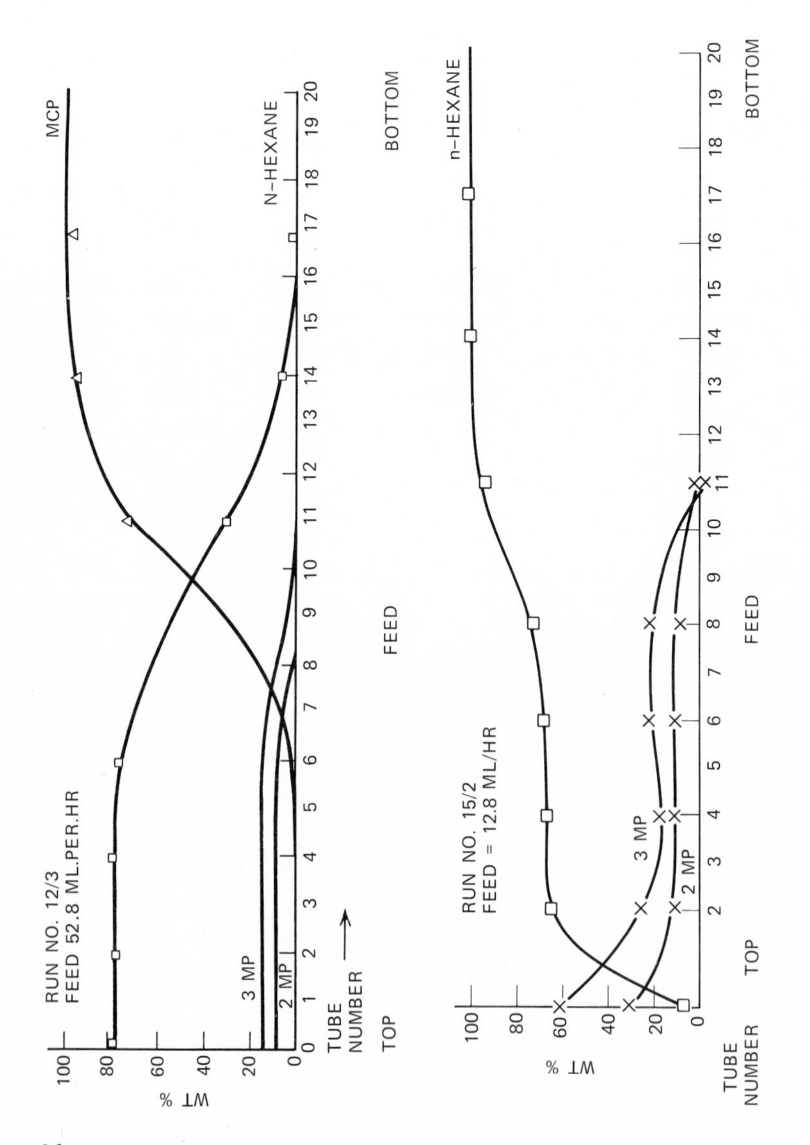

Figure 26a: Separation of n-hexane and methylcyclopentane.

b: Separation of pure n-hexane from 2 and 3 methylpentanes.

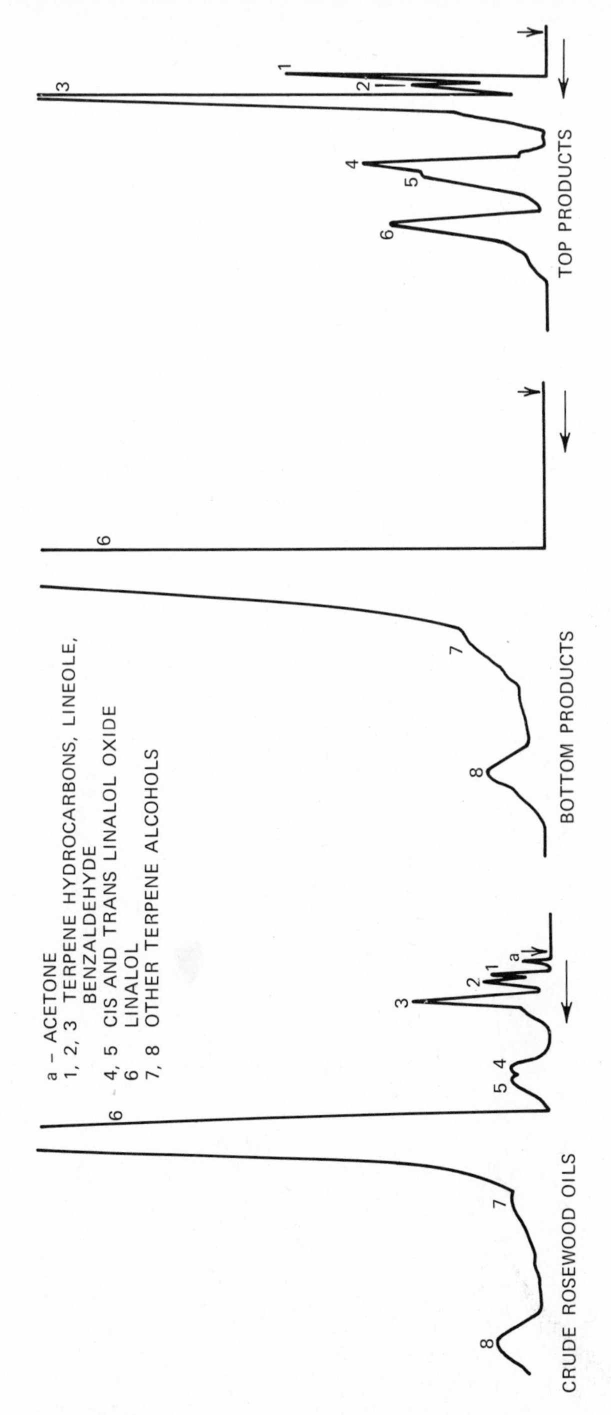

Figure 27: Separation of linalol from rosewood oils.

TABLE XXI

Separation of Linalol from Rosewood Oil

Run No.	Feed ml/hr	RPH of column	Ml stationary phase per hr	Carrier Gas at ambient conditions l/hr upper column offtake	lower column offtake	feed purge
1	8	0.1755	60.3	72	282	0
2	8	0.1755	60.3	102	300	0
3	3.9	0.476	164.4	165	402	0
4	9.7	0.2375	81.6	147	374	0

TABLE XXI (Cont)

Mean Column Pressures psig upper lower	column lower	F^R/S at mean column cond.	Column temp. °C	Upper offtake products	Lower offtake products
1	3	1373	110	light hydrocarbons, oxides, no linalol	some oxides, linalol & higher alcohols
1.6	4.6	1890	130	light hydrocarbons, oxides, a little linalol.	linalol & higher alcohols only
2.2	6.8	1046	130	light hydrocarbons, oxides only	linalol,higher alcohols
1.8	5.2	2006	135	light hydrocarbons, oxides,linalol, trace heavy alcohol	trace linalol, higher alco-hols

These experiments showed the ability of a compact chromatograph to handle high boiling heat sensitive liquid mixtures, giving products that had suffered no detectable decomposition effects. It is considered that this type of separation technique will find particular application in handling heat sensitive liquid mixtures, because the presence of the inert gas acts in a similar manner to low vacuum operation, while by operating primarily in the gaseous state, residence times within the equipment are small thereby reducing the chance for significant decomposition to occur. Materials of construction of the equipment obviously play an important role in determining decomposition effects. Currently 18/8 stainless steel and "Graflon" i.e., P.T.F.E. and carbon are used, although glass equipment is contemplated for those separations requiring it.

e. Separation of α-pinene from U.S. grade spirit of turpentine

Turpentine which contains primarily α and β pinenes together with a few per cent of camphene has been quoted frequently in the literature as a yard stick for separations by batch operated gas chromatographs. It is therefore of interest to compare the quoted batch operated throughput achievements with those for the compact circular chromatographs.

Abcor (43) have reported the separation of α-pinenes from β-pinenes using crude turpentine. They reported a maximum feed rate of 22 liters (41 lbs) per 24 hours when using a 4 inch diameter packed column 9 feet long fitted with special baffles to get maximum column efficiency. The reported recovery of 98.6% pure α-pinene was 90%, and the equivalent feed rate through a 1 inch diameter column corresponds to 55.4 ml/hr.

In the investigation by Barker and Al-Madfai reported in Table XXII and Figure 28, a 99% pure α-pinene at 91.3% recovery was obtained by C.C.R. techniques at a feed rate of 66.5 ml/hr.

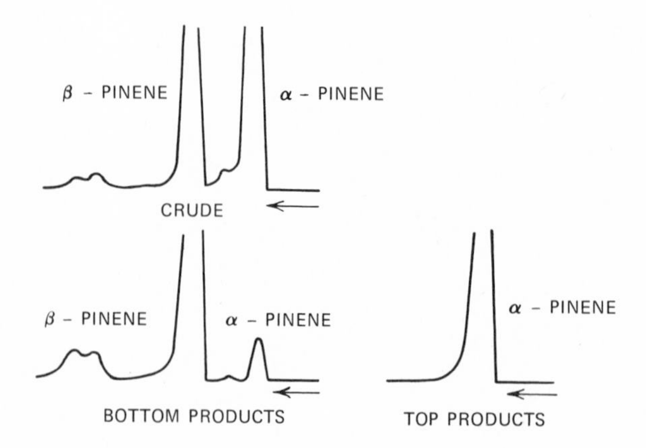

Figure 28: Separation of alpha-pinene from U.S. grade spirit of turpentine.

TABLE XXII

Separation of α- and β-Pinenes from U.S. Commercial Turpentine Crudes

Feed ml/hr	Carrier gas flow at ambient conditions L/hr		Oven Temperature	Average Column Pressure p.s.i.	% Purity	% Recovery	Postulated feed rate through 4 inch diameter column L/24 hr.
	Upper offtake	Lower offtake					
33.5	165	342	110°C	5	99+	98.3	12.91
66.5	165	264	113°C	4.8	99.5	91.3	21.65
109.8	189	240	120°C	5	99	62.0	42.17

This feed rate is 20% more than reported by Abcor for an equivalent 1 inch diameter column. With higher temperatures nearer to the boiling points of the pinenes (ca. 152°C). Barker and Al-Madfai anticipate even higher percentage feed differences over the static bed results.

The overall percentage recovery indicated in Table XXII refers to the product actually recovered by a two stage cooling process, which is not necessarily the maximum that could be recovered with more sophisticated equipment.

VI. OTHER EQUIPMENTS ACHIEVING CONTINUOUS CHROMATOGRAPHIC SEPARATIONS

A. Radial flow columns

An equipment due to L.C. Mosier (21) in which the feed travels from the center to the circumference of an annular packing is described in the literature. By relative rotation of the packing and the feed inlet the paths taken by different components of the feed will be different depending on the retention volume of the component, and so continuous separations can be achieved. The equipment shown in Figure 29 requires that either the feed injection and collection system can be rotated with the packing stationary, or the packing can be rotated with the feed system static, the latter method being preferred.

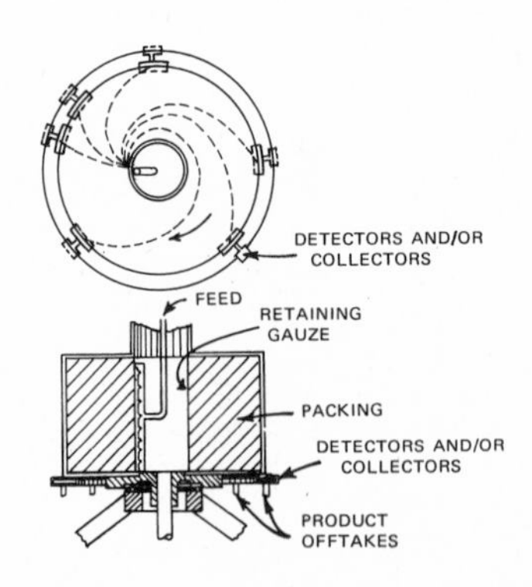

Figure 29: Design of a radial flow column.

B. Helical flow columns

This type of column first proposed by Martin (22) is based on an annular packed column, or a circular tube array which is rotated. The feed enters at the top, and the paths traveled by the different components are in the form of helices.

Dinelli et al. (23,24) have constructed such a column (see Figure 30 from 100 columns 6 mm diameter, 1.2 meters long), arranged in the form of a circle. The column bundle rotates at

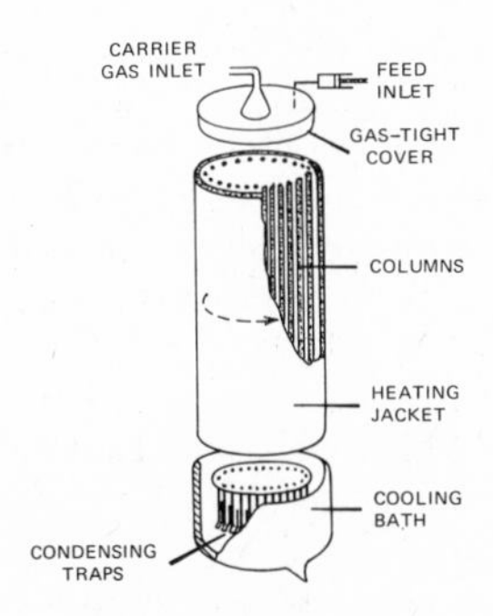

Figure 30: Rotating unit for preparative scale gas chromatography.

speeds of from 1 to 50 r.p.h. and both the feed inlet port and product offtake receivers are stationary.

The carrier gas travels down the tubes and due to the rotation, the different components of the feed, by virtue of their different retention times, will each follow a different path and so can be collected in separate traps. Knowledge of the speed of rotation, the carrier flow and retention times of the components, permits evaluation of the receiver at which the component should appear.

$$L = 2\pi rQt_R \qquad [45]$$

where

L = distance of receiver from feed position

r = cylinder radius

Q = speed of revolution

t_R = retention time of component.

Optimum operating conditions have also been determined (25). Using this column a mixture of heptane and toluene has been separated at 60 ml/hr, heptane appearing in receivers 8-22 and toluene in receivers 72-4 (compared with 9-24 and 70-3 as calculated).

As the feed rate is increased the number of receivers in which the component appears will increase, and eventually the components will begin to overlap, so there is a maximum feed rate that can be used to give products of a required purity. For separation of cyclohexane/benzene on tricresyl phosphate at 80°C at purities of more than 99.9% a theoretical maximum value of 220 ml/hr was calculated, although in practice it was found to be 200 ml/hr.

The unit was enlarged using 36 tubes of 2.4 m length in the form of an inverted 'U' and in this form has been used to separate isomers (26) and in the purification of hydrocarbon fractions for which higher throughputs and purities are obtainable than with an Oldershaw column.

The same type of column, but using a packed annulus of multiple tubes has also been proposed (27,28) although no operating data has been published. It is considered that the mechanical problems involved in making a perfectly symmetrical annulus would be greater than in the construction using multiple tubes, although the latter requires that the column must be matched to give the same elution characteristics (29).

Ingenious as these helical flow schemes are, they are still batch operated columns, suffering the inherent drawbacks outlined earlier in the chapter when attempting to separate large quantities of feed.

Turina et al. (30) have used an apparatus consisting of two parallel glass plates 1 mm apart containing the inert support. Carrier gas is introduced uniformly along one vertical edge and emerges at the opposite edge through twenty-six uniformly spaced exhaust ports. The non-volatile liquid flows through the packing vertically, and leaves at the base. If the feed mixture is introduced continuously into the packing at the corner situated between the entrance of both carrier and non-volatile liquid, the individual components will travel at an angle from the horizontal depending on the retention time, and appear at different positions in either the existing carrier gas or in the non-volatile liquid phase. This has been applied to continuous thin layer chromatography using triangular plates (31).

C. Sinusoidal flow columns

Thompson (32,33) developed a semi-continuous separation using a single analytical column in which the inlet stream varied from 100% carrier gas to 100% feed mixture in a continuous sinusoidal manner, so that the concentration of all the components in the feed mixture oscillated sinusoidally in phase with one another. As the concentration waves traveled through the column they were differentially retarded and attentuated by the stationary phase. Hence by adjustment of the frequency and flowrate selected, component waves can be caused to emerge from the column out of phase with one another so that the valves, operating at the same frequency as the wave generator, can direct the effluent to different condensing traps for collection of the material. Using a 1:1 feed mixture of ethane and propane with benzyl ether as the stationary phase and carbon dioxide as the carrier gas, Thompson was able to separate into products of 70% pure ethane 74% propane at a total feed rate of 82 ml/min and a frequency of 40 cycles/min.

D. Flowing Stationary phase columns

In all the vertical moving bed equipment described earlier, the solid support flowed under gravity down the column, but the same basic principle can be achieved by having a solid support held stationary, as in a conventional packed column, and allowing the non-volatile stationary phase liquid to flow down the column at a known rate over the support.

Tiley and co-workers (34) have investigated this principle. Using a 1 inch diameter column to separate 5 ml/hr of 1:1 feed mixture of diethylether and dichloromethane, with dinonyl phthalate as the non-volatile flowing liquid, they found that product purities in excess of 99.9% could be achieved although the efficiency was much less (H.E.T.P. about 2.5 mins) than using the moving bed column.

Although this scheme is simpler to construct than moving bed or circular columns, poorer efficiencies and low throughputs are in fact obtained. This is to be expected since the non-volatile liquids flowing over the packing pieces usually have high viscosities giving low diffusion rates, an effect which is counteracted in moving bed chromatographs by distributing the liquid as thin films over solid supports of very large surface area. Also packed columns suffer from liquid maldistribution effects, and have a much lower surface area/unit volume ratio than when using solid supports.

Kuhn et al. (35,36) have used the temperature dependence of the partition coefficients in flowing liquid columns for the separation of multicomponent mixtures. By having sections of the column at different temperatures and constant gas and liquid flowrates, the different components of a mixture will collect at different specific places in the column depending on the partition coefficient. The column used consisted of five sections each maintained at a different temperature, decreasing as the column was descended. The non-volatile liquid paraffin oil containing 10% stearic acid, flowed over steel spirals countercurrent to the flowing carrier gas. The ternary mixture of propionic, n-butyric and n-valeric acids was introduced into the top of the first section (hottest) at the base of the column. All three components rose through section 2 and the column temperature gradient was arranged so that n-valeric acid concentrated between sections 2 and 3, n-butyric acid between 3 and 4, and propionic acid between 4 and 5. The product could be continuously withdrawn and purities of 95% were achieved.

Other flowing liquid film equipment papers have been reported in the literature (37,38).

E. The continuous separation of non-volatile organic materials by liquid-solid chromatography

Whereas most of the published work using C.C.R. techniques relates to gas chromatographic type systems the liquid-solid chromatographic counterpart is in fact practiced commercially using batch techniques for separating glucose from fructose, reducing the molecular distribution of carbohydrate polymers such as Dextran, etc.

Using a smaller rotating circular column techniques to those described previously in the chapter for the separation of volatile organic substances, preliminary studies into the separation of carbohydrates has been completed (42) and work now continues into evaluating the technique for the separation of enzymes, antibiotics, hormones carbohydrate polymers, etc.

The equipment being used is a small laboratory model consisting of a circular bundle made up of 44 tubes, one foot long and 3/8 inch I.D. packed with, for example, porous silica beads in order to effect separations on the basis of molecular size (gel permeation chromatography).

With degassed deionized water as eluant, Dextran, the blood plasma substitute having a molecular weight range 40,000 to 200,000 has been fractionated continuously into samples with narrower molecular weight distributions.

By using a multihead micropump, the flow of liquid into and out of the apparatus have been controlled precisely, so that continuous night and day operation for periods of up to one month have been made.

F. Comparison between batch and continuous chromatographic refining

No experimental and economic data when processing identical systems exists in the literature to make a true comparison between production scale batch and continuous chromatographic processing. Abcor (41) have published some capital and processing cost data when batch processing, while Barker and Huntington (8) published cost data relating to two particular hydrocarbon separations assuming throughputs of 200 g.p.h. of feed. The more comprehensive data of Abcor gave operating costs varying from 3 cents/lb for relatively easy and volatile systems, (at the 5,000,000 lbs/yr rate) to 15$ per lb. for difficult separations and low volatile systems, (at the 8,000 lbs/yr rate). Barker (8) reported processing costs for volatile systems varying from easy to average difficulty of from 1.5 to 3.5 cents/lb. (at the 14,000,000 lbs/yr rate). At the one end of the scale, processing costs of a similar order of magnitude are being quoted, but the accuracy is inadequate to give a clear decision. Further work is obviously needed in this area.

With most new developing techniques it is impossible to say with certainty whether batch or continuous processing will finally predominate, although the author and his colleagues who have worked for fifteen years on the latter favor the continuous process.

VII. NOMENCLATURE

ϵ_1	extraction factor z/y
F_1^R	gas flow rate in rectification section of column M,N (Fig. 10)
F_2^R	gas flow rate in section of column P (Fig. 10)
f(F)	function (f) of F, the gas flow rate
F^R	gas flow rate in rectification section of column
F^S	gas flow rate in stripping section
H_{OG}	height of an overall gas phase transfer unit
i	a component i in a mixture
I	product impurity fraction
K^F	partition coefficient of a solute at finite concentration on a solute free basis
K^R	partition coefficient of a solute at infinite dilution in the rectification section of the column
K_S	partition coefficient of a solute at infinite dilution in the stripping section of a column
K^∞	partition coefficient of a solute at infinite dilution
$K_G a$	overall gas phase mass transfer coefficient
$\frac{1}{k_g a}$	gas phase mass transfer coefficient
$\frac{1}{k_L a}$	liquid phase mass transfer coefficient
L	column length
M	solute flow rate
N_{OG}	number of overall gas phase transfer units $\int_{y_2}^{y_2} \frac{dy}{y-y_\epsilon}$
N_c	number of theoretical plates (steady state conditions)
n,m	number of stages above or below the feed respectively
Q	speed of revolution

R_A, R_B	rate constants of desorption of solutes A & B respectively
r	cylinder radius
S	solvent rate
S_1^R	solvent flow rate in rectification section of column M,N (Fig. 10)
S_2^R	solvent flow rate in section of column P, (Fig. 10)
SF	separation factor = K_B/K_A
t_R	retention time of component
W_A, W_B	feed rate of components A and B respectively
x	parts of component i removed in the gas stream
y	parts of component i passing to the $(m+1)^{th}$ stage
Y_A, X_A	weight of component A per unit volume of carrier phase and solvent phase respectively
z	parts of component i passing to the $(m-1)^{th}$ stage

Subscripts

A,B, etc., refer to solutes A, B, etc.
T,B, etc., refer to Top and Bottom products respectively

Superscript	S refers to stripping section
Superscript	R refers to rectification section
γ_A, γ_B	factors accounting for effect of finite concentration on the partition coefficient
δ_A, δ_B	factors accounting for effect of finite column length on partition coefficient
$(\mu_z)_A$	mass ratio of component A in bottom/feed
$(\mu_z)_B$	mass ratio of component B in top/feed
ϕ_{max}	operating ratio a maximum as defined by equation 38
ϕ_{min}	operating ratio a minimum as defined by equation 35
Ψ	fraction of component i not extracted

VIII. REFERENCES

1. Barker, P.E. & Lloyd, D.I. J. Inst. Petr. 49, 73 (1963)
2. Fitch, G.R., Probert, M.E., Tiley, P.E. J. Chem. Soc. 4875 (1962)
3. Sciance, C.T., Crosser, O.K. J.A.I.Ch.E. 12, 100 (1966)
4. Schultz, H. "Gas Chromatography 1962" (ed. Van Swaay, M.), Butterworths 1963, p. 225
5. Scott, R.P.W. "Gas Chromatography 1958" (ed. Desty, D.H.) Butterworths 1958, p. 189
6. Chilton, T.H. & Colburn, A.P. Ind. Eng. Chem. 27, 255 (1935)
7. Alders, L. Liquid-Liquid Extraction 2nd Ed., Elsevier 1959, Chap. V
8. Barker, P.E., Huntington, D.H. Dechema Monographien Vol. 62 p. 153 (1968), Published Dechema
9. Huntington, D.H. Ph.D. Thesis, Birmingham University (1967)
10. Barker, P.E. & Lloyd, D.I. U.S. Patent 3,338,031
11. Barker, P.E. & Huntington, D.H. "Advances in Gas Chromatography 1965", ed. A. Zlatkis & Ettre, L.S., Preston Technical Abstracts Co., Evanston 1966, p. 162.
12. Pichler, H. & Schultz, H. Brennstoff Chemie 39, 48 (1958)
13. Gulf Research & Development Co. U.S. Patent 2,893,955 (1959)
14. Mine Safety Appliances Co. (Luft, L.) U.S. Patent 3,016,107 (1962)
15. Glasser, D. "Gas Chromatography 1966" (ed. Littlewood, A.B.) Institute of Petroleum, p. 119 (1967)
16. Barker, P.E. & Universal Fisher Eng. Co. Ltd. Crawley, Sussex, Patent Application No. 33630/65 and 43629/65. Patent issued under no. 114596, also foreign patents.
17. Willmott, F.W. & Littlewood, J.B. J. Gas Chromat. 4, 401 (1966).
18. Barker, P.E., Huntington, D.H. Proceedings of the Sixth International Symposium on Chromatography, Rome 1966, Published by Inst. of Petroleum
19. Barker, P.E. & Universal Fisher Eng. Co. Ltd. Patent Application No. 5764/68 and 44375/68
20. Barker, P.E. & Al-Madfai, S. Proceedings of the Fifth International Symposium on Advances in Chromatography, Las Vegas, 1969. Published by Preston Technical Abstracts, pg. 123
21. Cities Service Research & Development Co. (Mosier, L.C.) U.S. Patent 3,078,647, 1963
22. Martin, A.J.P. Disc Farad. Soc. 7, 332 (1949)
23. Dinelli, D., Polezzo, S., Taramasso, M. J. Chromatog. 7, 477 (1962)
24. Dinelli, D., Taramasso, M., Polezzo, S. U.S. Patent 3,187,486 (1965)
25. Polezzo, S., Taramasso, M. J. Chromatog. 11, 19 (1963)
26. Taramasso, M., & Dinelli, D. J. Gas Chrom. 2, 150 (1964)
27. Deutsche - Erdol - Aktiengeselischaft. G. Patent 1,033,683 (1958)

28. Deutsche - Erdol - Aktiengeselischaft. B. Patent 810,767 (1959)
29. Anon. Chem & Eng. News 40, 74 (1962)
30. Turina, S., Krajovan, V., Kostomaj, T. Z. Anal. Chem. 189, 100 (1962)
31. Turina, S., Krajovan, V., Obradovic, M. Anal. Chem. 36, 1905 (1964)
32. Canadian Patents & Developments Ltd. (Thompson, D.W.) U.S. Patent 3,136,616 (1964)
33. Thompson, D.W. Trans. Instn. Chem. Engrs. 39, 19 (1961)
34. Bradley, B.J. and Tiley, P.F. Chem & Ind 18, 743 (1963)
35. Kuhn, W., Narten, E. and Thurkauf, M. 5th World Petr. Congress 1958. Section V, paper 5, p. 45
36. Kuhn, W., Narten, E. and Thurkauf, M. Helv. China Acta 41, 2135 (1958)
37. Grubner, O. and Kucera, E. Collection Czech. Chem. Commun. 29, 722 (1964)
38. Phillips Petroleum Co. (Ayers, B.O.). U.S. Pat. 3,162,036 (1964)
39. Giddings, J.C. and Eyring, H. J. Phys. Chem. 59, 416 (1955)
40. Barker, P.E., Lloyd, D.I. Proceedings of the Symposium on "Less Common Means of Separation". April 1963. Published by the "Institution of Chemical Engineers", London, p. 68
41. Ryan, J.M., Timmins, R.S., O'Donnell, J.F. Chemical Eng. Progress, 64, 53 (1968)
42. Barker, P.E., Barker, S.A., Hatt, B.W., Somers, P.J., Chemical Processing Eng. vol. 52, No 1, (1971) ps. 64-66
43. Anon. Chem. & Eng. News, May 23 (1966)
44. Kremser, A. National Petroleum News 22, 43 (1930)

Index